OPTICS: A Short Course for Engineers & Scientists

WILEY SERIES IN PURE AND APPLIED OPTICS

Advisory Editor

STANLEY S. BALLARD University of Florida

Introduction to Laser Physics, BELA A. LENGYEL
Laser Receivers, MONTE ROSS
The Middle Ultraviolet: its Science and Technology, A. E. S. GREEN, *Editor*
Optical Lasers in Electronics, EARL L. STEELE
Applied Optics, A Guide to Optical System Design/Volume 1, LEO LEVI
Laser Parameter Measurements Handbook, HARRY G. HEARD
Gas Lasers, ARNOLD L. BLOOM
Infrared System Engineering, R. D. HUDSON, JR.
Laser Communication Systems, WILLIAM K. PRATT
Optical Data Processing, A. R. SHULMAN
The Applications of Holography, H. J. CAULFIELD AND SUN LU
Far-Infrared Spectroscopy, KARL D. MOLLER AND WALTER G. ROTHSCHILD
Lasers, second edition, BELA A. LENGYEL
Polarization Interferometers, M. FRANCON AND S. MALLICK
Optics, CHARLES S. WILLIAMS AND ORVILLE A. BECKLUND

OPTICS: A Short Course for Engineers & Scientists

CHARLES S. WILLIAMS
and
ORVILLE A. BECKLUND
Texas Instruments Incorporated

WILEY-INTERSCIENCE
a division of John Wiley & Sons, Inc.
New York London Sydney Toronto

Library of Congress Cataloging in Publication Data

Williams, Charles Sumner, 1911–

Optics: a short course for engineers and scientists.

(Wiley series in pure and applied optics)

Includes bibliographical references.

1. Optics. I. Becklund, Orville A., 1915– joint author. II. Title.

QC357.2.W55 535 73-39046

ISBN 0-471-94830-6

Printed in the United States of America

10 9 8 7 6 5 4

To Our Wives,
Dorothy and Alma

Preface

Several recent trends show the need for a particular kind of optics book. First, many college-trained industrial employees—trained in electrical engineering, mechanical engineering, chemistry, mathematics, or physics with scant optics background—find themselves working on projects that are primarily optics. To contribute to the development of optical devices, these people need a concise, understandable, and fundamental discussion of optics. Second, the application of Fourier transform theory, statistical analysis, complex number theory, and matrix algebra to optical system analysis and design, often programmed on electronic computers, has dramatically changed the mathematical background needed by people working in optics. Several subjects in this optics book—coherence theory, image evaluation, aberration theory, and interference spectroscopy—are best developed in terms of these mathematics rather than by references and analogies to communication theory and systems analysis. Third, technological trends—the advent of color television as an example—have changed the relative importance of certain optics-related topics; image recording processes, theory of color, and vision have become much more significant than they were a decade ago.

Though we could not fulfill all of the indicated needs in one short book, their satisfaction was our objective. We hope the style is tutorial, but considering our intended reader, an experienced technical man, we could not write at so elementary a level as to become nonmathematical. Although we have striven for conciseness, we hope that our developments are complete enough to be understood on first reading.

With a mature reader in mind, we have attempted to make each chapter reasonably independent so that most topics can be studied with only occasional reference to other material in the book. This may be particularly helpful to the phenomenologically inclined reader who might want to bypass the more lengthy mathematical discussions. For instance, if optical imaging systems are of particular interest, Chapters 6–9 cover most of what we say on this subject; Chapters 10 and 11 are on spectroscopy; Chapters 14 and 15 deal with vision and color. Other groupings will no doubt be suggested by the index.

Many topics could not be included, and others are treated so briefly as only to pique the reader's interest. But this is probably just what we wish to do, that is, to provide just enough background that the interested reader can thoroughly pursue each topic elsewhere in the references cited.

No book as broad in scope as the present one can be completed without the help of many people. Credits have been expressed wherever we could trace our ideas or ways of presentation back to their sources, but undoubtedly many more significant contributors go unrecognized because their ideas passed through too many minds before they reached ours.

The semiconductor material of Chapter 12, "Radiant Energy Detectors," is based on unpublished notes of George R. Pruett, who supplemented his excellent written material with many informal discussions. Sebastian R. Borrello also contributed generously to our understanding of semiconductor detectors in the lengthy conversations that we had with him.

For professional polish, we depended on our series editor, Dr. Stanley S. Ballard, who contributed much more to the final revisions of our material than his responsibilities required.

Our publisher was personified for us in the person of Miss Beatrice Shube, whose editorial efficiency and understanding made production from manuscript to finished book the easiest part of our task.

Perhaps the most sustained help that we have received has come from our associates at Texas Instruments, who have supported us with their encouragement and good wishes. We are especially grateful for the dedicated attention given our manuscript by two secretaries, Irene (Mrs. John F.) Robbins and Elaine (Mrs. Billy R.) Irish.

Dallas, Texas

CHARLES S. WILLIAMS
ORVILLE A. BECKLUND

Contents

OPTICS: A Short Course for Engineers & Scientists

1

The Nature of Radiant Energy

Waves and Photons

Radiant energy propagates through empty space independently of matter. Although this kind of energy can be a stimulus to the eyes and can elicit a response—that is, cause the observer "to see" his environment—radiant energy is unobservable in the sense that a person can see a water wave and watch its progress. We comprehend the nature of radiant energy only by putting together the results of experiments and measurements, describing assumed physical processes, and then carefully checking the harmony and consistency between the descriptions and the experimental facts.

In the study of radiant energy, we note the following: A number of simple and complex scalar and vector wave functions satisfy Maxwell's electromagnetic field equations. These wave functions appear to describe various physically real, propagating electromagnetic disturbances, and we can show mathematically that energy would accompany each disturbance. Radiant energy then could well be electromagnetic energy. The explanations of numerous experiments involving interference and diffraction phenomena also influence us to believe that radiant energy has a wave nature, the waves comprising periodic fluctuations of an electromagnetic field quantity. Furthermore, polarization experiments lead us to believe that there are two kinds of radiant energy, polarized and unpolarized, and the wave functions representing the former must, to explain the experiments, be vector wave functions. Other phenomena, such as the photoelectric effect, compel us to believe that light has a particle-like nature. Happily, quantum mechanics unifies these two seemingly incompatible natures, wave and particle.

Wave Nature of Radiant Energy

A plane wave. The wave nature of radiant energy is illustrated by using a simple example: the plane, polarized, electromagnetic wave. At any point in a plane determined by $y = y_1$ (Fig. 1.1), there exists an electric field in the plane that varies according to

$$\mathbf{E}_1 = \mathbf{E}_0 \sin (2\pi\nu t + \theta_1). \tag{1-1}$$

The quantity in the parenthesis is the phase; ν is the frequency, t is the time, and θ_1 is the initial phase (when $t = 0$) associated with y_1. In mechanics, the sine function occurs in the familiar expression for a traveling wave on a string; however, in the light wave, there is no vibratory motion of matter but simply the changing of the electric field represented by a vector that changes in magnitude and sign. (Field quantities are discussed in Chapter 4.) In a plane determined by another value of y, such as $y = y_2$, there will be an electric vector $\mathbf{E}_2$ parallel to $\mathbf{E}_1$. $\mathbf{E}_2$ is described by the same equation as $\mathbf{E}_1$ except that there will be a different initial phase θ_2. In general, the vector $\mathbf{E}$ will have a y-dependent phase term, $\theta(y)$, so that

$$\mathbf{E} = \mathbf{E}_0 \sin 2\pi[\nu t + \theta(y)/2\pi]. \tag{1-2}$$

One y-dependent phase term that satisfies wave theory is

$$\theta(y) = \theta_0 - 2\pi y/\lambda \tag{1-3}$$

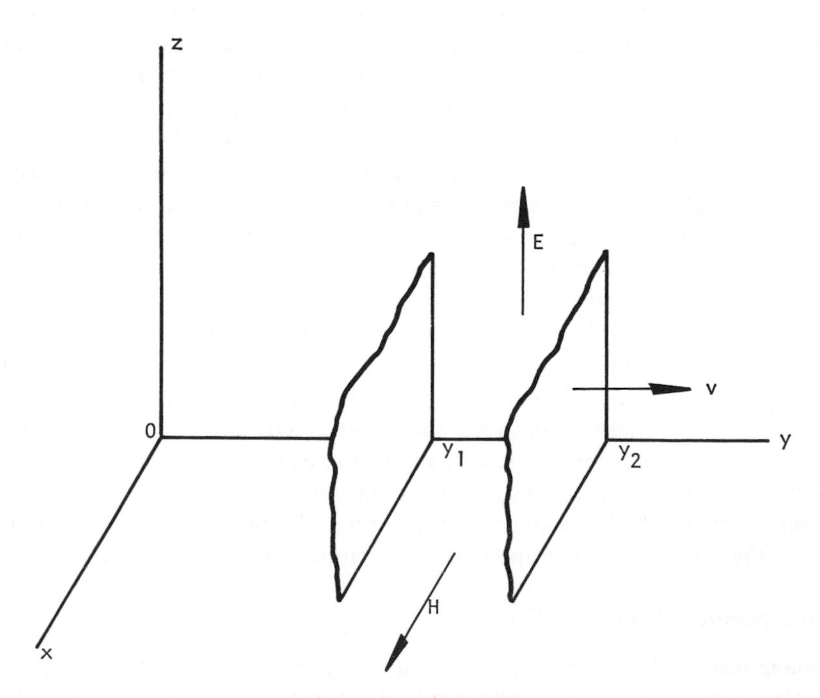

Figure 1.1 Plane wave.

in which θ_0 and λ are constants. When the phase function is written explicitly and the reciprocal of a constant called the period, $\mathscr{T}$, is substituted for the frequency, the expression for **E** becomes

$$\mathbf{E} = \mathbf{E}_0 \sin 2\pi\left[\frac{t}{\mathscr{T}} - \frac{y}{\lambda} + \frac{\theta_0}{2\pi}\right]. \tag{1-4}$$

If y is held constant, **E** is seen to go through a complete cycle of values when t increases by $\mathscr{T}$. Similarly, if t is held constant, **E** is seen to go through a complete cycle of values when y increases by λ, which is called the wavelength. When the algebraic value of **E** is plotted against y for successive values of t, one sees that the equations describe a traveling wave; furthermore, this wave goes a distance λ in time $\mathscr{T}$. Hence, the wave propagates at a speed

$$c = \lambda/\mathscr{T}. \tag{1-5}$$

Wavelengths of "visible" radiant energy range from about 0.4 μm to 0.7 μm corresponding to frequencies of 7.5×10^{14} to 4.29×10^{14} Hz. Infrared is the same kind of radiant energy, having wavelengths extending from 0.7 μm on out to several hundred micrometers. The wavelengths of ultraviolet "light" extend from 0.4 μm, or 4000 Å, all the way down to the neighborhood of 100 Å (Å indicates the angstrom unit, which is 10^{-10} m). For all electromagnetic waves, the speed c in free space is 2.997925×10^8 m/sec.

Associated with **E** will be a magnetic field vector **H**, which varies in the same manner as **E**,

$$\mathbf{H} = \mathbf{H}_0 \sin 2\pi\left[\frac{t}{\mathscr{T}} - \frac{y}{\lambda} + \frac{\theta_0}{2\pi}\right]. \tag{1-6}$$

In free space, the magnitudes of **E** and **H** are related by

$$|\mathbf{H}| = (\epsilon/\mu)^{1/2}\,|\mathbf{E}|, \tag{1-7}$$

where ϵ is the dielectric constant and μ is magnetic permeability. The values of these constants depend upon the system of units used [Ref. 1]. **E** and **H** are perpendicular to each other and to the direction of propagation, which in our coordinate system is the positive y direction. Because all points where **E** (or **H**) is in the same phase lie in a plane, the wave is said to be a plane wave. That is, the wave fronts are planes. From electromagnetic theory one learns that energy is traveling with the wave (in this example, in the y direction).

Elliptically polarized waves. Suppose that there are two ideal wave trains, each represented by Eq. (1-4), of equal amplitude, of the same frequency, and traveling in the same direction with the electric field of each parallel to the x–z plane. Let the electric field vector of one be oriented at an angle of $\pi/2$ (90 deg) with that of the other; therefore at a given point in space the electric field vectors will add instantaneously as two vectors with an included angle of $\pi/2$. If the two waves are precisely in phase, the superposition of the two gives a resultant polarized at an angle of $\pi/4$ to the individual waves and having an amplitude greater by the factor $\sqrt{2}$ than either of them. Thus, we visualize the superposition of two waves, but we have no way of determining experimentally that there are two waves present instead of the resultant, which can be measured. This kind of wave is said to be linearly polarized in the direction of the **E** vector.

Next, suppose that one of the wave trains is delayed in phase with respect to the other. Let the delay be $\pi/2$, a quarter wave. Then at any point in space the electric field vectors are timed so that when one is zero the other has its maximum amplitude. If we study the instantaneous sum of the electric field vectors, we find that the resultant has a constant amplitude but rotates with time. The rotation is in a plane parallel to the x–z plane, and the time interval required for one complete cycle of rotation is the period $\mathscr{T}$. There are two senses in which the vector can rotate depending upon which wave is delayed $\pi/2$ with respect to the other. If the rotation is clockwise when one looks toward the source of wave trains, the sense of rotation is said to be right-handed. The rotation is "seen" only by an observer in a fixed plane through which the wave trains are passing. An observer traveling along with the waves would see a resultant field vector of fixed amplitude and direction. The superposition of the two waves in this example gives a wave that is circularly polarized. Again, we cannot determine experimentally whether there are two waves or the one resultant.

When the two individual wave trains have different amplitudes or are neither precisely in phase nor out of phase by precisely $\pi/2$, the polarization of the resultant is neither linear nor circular but elliptical. In the simplest type of elliptically polarized radiant energy wave train, the electric field vector, as seen by an observer in a fixed plane, rotates and changes amplitude, completing both the cycle of changing amplitude and a cycle of rotation in the period $\mathscr{T}$. Linear and circular polarizations are special cases of elliptical polarization. Also, the process of "building" the elliptically polarized wave train from two linearly polarized wave trains is actually unnecessary in our concept of the nature of radiant energy; experimentally and conceptually an elliptically polarized wave train can be as basic and as ideal as a linearly polarized wave train.

Figure 1.2 illustrates graphically the several types of polarized wave trains. The quantity represented is the electric vector at a point in space.

If two wave trains, each represented by Eq. (1-4), were to exist at the same time, differing only in phase and amplitude, the two would combine to form a

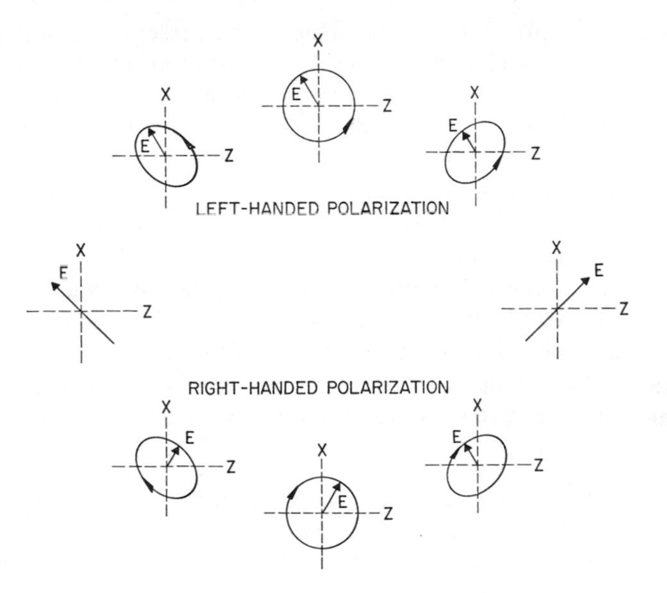

Figure 1.2 Representations of several types of polarized light traveling toward the observer.

single wave train with amplitude and phase different from the amplitudes and phases of the components. As already suggested for a particular example, we would, in general, be unable to distinguish experimentally between a single wave and a resultant made up of two component waves.

A radiant energy field of many waves. Let us suppose there is a myriad of plane wave trains present within the same region of space during a particular time interval. Whether these linearly polarized plane waves are treated as existing individually or as components of elliptically polarized waves does not affect any conclusions about the system.

A number of questions can be asked about the resulting radiant energy field. In how large a region of space and during what period of time will the field conditions be reasonably stable? What different frequencies are present among the several wave trains? What are the directions of the several electric vectors? What are the directions of propagation? Is the wavelength at a given frequency different in different directions? What are the phase differences among the wave trains at the same frequency? Will these quantities remain constant or vary in a "well behaved" manner throughout the region and time interval considered? What can be learned experimentally about these waves? What experiments can we perform? What terms need to be defined for describing the field or a radiant energy beam?

Two important principles apply. One is the principle of superposition, first stated clearly by Thomas Young in 1802, that at any point and at any instant the resultant field vector is the vector sum of the instantaneous field vectors that would have been produced by the individual wave trains had each been present alone.

The second principle is that no detector of radiant energy has a response time short enough to determine the instantaneous magnitude of the electric vector, even if a single, long wave train could be isolated. The quantity that can be detected is called radiant flux density, which is proportional to the square of the amplitude of the electric field vector and is also proportional to (and sometimes defined as) the time average of the energy that crosses a unit area perpendicular to the direction of energy flow in unit time. The radiant flux density of a beam consisting of several wave trains is the sum of the individual flux densities.

Isolating a single wave train. If a radiant energy field does exist that can be described as the total of many wave trains individually represented by Eq. (1-4), then there are experimental methods for isolating certain groups. For example, the following could be isolated:

(1) Those traveling close to a given direction (by using an optical collimator).
(2) Those, or their components, having electric vectors oriented near a given direction (by using a linear polarization analyzer).
(3) Those having their frequencies near a given frequency (by using a narrow bandpass optical filter or a monochromator).

Let us suppose that a succession of experiments has been performed using these three methods in turn, each operating on the group of wave trains isolated by the preceding method. Nearly perfect isolation by the three methods is assumed so that a "bundle" of a few aligned wave trains, of indefinite lengths, with all frequencies near a median frequency ν_0 results. An alternate and equivalent way to visualize this "beam" of radiant energy comes out of applying the principle of superposition to the "bundle." Figure 1.3 indicates how one assumed group of constant amplitude waves can add to a disturbance, or a sequence of disturbances, consisting of a wave pulse at the nominal frequency ν_0 that increases gradually from a minimum amplitude to a maximum and then decreases again to a minimum. For the component wave trains, we may think of improving our method of isolation so that the range of frequencies about ν_0 becomes narrower, the limit being a single, never-ending wave train of frequency ν_0. For the pulse equivalent of the wave trains, the corresponding process is the lengthening of the pulse, the change in amplitude becoming more and more gradual with the same never-ending, single-frequency wave train as the limit.

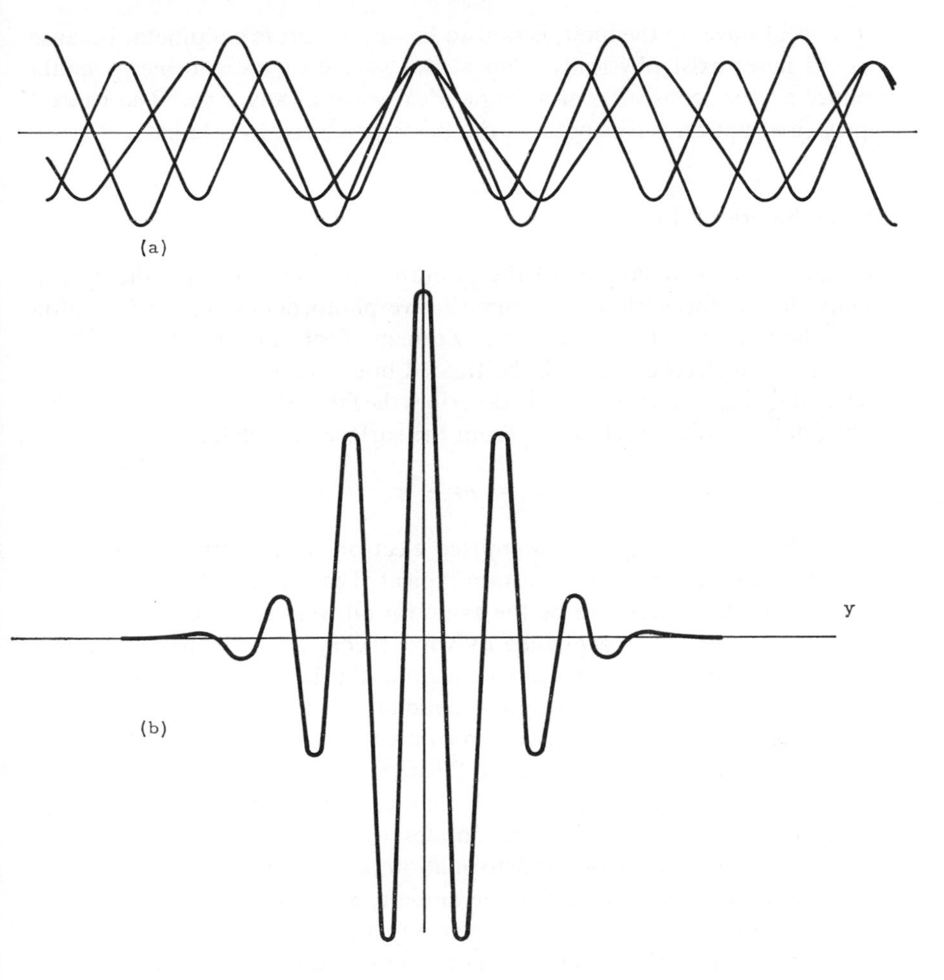

Figure 1.3 Two equivalent ways to represent a group of wave trains.

This ultimate ideal wave is said to be coherent both spatially and temporally. It is spatially coherent because we could choose any two points in the space traversed by the wave and determine the fixed difference in phase of the wave at these two points. It is temporally coherent because at a point in space there is a fixed change in phase corresponding to a given time interval. Temporal coherence is also sometimes called longitudinal coherence because there is a fixed phase for a given distance along (the direction of propagation of) the wave. The ideal wave is said to be monochromatic because there is only one frequency present. Also, conversely, a truly coherent beam of radiant energy is both polarized and monochromatic.

The ideal wave, in the limit, is said to be one of nature's ultimates because it could never exist physically. No actual source of radiant energy could produce a beam consisting of a single ideal wave of any type. The nearest approach at optical wavelengths is the beam produced by a laser.

Photon Nature of Light

Strong arguments in support of the photon theory are that this theory can explain the photoelectric effect, quantitative photochemistry, the Compton effect, the x-ray absorption edge, the Zeeman effect, and the Raman effect. Only the photoelectric effect will be treated briefly here.

The following empirical formula describes the facts observed experimentally in the photoemission of electrons from the surface of metals:

$$U_e = h\nu - \psi. \tag{1-8}$$

U_e is the kinetic energy of an emitted electron, ν is the frequency of the incident radiant energy, ψ is a constant typical of each particular metal, and h is a constant which seems to be the same for all frequencies and all metals. The facts that cannot be explained by wave theory are: (1) that the kinetic energy of an emitted electron is independent of the flux density of the incident radiant energy beam, and (2) that the radiant energy of frequencies lower than a threshold frequency produce no emitted electrons for any flux density.

To explain the photoelectric effect Einstein postulated that the energy in an electromagnetic wave is quantized, that is, it exists in discrete packets of energy called photons. Photons are manifest during absorption and emission of radiant energy, the energy transformation taking place in multiples of an amount of energy that is precisely the amount of energy ascribed to a single photon. During photoemission, the energy of one photon increases the energy of the absorbing electron. The energy packets are, in the present context of the photoelectric effect, indivisible; either the total energy of a photon is absorbed when the electron is emitted, or none of the energy is absorbed, in which instance the photon remains unchanged and the electron remains with its atom. Generally, two or more photons do not pool their energy; they must each act independently. The experimental results indicate that the energy of each photon in high frequency radiant energy is greater than the energy of each photon at lower frequencies. Thus, an electron may be emitted from the metal only if the energy of a single photon is sufficient to release the electron from its atom. A photon of radiant energy below a certain threshold frequency does not have enough energy to free the electron. Increasing the flux density of the beam increases the number of photons but does not supply more energy for each individual encounter between a photon and an electron, and therefore does not increase the energy of an emitted electron.

The amount of energy in each photon is, according to the Bohr equation, found to be

$$U_p = hc/\lambda = h\nu, \tag{1-9}$$

where h is a universal constant, the same as the onc that was used in the photoelectric equation; it is known as Planck's constant and has the value 6.62559×10^{-34} J sec. The energy of a photon corresponding to yellow-green light at 0.6 μm, for example, is 3.3×10^{-19} J. As suggested in the discussion above, the photon is an all-or-nothing entity. It has energy U_p and velocity c, in a vacuum, or it does not exist. Its momentum is said to be

$$h/\lambda = h\nu/c, \tag{1-10}$$

and from energy considerations it has an apparent mass of

$$h/\lambda c = h\nu/c^2. \tag{1-11}$$

It must be remembered, however, that a photon can have no rest mass because a photon at rest is nonexistent. In spite of the particle nature of photons, we cannot accept a simple corpuscular theory of light. Corpuscles in the form of mass particles cannot travel at the speed of light; and such particles, unlike photons, would have a rest mass.

The conceptual difficulties between the two natures of light are resolved by "quantum mechanics," but we are not prepared here to study its experimental and mathematical foundations. It is interesting to note, however, that quantum mechanics is also applied to an electron beam and that the beam can produce a diffraction pattern. This implies that the electron, often thought of as a particle, must have a wave associated with it. The wave of a particle in motion can be assigned, consistently with Eq. (1-10) above, the wavelength

$$\lambda = h/mv, \tag{1-12}$$

where mv is, of course, the momentum of the particle.

It is also necessary to conceive of photons, which do not collide, traveling along paths that are not necessarily straight lines. In fact if photons travelled in precisely straight lines, an interference pattern would not be formed. Instead, quantum mechanics predicts the motion of photons by the following hypothesis: First, assume the propagation of waves from a source of radiant energy as prescribed by the classical theory. Further, assume that these waves do not carry energy uniformly distribution over the wave front, but, instead, that the energy is concentrated in quanta $h\nu$. The waves serve only the mathematical function of describing the probability that photons are traveling through any given region of space. In an interference pattern, for instance, a study of the flux density distributions by wave theory leads to certain maxima and minima at definite positions in space, indicating that photons are traveling through these positions with a high or low probability, respectively, and so produce bright and dark fringes on a screen. Specifically, the hypothesis is made that the flux density, computed for any spot in the fringe pattern, is proportional to the probability of finding photons traveling through the spot.

The wave associated with a particle is not a long wave train but a short wave packet. The physical significance of this packet is that the flux density at any point in space represents the probability of finding the particle at that point. The theory yields the statistical distribution of the particles, and it denies the possibility of going further than this. So, for radiant energy, wave theory gives us the statistical or average distribution of photons as the flux density in the electromagnetic wave.

The General Beam of Radiant Energy

Since wave theory is emphasized more than photon theory in this book, a mathematical and conceptual representation of a light beam that is more complex than the ideal wave already discussed is needed.

A beam of radiant energy originating from one of the various possible sources may be one of a myriad of types. One's visualization of the beam depends upon the source, the experiments to be performed, and even upon the personal bias of the individual concerned. We like to picture the beam as many wave trains superposed to form the resulting beam. Enough wave trains are assumed present at any instant to account for the radiant energy being transported. Each wave train has a finite length and the wave-train length from wave to wave varies appreciably, although the average length in a given beam will be almost constant. With so many wave trains, the presence or absence of any particular one is a chance event. Since each has a finite length, it will include frequencies besides the nominal frequency. The polarization of any individual wave train may ordinarily take any form (linear, elliptical, or circular); experimentally no particular directional preference can be found for the electric field in a plane perpendicular to the direction of propagation. The beam as a whole is said to be unpolarized. At a point in space we assume that some property of the beam can be represented by a scalar quantity that satisfies the wave equation. Let $V^r(r,t)$ represent this scalar quantity, a function of position r and time t, which may be, for example, the component of the electric field vector parallel to the x-axis. (The superscript r notation anticipates later expressions where this becomes the real component of a complex number.)

The conglomerate nature of the field disturbance represented by $V^t(r,t)$, which for any actual beam will be a fluctuating quantity, should be emphasized. The conglomerate nature comes from the variety of possible frequencies, phases, polarizations, and directions that exist at the same time and that change abruptly with time. For instance, a wave train may suddenly and unpredictably be added to a beam of radiant energy. The presence of the new wave train alters the amplitudes, phases, and directions of the field quantities and the state of polarization. Electromagnetic wave disturbances discharged from atoms are wave trains of finite length. Since the event of emitting a wave train and the length of a wave train are chance variables, the superposition of the wave trains is describable only in statistical terms, and $V^t(r,t)$ may, therefore, be treated as a statistical random variable. The nature of the radiant energy beam makes $V^t(r,t)$ both stationary and ergodic. It is stationary because the statistical properties are independent of the particular choice of zero time. It is ergodic because the statistical properties obtained by taking the ensemble average—i.e., by sampling at a given time from a number of separate records of $V^t(r,t)$—are the same as found by taking the time average—i.e., sampling along one individual record [Ref. 2, pp. 5–6].

In practice, a light beam is neither the single ideal wave train nor the completely general case just described. Beams from most sources are somewhere in between these two extremes. The amplitudes of the wave trains present will usually have significant values only within some finite frequency range from ν_1 to ν_2. The bandwidth $\nu_2 - \nu_1 = \Delta\nu$ is frequently very small compared with ν,

$$\Delta\nu \ll \nu. \tag{1-13}$$

Nearly every beam tested has electric field vectors at least slightly stronger in one direction so that there is a degree of polarization, considering the beam as a whole; also, there is usually some coherence.

Analytic Signal Representation of the General Beam

Let $\xi(t)$ represent some scalar quantity (such as a component of the electric field vector) associated with the electromagnetic disturbance at a point in space. In its simplest form, $\xi(t)$ could be expressed by a cosine function, a contribution from a single wave train:

$$\xi(t) = \xi_0 \cos(2\pi\nu t - \varphi). \tag{1-14}$$

The total disturbance will be the sum of a large number of individual components, each represented by an expression like Eq. (1-14):

$$\xi_i(t) = \xi_{i0} \cos(2\pi\nu_i t - \varphi_i), \tag{1-15}$$

where $i = 1,2,3,4,5,\ldots$. The frequencies will be near a median frequency ν_0. Let the sum of all these components be $V^r(r,t)$:

$$V^r(r,t) = \left[\sum_{i=1} \xi_{i0} \cos(2\pi\nu_i t - \varphi_i)\right]_r. \tag{1-16}$$

In accordance with our comments about the finite length of a wave train, each contribution to the disturbance $\xi_i(t)$ will exist only for a finite time interval. Consequently, because of this finite time interval and other reasons that will be discussed in a later chapter, ξ_{i0} will have a finite and significant value within a continuous range of frequencies in the neighborhood of ν_i. If the nominal frequencies ν_i of the several wave trains contributing to the disturbance are reasonably near each other, ν will be a continuous function instead of occurring at only discrete frequencies. Then the function $V^r(r,t)$ can be represented as an integral instead of a sum:

$$V^r(r,t) = \left[\int_{\nu_1}^{\nu_2} \xi_0(\nu) \cos[2\pi\nu t - \varphi(\nu)]\, d\nu\right]_r. \tag{1-17}$$

The contributions are considered to be negligible outside the range of frequencies from ν_1 to ν_2. All of the quantities, $V^r(r,t)$, ν, $\xi_0(\nu)$, and $\varphi(\nu)$, are real quantities. Let us also consider the real function $V^i(r,t)$ obtained by subtracting $\pi/2$ from the phase of each component in $V^r(r,t)$:

$$V^i(r,t) = \left[\int_{\nu_1}^{\nu_2} \xi_0(\nu) \sin[2\pi\nu t - \varphi(\nu)]\, d\nu\right]_r. \tag{1-18}$$

We will now drop the notation that expresses explicitly the dependence of the function on position r; all the functions considered in the following discussion will be understood to be point functions of position in space.

A complex quantity, $\hat{V}(t)$, can be defined by using $V^r(t)$ and $V^i(t)$:

$$\begin{aligned}\hat{V}(t) &= V^r(t) + iV^i(t)\\ &= \int_{\nu_1}^{\nu_2} \xi_0(\nu) \exp i[2\pi\nu t - \varphi(\nu)]\, d\nu.\end{aligned} \tag{1-19}$$

Then

$$\hat{V}(t) = \int_{\nu_1}^{\nu_2} \hat{v}(\nu) \exp 2\pi i\nu t\, d\nu \tag{1-20}$$

is obtained by setting $\hat{v}(\nu) = \xi_0(\nu) \exp[-i\varphi(\nu)]$. Since $\xi_0(\nu)$ is considered to be zero outside the range ν_1 to ν_2, the range of integration can be extended over all frequencies:

$$\hat{V}(t) = \int_0^{+\infty} \hat{v}(\nu) \exp 2\pi i\nu t\, d\nu. \tag{1-21}$$

Let us assume that $V^r(t)$ is derivable from a function $\hat{v}_1(\nu)$ by the inverse Fourier transform:

$$V^r(t) = \int_{-\infty}^{+\infty} \hat{v}_1(\nu) \exp 2\pi i\nu t \, d\nu. \tag{1-22}$$

Then by the inversion theorem of Fourier integrals:

$$\hat{v}_1(\nu) = \int_{-\infty}^{+\infty} V^r(t) \exp(-2\pi i\nu t) \, dt. \tag{1-23}$$

It is known from Fourier integral theory that since $V^r(t)$ is real, $\hat{v}_1(-\nu)$ is the complex conjugate of $\hat{v}_1(\nu)$ [Ref. 3, p. 11].

$$\hat{v}_1^*(\nu) = \hat{v}_1(-\nu) \tag{1-24}$$

The asterisk denotes the complex conjugate. There is, therefore, no information contained in the negative frequencies that cannot be obtained in the positive frequencies; so the negative frequencies can be suppressed during computations. Dividing the integral in Eq. (1-22) into the sum of two integrals and reversing the limits in the first integral gives

$$V^r(t) = \int_0^{-\infty} \hat{v}_1(\nu) \exp 2\pi i\nu t(-d\nu) + \int_0^{+\infty} \hat{v}_1(\nu) \exp 2\pi i\nu t \, d\nu. \tag{1-25}$$

Then changing ν to $-\nu$ in the first integral, which, in general, does not affect the value of the integral, gives

$$V^r(t) = \int_0^{\infty} \hat{v}_1(-\nu) \exp(-2\pi i\nu t) \, d\nu + \int_0^{\infty} \hat{v}_1(\nu) \exp 2\pi i\nu t \, d\nu. \tag{1-26}$$

The first integral is now recognized as the complex conjugate of the second. The following is known from the theory of complex numbers:

$$Z + Z^* = 2 \,\mathcal{Re}\, Z$$

where $\mathcal{Re}$ signifies "the real part of." Hence the expression,

$$V^r(t) = 2\mathcal{Re}\int_0^{\infty} \hat{v}_1(\nu) \exp 2\pi i\nu t \, d\nu, \tag{1-27}$$

results from Eq. (1-26). The equality,

$$\int_0^{\infty} \xi_0(\nu) \cos[2\pi\nu t - \varphi(\nu)] \, d\nu = 2 \,\mathcal{Re}\int_0^{\infty} \hat{v}_1(\nu) \exp 2\pi i\nu t \, d\nu, \tag{1-28}$$

is satisfied provided

$$\hat{v}_1(\nu) = \tfrac{1}{2}\xi_0(\nu) \exp[-i\varphi(\nu)] = \tfrac{1}{2}\hat{v}(\nu). \tag{1-29}$$

Finally $\hat{V}(t)$ is found from Equations (1-21) and (1-29):

$$\hat{V}(t) = V^r(t) + iV^i(t) = 2\int_0^\infty \hat{v}_1(\nu) \exp 2\pi i \nu t \, d\nu. \tag{1-30}$$

The function $\hat{V}(t)$ is called the analytic signal associated with $V^r(t)$. As long as the operations on $\hat{V}(t)$ are linear, we can operate mathematically on this function instead of on $V^r(t)$, using the methods of complex numbers; the real part of the result will be the same answer that we would get by operating on $V^r(t)$. Also, the analytic signal avoids integrating over negative frequencies. There are often significant mathematical advantages in using the analytic signal to represent some quantity of the beam in radiant energy problems.

In summary, $V^r(t)$ adequately represents, with considerable generality, the properties of a general radiant energy beam. The convenient complex analytical signal $\hat{V}(t)$ includes $V^r(t)$ as its real part and is often used instead of $V^r(t)$ in mathematical operations. [See Ref. 4, p. 128 for further significance of $\hat{V}(t)$ as a mathematical function.] The absence of negative frequency components ensures that each component of $\hat{V}$, considered as a function of complex t, will be analytic and regular in the lower half of the complex t-plane. For some mathematical operations, it is significant that the two functions $V^r(t)$ and $V^i(t)$ form a Hilbert transform pair [Ref. 5, p. 15] which are related by the equations

$$V^r(t) = -(1/\pi)P_c\int_{-\infty}^{+\infty} [V^i(t')/(t' - t)] \, dt' \tag{1-31}$$

$$V^i(t) = (1/\pi)P_c\int_{-\infty}^{+\infty} [V^r(t')/(t' - t)] \, dt', \tag{1-32}$$

where P_c signifies that the Cauchy principal value is to be taken at $t' \to t$.

Energy Flow in a Radiant Energy Beam

The rate at which energy flows in an electromagnetic wave depends upon the velocity of propagation, hence upon the index of refraction, and in general upon the nature of the medium. Here we shall simply state that the flow of energy per second through a unit cross section area is proportional to the square of the electric field vector in a propagating electromagnetic field. The average flow is proportional to the time average, over a time $2T$ long compared with a mean period $\mathcal{T}$, of the field quantity $\xi^2(t)$ of Eq. (1-14). We shall ignore a constant of proportionality (i.e., assume it to be unity) which depends partially upon the characteristics of the medium and partially upon the specific quantity represented by ξ. Except for this constant, the radiant flux density W in the wave represented by Eq. (1-14) is

$$W = \lim_{T\to\infty} \frac{1}{2T}\int_{-T}^{T} \xi^2\, dt$$

$$= \lim_{T\to\infty} \frac{1}{2T}\int_{-T}^{T} \xi_0^{\,2} \cos^2 (2\pi\nu_0 t - \varphi)\, dt$$

$$= \lim_{T\to\infty} \frac{\xi_0^{\,2}}{2T}\int_{-T}^{T} [\tfrac{1}{2} - \tfrac{1}{2}\cos (4\pi\nu_0 t - 2\varphi)]\, dt. \tag{1-33}$$

$$W = \xi_0^{\,2}/2. \tag{1-34}$$

If the beam is represented by the analytic signal, $V^r(t)$ given by Eq. (1-17) represents the resultant field quantity produced by all its wave trains. The radiant flux density of the beam can be found by operating on the analytic signal; the assumptions concerning $V^r(t)$ determine the limitations on the results. The real quantity $V^r(t)$, as defined earlier, represents only one component of the beam electric field vector. A wave train linearly polarized perpendicular to this component, for instance, is not accounted for in our computation. However, it is assumed that the probability of a wave train having a given direction of linear polarization is the same for any other direction. Thus, whatever value or statistical property found for one component of the field vector would hold for any other. Also, the general assumption is made that whatever can be said statistically about the scalar quantity represented can also be said about any other scalar quantity of the same beam that can be represented by $V^r(t)$. So, when the radiant flux density is expressed in terms of the analytic signal, a quantity is found that is proportional, rather than equal, to the flux density. If certain additional assumptions are made, for example that the beam is linearly polarized in the direction represented by $V^r(t)$ or that the beam has a degree of coherence, then the quantity representing the flux density would more nearly approach the true value.

To derive the expression for the beam radiant flux density, we begin with the following expression, which defines a convenient notation in terms of an integral:

$$\hat{V}_1 * \hat{V}_2{}^* \equiv \int_{-\infty}^{+\infty} [\hat{V}_1(t - \theta)\hat{V}_2{}^*(t)]\, dt. \tag{1-35}$$

Here $\hat{V}_1$ and $\hat{V}_2$ are the analytic signals representing two radiant energy beams, and θ is an arbitrary time delay that is introduced in one of them. Expanding the complex quantities in terms of their real and imaginary components, one obtains

$$\begin{aligned}\hat{V}_1 * \hat{V}_2^* &= (V_1^{r} + iV_1^{i}) * (V_2^{r} - iV_2^{i}) \\ &= (V_1^{r} * V_2^{r}) + (V_1^{i} * V_2^{i}) \\ &\quad + i[(V_1^{i} * V_2^{r}) - (V_1^{r} * V_2^{i})].\end{aligned} \tag{1-36}$$

Either member of Eq. (1-35) is known as the convolution of $\hat{V}_1$ with $\hat{V}_2^*$. Similarly, expressions like

$$(V_1^{r} * V_2^{r}) \tag{1-37}$$

also are convolutions. (See Chapter 10 under the *Convolution of Instrument Function and True Spectrum* for an explanation, in one dimension, of a physical meaning of the convolution integral.) A property of Hilbert transforms can be used to advantage here since $V^{r}(t)$ and $V^{i}(t)$ turn out to be a Hilbert-transform pair. The convolution of two real functions is equal to the convolution of their Hilbert transforms in the same order [Ref. 5, p. 19]. So, using the transforms defined in Equations (1-31) and (1-32), we get

$$(V_1^{i} * V_2^{r}) = -(V_1^{r} * V_2^{i}), \tag{1-38}$$

$$(V_1^{r} * V_2^{r}) = (V_1^{i} * V_2^{i}). \tag{1-39}$$

From (1-36)

$$\hat{V}_1 * \hat{V}_2^* = 2(V_1^{r} * V_2^{r}) - 2i(V_1^{r} * V_2^{i}). \tag{1-40}$$

This two-beam expression can be reduced to a single-beam expression by setting $\hat{V}_1 = \hat{V}_2 = \hat{V}$ and letting $\theta = 0$. Then

$$\hat{V} * \hat{V}^* = 2(V^{r} * V^{r}) + 2i(V^{r} * V^{i}). \tag{1-41}$$

But from Eq. (1-38),

$$(V^{r} * V^{i}) = 0 \tag{1-42}$$

so that

$$\int_{-\infty}^{+\infty} [\hat{V}(t)\hat{V}^*(t)]\, dt = 2\int_{-\infty}^{+\infty} [V^{r}(t)]^2\, dt. \tag{1-43}$$

The product of a complex number and its complex conjugate gives

$$\int_{-\infty}^{+\infty} [\hat{V}(t)\hat{V}^*(t)]\, dt = \int_{-\infty}^{+\infty} |V(t)|^2\, dt, \tag{1-44}$$

where $V(t)$ is the modulus (sometimes called the "absolute value") of the complex number $\hat{V}(t)$. The integrand on the right of Eq. (1-43) is the square of a quantity representing the electric field; quantitatively, it is power and the integral must represent the total energy in the beam. Also one recognizes, from the theory of convolutions, that the right member of Eq. (1-44) represents total energy. Since the function represents a light beam, the integral is proportional to the total energy passing through an arbitrary unit area. This total energy cannot be finite unless $V^r(t)$ becomes vanishingly small as $|t|$ becomes large. However, without imposing this condition on $V^r(t)$, an expression for average power, the flux density, can be found by integrating from $-T$ to $+T$, dividing by $2T$, and then letting T approach infinity:

$$\begin{aligned} W &= 2 \lim_{T\to\infty} \frac{1}{2T} \int_{-T}^{T} [V^r(t)]^2 \, dt \\ &= \lim_{T\to\infty} \frac{1}{2T} \int_{-T}^{T} |V(t)|^2 \, dt \\ &= \lim_{T\to\infty} \frac{1}{2T} \int_{-T}^{T} [\hat{V}(t)\hat{V}^*(t)] \, dt. \end{aligned} \tag{1-45}$$

In practical evaluations, $|T|$ need be increased only until it is very large compared with a mean period $\mathscr{T}_m$ and with the coherence time (which will be discussed later).

Polarization

If there is no permanent ("permanent" over a relatively long Δt, the coherence time) phase relation between the various components of a beam of radiant energy, and if the orientation of the electric field vectors with respect to each other is random following no particular pattern, the beam is said to be unpolarized. An unpolarized beam can be polarized by any of several procedures. (The analytic signal can still be used to represent a beam of polarized radiant energy as well as the more general unpolarized beam.)

There are linear sheet polarizers which transmit the electric field components oriented in a particular direction much better than the orthogonal components of the same wave trains. The transmitted beam is said to be linearly polarized although the actual polarization is never complete. An unpolarized beam can be linearly polarized also by reflection from or transmission through certain simple dielectric materials such as glass. This phenomenon is discussed further in Chapter 4.

When the material is anisotropic because of its crystalline structure or its stressed condition, the speed of light depends upon the orientation of the electric vector with respect to the transmitting material. Such materials are said to be doubly refracting or birefringent. A birefringent material is often made up into what is called a wave plate and is described more specifically by giving a wave fraction (for example, a quarter wave plate or a half wave plate). Such a wave plate can be used to separate the electric field vector of a linearly polarized beam into two orthogonal components. The difference in velocity causes one component to be delayed with respect to the other, and the superposition of the two components results in an elliptically polarized beam.

Generally the radiant energy emitted by most sources can be shown experimentally to have at least a small amount of polarization. There are several procedures (the Poincaré Sphere, Stokes Vectors, Jones Vectors, and quantum mechanics wave functions) for quantitatively specifying the degree and state of polarization of a radiant energy beam. Though these will not be discussed further here, they may be reviewed in Refs. 6–13.

Coherence

A formal definition of coherence is given in Chapter 11. Here "degree of incoherence" is explained in such a way that coherence and incoherence have meaning as limiting cases. Partial coherence, or degree of coherence, is closely associated with the "visibility" of interference fringes; this association is discussed in Chapter 11.

Radiant energy that reaches a given point in space, where observations are being made, may originate from separated, discrete points on a source, from points on separate sources, or from a common point but over different paths. Let us think of just two beams superposed in the region near the point of observation. Degree of coherence depends upon how much the phase in the second beam depends upon the phase in the first. If there is no dependence at all, the beams are incoherent. But the independence, or dependence, must continue in time. At each instant, for both coherent and incoherent beams, there is a precise phase difference. Complete phase independence implies that the phase difference at a time t_2 is completely independent of the phase difference that existed at an earlier time t_1.

A composite wave, formed by the superposition of the two beams, exists at each instant in the region near the point of observation. Incoherence implies that the composite wave existing at a time t_2 is independent of the composite wave that existed at an earlier time t_1. Degree of coherence depends upon the time interval $\Delta t = t_2 - t_1$ required for the composite waves to be independent. If the interval is long compared with the time constant of the eye and other detectors that might be involved in the observation, there is a high degree of coherence and we are able to "see" the flux density pattern produced by one of the persisting composite waves. On the other hand, near incoherence implies that the composite wave varies within such a short time interval that a picture of any one of the many different density patterns appearing successively cannot be seen. Only an average pattern can be observed, an average flux density at each point taken over a time interval at least as long as the time constants of the detectors.

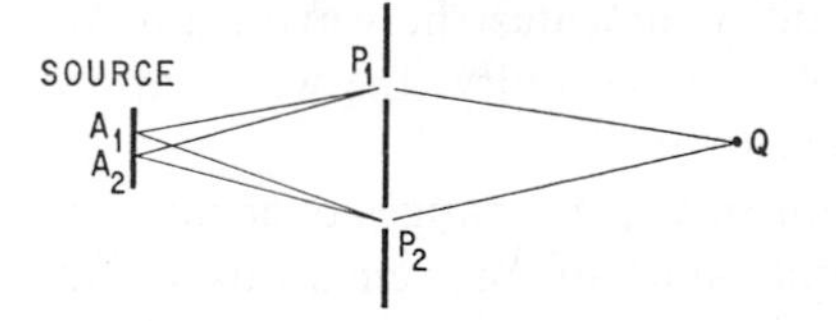

Figure 1.4 Schematic of an arrangement for producing fringes.

If a beam of radiant energy is divided by suitable apparatus into two beams which are then superposed after one has been delayed with respect to the other, an interference pattern may be produced in the region of superposition. Within an interference pattern the radiant flux density varies from point to point in the space from the maximum W_a, which can exceed the sum of the flux densities of the individual beams, to a minimum W_i, which can be as small as zero. The bright and dark regions are called interference fringes. The fringe visibility $\mathscr{V}$ is defined quantitatively by the following equation, in which W_a and W_i are determined in a bright fringe and in an adjoining dark fringe:

$$\mathscr{V} = (W_a - W_i)/(W_a + W_i). \tag{1-46}$$

A possible experimental setup is shown schematically in Fig. 1.4. A suitable optical system is used to direct a portion of the radiant energy from the element of area Δa_1 in a source at A_1 through the pinhole P_1 and another portion from the same element of area through another pinhole P_2. A second optical system collects the radiant energy from both P_1 and P_2 to illuminate an element of area $\Delta a'$ at Q, where the two beams superpose. The phase difference at Q is

$$\Delta\phi_1 = (2\pi\delta_1/\lambda) \tag{1-47}$$

where $\delta_1 = A_1P_1Q - A_1P_2Q$. Similarly for another element of area Δa_2 at A_2, the phase difference would be

$$\Delta\phi_2 = (2\pi\delta_2/\lambda) \tag{1-48}$$

where $\delta_2 = A_2P_1Q - A_2P_2Q$.

When the source is small, the phase differences $\Delta\phi_i$ for the elements of area Δa_i ($i = 1,2,3, \ldots$) are nearly equal, and a mean $\Delta\phi_m$ can be considered to be the phase difference. When we speak of the coherence length $\Delta s = c\,\Delta t$ for the beam, the coherence time Δt is the time required for an individual wave train of average length to pass through the pinhole. If the total phase change from one end to the other of an individual wave train is ϕ_m, the visibility of fringes in the neighborhood of Q will be high, approaching unity, if $\phi_m \gg \Delta\phi_m$ (where $\Delta\phi_m$ is the phase difference caused by the difference in path length) and if the flux densities in the two beams are nearly equal.

Conversely, a high visibility indicates the wave trains to be relatively long; in the limit of completely unit visibility, the waves approach the ideal wave discussed earlier in the chapter.

When the source is not small, $\Delta\phi_i$ ranges over many values. If the differences between the several values of $\Delta\phi_i$ remain fixed for a time equal to the coherence time, interference fringes will still be produced at Q and the visibility may still be high. If, however, the differences between the several $\Delta\phi_i$ are varying at random so that there is no correlation of the phase difference of beams from A_i with the phase difference of beams from A_j, fringes cannot be produced and the visibility is zero.

The degree of coherence of a source, or the beam produced by a source, is defined as the visibility of the fringes when the fringes are obtained under the most favorable conditions of small path difference and equal flux densities. Thus, incoherence and coherence are the two limiting extremes of partial coherence.

An effect of spatial coherence can be demonstrated by using the set-up shown schematically in Fig. 1.4 in which the source is assumed to be an incoherent, distributed source. While the separation between the two pinholes is varied, with the other parameters fixed, the fringe visibility is measured at several separations. As the separation increases, visibility of fringes gradually decreases until a separation is reached for which no fringes are formed. Separating the pinholes still further results in a reappearance of fringes of low visibility. Still further separation results in an increasing, then a decreasing visibility. A typical plot of visibility as a function of separation, for a circular source, is shown in Fig. 1.5. A similar plot can be made

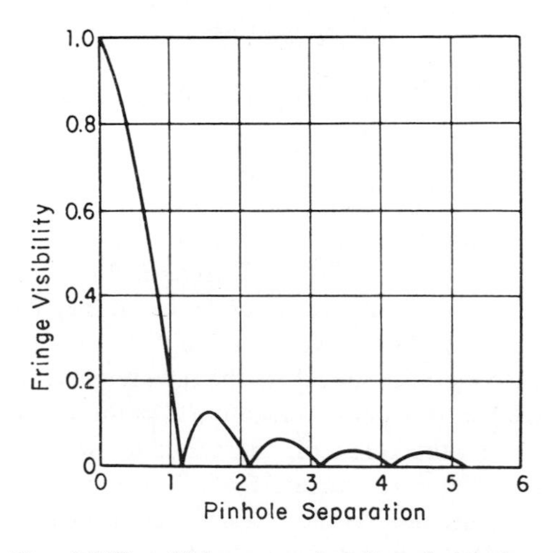

Figure 1.5 The visibility of fringes as a function of pinhole separation [Ref. 5].

while varying one of the other parameters. For example, an almost identical plot applies for the visibility of fringes as a function of source diameter.

Our discussion of coherence has necessarily been simplified. Coherence phenomena—one of the simplest being the interference effects with which optical coherence has long been associated in our experience—are manifestations of correlations between the fluctuations in the field quantities represented by the real part $V^{r}(t)$ of the analytic signal. Coherence theory, in general, is concerned with statistical descriptions of beams and their fluctuations. We discuss this topic again in Chapter 11. (For more complete and rigorous treatments of coherence, the reader is referred to Refs. 5 and 14–20.)

REFERENCES

1. SUN (Commission for Symbols, Units and Nomenclature of the International Union of Pure and Applied Physics), "Symbols, Units and Nomenclature in Physics," *Phys. Today*, **15**, 20 (June, 1962).
2. J. S. Bendat, *Principles and Applications of Random Noise Theory*. John Wiley & Sons, Inc., New York, 1958.
3. A. Papoulis, *The Fourier Integral and Its Applications*. McGraw-Hill Book Company, Inc., New York, 1962.
4. E. C. Titchmarsh, *Introduction to the Theory of Fourier Integrals*. Second Edition. Clarendon Press, Oxford, 1948.
5. M. J. Beran and G. B. Parrent, Jr., *Theory of Partial Coherence*. Prentice-Hall, Inc., Englewood Cliffs, N.J., 1964.
6. Wm. A. Shurcliff, *Polarized Light*. Harvard University Press, Cambridge, Mass., 1962.
   Wm. A. Shurcliff and S. S. Ballard, *Polarized Light*. D. Van Nostrand Co., Inc., Princeton, N.J., 1964.
7. G. N. Ramachandran and S. Ramaseshan, "Crystal Optics," in *Handbuch der Physik*, Vol. XXV/1. Springer-Verlag, Berlin, 1961.
8. M. J. Walker, "Matrix Calculus and the Stokes Parameters of Polarized Radiation," *Am. J. Phys.*, **22**, 170 (1954).
9. W. H. McMaster, "Polarization and the Stokes Parameters," *Am. J. Phys.*, **22**, 351 (1954).
10. H. G. Jerrard, "Transmission of Light through Birefringent and Optically Active Media: The Poincaré Sphere," *J. Opt. Soc. Am.*, **44**, 634 (1954).
11. R. Clark Jones, "New Calculus for the Treatment of Optical Systems: I. Description and Discussion of the Calculus," *J. Opt. Soc. Am.*, **31**, 488 (1941).
12. Wm. H. McMaster, "Matrix Representation of Polarization," *Rev. Mod. Phys.*, **33**, 8 (1961).
13. J. M. Jauch and F. Rohrlich, *Theory of Photons and Electrons*. Addison-Wesley Publishing Company, Inc., Reading, Mass., 1955.
14. L. Mandel and E. Wolf, "Coherence Properties of Optical Fields," *Rev. Mod. Phys.*, **37**, 231 (1965).
15. M. Born and E. Wolf, *Principles of Optics*, Chapter X. Pergamon Press, Oxford, 1965.
16. R. J. Glauber, "Coherent and Incoherent States of the Radiation Field," *Phys. Rev.*, **131**, 2766 (1963).

17. B. Karczewski, "Coherence Theory of the Electromagnetic Field," *Nuovo Cimento*, **30**, 906 (1963).
18. Roy J. Glauber, "The Quantum Theory of Optical Coherence," *Phys. Rev.*, **130**, 2529 (1963).
19. L. Mandel and E. Wolf, "Some Properties of Coherent Light," *J. Opt. Soc. Am.*, **51**, 815 (1961).
20. M. Francon, *Optical Interferometry*, Chapters I, VIII, and IX. Academic Press, New York, 1966.

2

The Measurement of Radiant Energy

As we undertake the study of radiometry and then of photometry, we assume the existence of an instrument called a radiometer. If the radiometer is calibrated in photometric units, it is called a photometer. The radiant energy is assumed to be incident upon a surface of known area associated with the radiometer. Also known are the area and orientation of the surface relative to the direction from which radiant energy is received. This energy will produce a response, for example a voltage produced by a radiant energy detector in the instrument, that is quantitatively related to the amount of radiant energy incident upon the area. Radiometers are discussed briefly in Chapter 7, and various detectors and their response characteristics are discussed in Chapter 12. In the present chapter we concentrate on the measurement of radiant energy from a source and extend this fundamental measurement to the following: the rate at which energy is radiated (power), the power either emitted from or incident upon a unit area, the power radiated within a small solid angle about a given direction, and the variation of these quantities as the direction and position on the surface vary.

We will confine our study to incoherent radiant energy beams. Careful consideration must be given to localized interference effects whenever the measurement concepts of this chapter are extended to coherent beams [Ref. 1].

About Radiant Energy Quantities and Units

Because the concepts of certain physical quantities pertaining to optics are quite different from the corresponding physical quantities encountered in other sciences, we will take particular pains to develop definitions of physical quantities and definitions pertaining to the behavior of radiant energy; these definitions are fundamental to all other optical topics.

A discussion of any system of units, symbols, and nomenclature usually evokes an argument. Not only do individual scientists and engineers prefer certain systems over others, but a reader discovers that several learned societies and their journals differ in their preferences. Moreover, a good argument can usually be made for each of several mutually discordant units and terms that might be included within a system. In this book, we have chosen units, symbols, and terms that are recommended for the most part by the Optical Society of America and by several other scientific societies.

Loose terminology seems to abound in optics. Perhaps the most difficult task for most of us will be to break the habit of using certain familiar terms and units that are vaguely defined or that are no longer recommended [Refs. 2–9] but which are still commonly found in the new as well as the old literature. For instance, one should try to use words ending in "tion" to mean processes rather than quantities. Thus we can speak of *radiation of radiant energy* but not of *radiation of radiation*! A certain amount of risk is taken in our dedication to backing the new just because it is logical; the time-honored "micron," for instance, may never give way to the more explicit "micrometer!"

We build upon the terms defined in the last chapter: radiant energy, frequency ν, wavelength λ, and period $\mathscr{T}$. To continue, a closely related term, wavenumber, is defined as

$$\sigma = 1/\lambda = \nu/c. \tag{2-1}$$

By this definition, wavenumber simply tells the number of wavelengths or cycles that occur in a unit of length, measured in the direction of propagation. The velocity of propagation will differ in different media, but frequency is an invariant with the change in velocity. So, if

$$\nu = c\sigma = c/\lambda, \tag{2-2}$$

TABLE I FUNDAMENTAL WAVE QUANTITIES AND UNITS

Quantity		Unit	
Name	Symbol	Symbol	Name
Frequency	ν	Hz	Hertz (cycle per second)
Wavelength	λ	μm or nm	Micrometer or nanometer
Period	$\mathscr{T}$	sec	Second
Wavenumber	σ	cm^{-1}	Reciprocal centimeter

both wavenumber and wavelength must vary from medium to medium because of the velocity change. When values of wavenumber or wavelength are given without reference to the propagating medium, a perfect vacuum (sometimes referred to as "free space" or "vacuo") is assumed; and the symbols are modified to σ_0 and λ_0. In most gases, the vacuum values are only a few hundredths of a percent in error (e.g., $c_0 = 1.00029\ c_{\text{air}}$).

Table I summarizes the quantities that have been discussed thus far, which we will call the fundamental wave quantities.

Radiometry

Radiometric quantities are physical quantities; the definitions are quite straightforward, being expressed in terms of energy and geometrical units [Ref. 10]. They are listed systematically in Table II.

Starting with energy U as the basic quantity, we choose the erg or the joule (10^7 ergs) as the unit. Energy density u, then, can be expressed for example as joules per centimeter^3. Although energy or energy density can be associated with a storage mechanism like the accumulation of energy in the image on a photographic film, a more general concern in optics is with radiant energy being transferred, which implies a rate, that is, radiant power Φ_e. The usual unit is the watt, a joule per second. (The subscript will be explained later in the chapter.) Radiant power generally is distributed in several ways, with wavelength, position, direction, time, and polarization. Also, several of these distributions may vary simultaneously.

Radiant Flux Density

Absorption and emission suggest the interaction of energy with surfaces. If one thinks of the flow of energy toward or away from a surface as flux, a new term, radiant flux density, can be defined as the radiant power Φ_e incident upon or leaving a surface, divided by the area s of that surface. Then, when the flux is uniform, the radiant flux density W is

$$W = \Phi_e/s. \tag{2-3}$$

However, if the flux is not uniform, the differential form should be used:

$$W = \partial\Phi_e/\partial s. \tag{2-4}$$

Irradiance and Radiant Exitance

Flux density can be more specifically defined as irradiance (sometimes called radiant incidance) E_e if the energy flows toward the surface, or as radiant exitance (formerly called radiant emittance) M_e if the energy flows away

TABLE II RADIOMETRIC UNITS: PREFERRED SYMBOLS, NAMES, AND UNITS

Symbol and Defining Equation	Name	Description	Unit
U	Radiant energy		joule
$u = dU/dv$	Radiant energy density		joule cm^{-3}
$\Phi_e = dU/dt$	Radiant power (flux)	Rate of transfer of radiant energy	watt
$W = \partial\Phi/\partial s$	Radiant flux density	Radiant power incident upon or leaving a surface, divided by the area of the surface	watt cm^{-2}
$M_e = \partial\Phi/\partial s$	Radiant exitance	Radiant flux leaving an infinitismal element of surface divided by the area of that surface	watt cm^{-2}
$E_e = \partial\Phi/\partial s$	Irradiance	Radiant power per unit area incident upon a surface	watt cm^{-2}
$I_e = \partial\Phi/\partial\omega$	Radiant intensity	Radiant power leaving a point source per unit solid angle	watt sr^{-1}
$L_e = \partial^2\Phi/(\partial s\, \partial\omega \cos\theta)$	Radiance	Radiant power leaving or arriving at a surface, at a given point in a given direction, per unit solid angle and per unit of surface projected orthogonal to that direction	watt sr^{-1} cm^{-2}
ϵ	Radiant emissivity	Ratio of emitted radiant power from a surface to the radiant power from a blackbody surface at the same temperature	
α	Radiant absorptance	Ratio of absorbed radiant power to incident radiant power	
ρ	Radiant reflectance	Ratio of reflected radiant power to incident radiant power	
τ	Radiant transmittance	Ratio of transmitted radiant power to incident radiant power	

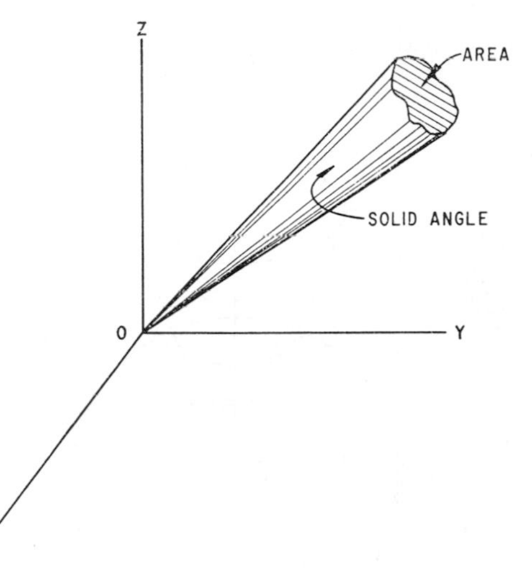

Figure 2.1 Solid angle.

from the surface. In the latter situation, the radiant flux (energy flow) may be emitted by (that is, generated essentially at) the surface, reflected at the surface, or may leave the surface having been transmitted through the material before arriving at the surface. Thus, in the same order, the terms emitted exitance, reflected exitance, or transmitted exitance apply.

Solid Angle

As radiant energy leaves an infinitesimal area on a surface, it may take any radial direction in an imaginary hemisphere centered on the area. To describe in numbers or equations what is going on, we need the geometric concept of the solid angle.

In many optics developments, it is convenient to use a solid angle subtended by a surface that is viewed from a point called the vertex (0 in Fig. 2.1). The solid angle is defined as the cone generated by a line that passes through the vertex and a point that is moved along the periphery of the surface. The size of the angle in steradians is equal to the area intercepted by the cone on an imaginary unit sphere that has its center at the vertex. From this definition, it is evident that the total angle about a point in space is 4π steradians. If we return now to the surface that subtends the solid angle, the size of the angle in steradians can be related to the area of the surface. Suppose that an element ds of the surface is at a distance r from the vertex and that its normal makes an angle θ with a line between the surface and the vertex as shown in Fig. 2.2. Then, starting with the definitions and geometry given above, we

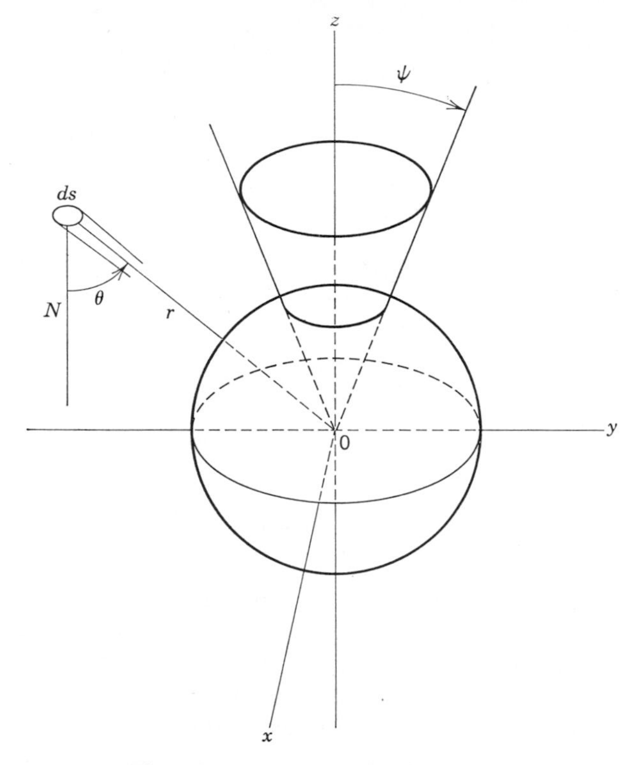

Figure 2.2 Geometry of solid angles.

can show that the differential $d\omega$ of solid angle is related to ds as follows:

$$d\omega = \frac{ds \cos \theta}{r^2}. \tag{2-5}$$

Expressing the total solid angle ω in terms of the surface s now becomes an integration problem, which requires a specific description of the surface s. Because practical difficulties often occur when we try to express the description mathematically and when we attempt the integration, certain approximations are frequently used to simplify (sometimes to make possible!) a solution.

A table of solid angles is available from the Office of Technical Services [Ref. 11]. We will consider here a number of approximations for solid angles defined by right circular cones (which may, in turn, be approximations of the true solid angles).

If ψ is the cone half-angle of a right circular cone (Fig. 2.2), the solid angle at the apex (measured, according to the definition above, by the area of the cap cut out of the unit sphere) is

$$\omega = 2\pi(1 - \cos \psi) = 4\pi \sin^2 (\psi/2). \tag{2-6}$$

A cone half-angle approximation for a unit solid angle ($\omega = 1$ sr) is sometimes useful [Ref. 12]. Solving for ψ in the above expression for the solid angle, one obtains

$$\psi = \arc\cos\,(1 - \omega/2\pi). \tag{2-7}$$

The cone half-angle corresponding to a unit solid angle is then

$$\begin{aligned}\psi_1 &= \arc\cos\,[1 - (1/2\pi)] \\ &= 0.571988 \text{ rad} = 32 \text{ deg, } 46 \text{ min, } 14 \text{ sec} \\ &= [(\pi/2) - 0.998808] \text{ rad} = 90 \text{ deg} - 57 \text{ deg, } 13 \text{ min, } 46 \text{ sec,} \\ &\approx [(\pi/2) - 1] \text{ rad,}\end{aligned} \tag{2-8}$$

with an error of approximately 0.2%. If the half cone angle is set exactly equal to $[(\pi/2) - 1]$, the corresponding solid angle is one steradian with an error of approximately 0.4%.

For small angles, a simple relationship can be used between the solid angle and the cone half-angle with good accuracy. A common trigonometric approximation for a small angle is

$$\cos\psi \approx 1 - \psi^2/2. \tag{2-9}$$

When this value is substituted in the expression for the solid angle,

$$\omega \approx \pi\psi^2. \tag{2-10}$$

The approximation is within 1% for conc half-angles less than about 20 deg; it is within 0.1% for cone half-angles less than about 6 deg.

Radiant Intensity

Radiant intensity I_e in watts per steradian is the radiant flux leaving a "point source" per unit solid angle:

$$I_e = \partial\Phi_e/\partial\omega, \tag{2-11}$$

where $\partial\omega$ is the element of solid angle through which the radiant energy passes. "Point source" here is a source of dimensions negligibly small compared with the distance to the point where radiant intensity is determined.

Radiance

Radiance L_e is the radiant flux leaving from or arriving at a given point on a surface, per unit solid angle and per unit projected area of surface:

$$L_e = \partial^2\Phi_e/(\partial\omega\,\partial s\cos\theta), \tag{2-12}$$

where θ is the angle between the normal to the surface and the given direction determined by the element of solid angle.

Although the definition of radiance involves a surface, the term can be extended to a small region of space by constructing an imaginary surface in the region so that the power is incident on one face and leaves from the other face.

Unfortunately, another quantity, which omits projecting the surface, is also frequently called radiance. So when the reader encounters this term, he must examine the context to determine which concept is to be understood.

As the differential notation in the radiance definition suggests, we must be able to assume that the dimensions of the radiating area are negligibly small when compared with the distance to the point at which we are making observations. Otherwise, the area, the direction, or the solid angle will be ambiguous.

Generally, the radiance varies with the orientation of $\partial\omega$ about ∂s; so the direction has to be specified. However, if the radiance is the same in all directions from the surface, the surface is said to be a diffuse, or a Lambertian, radiator. The radiant intensity at a point so distant from a diffuse radiator that the radiator may be considered a point source follows the Lambert cosine law,

$$I_e(\theta) = I_e(0) \cos \theta. \tag{2-13}$$

In geometrical optics, radiance is invariant, with certain restrictions, along any ray in a single medium [Ref. 13]. When the ray passes from one medium to another, the radiance must be modified by the change in index of refraction by

$$L_{e2} = (n_2/n_1)^2 L_{e1}, \tag{2-14}$$

where n is the index of refraction and the subscript 1 denotes the first medium and 2 the second. This principle is useful in evaluating the radiance of object and image in an optical system. However, the following restrictions have to be observed: We must not consider truly point sources or truly parallel beams, and the principle does not take into account attenuation of the beam intensity due to absorption and scattering.

Wavelength-Dependent Quantities

The quantities W, E, M, L, and I, which have already been defined in this chapter, are generally functions of wavelength; thus it is convenient to define derivative quantities called spectral radiant flux density, spectral irradiance, spectral radiant exitance, spectral radiance, and spectral radiant intensity, respectively. For the sake of simplicity we shall temporarily omit the subscript *e*. The subscript λ is added to each symbol to denote the spectral quantity. For instance, L_λ is spectral radiance typically in watts sr^{-1} cm^{-2} cm^{-1}. The differential of radiance is

$$L_\lambda \, d\lambda,$$

and the total radiance between wavelengths λ_1 and λ_2 is

$$L = \int_{\lambda_1}^{\lambda_2} L_\lambda \, d\lambda, \tag{2-15}$$

in watts/cm^2 sr. The spectral quantities may alternatively be expressed as functions of wavenumber or frequency, instead of wavelength. Various spectral quantities are summarized, with their units, in Table III.

Emissivity, Reflectance, Absorptance, and Transmittance

Emission, reflection, absorption, and transmission of radiant energy are often treated as surface phenomena. This is not strictly true because, as we shall see later, all of these phenomena must take place within matter; so there must be some penetration of the radiant energy to the medium under the surface. Even reflection at the surface of a metal involves penetration, though extremely shallow, to a finite depth into the metal. Emission, the transformation of molecular or atomic energy into radiant energy to cause radiation, must also take place within the medium below the surface. Nevertheless, for simplicity we shall, for the present, treat these phenomena as essentially taking place on a surface.

A true blackbody, by definition, absorbs all incident radiant energy. In actual surfaces, however, the absorption is never complete [Ref. 14]. The ratio of the absorbed energy to the incident energy is called the absorptance.

Let us assume that an objective having an absorptance $\alpha(\lambda)$ less than unity (therefore, not a blackbody) is placed within an ideal blackbody cavity. According to the principles of thermodynamics, the object will reach the temperature T of the cavity and remain at this temperature. At this equilibrium condition, the spectral irradiance at the object surface is equal to the spectral emitted radiant exitance $M_\lambda(T)$ of a blackbody surface. Blackbodies are discussed in the next chapter, in which we explain that $M_\lambda(T)$ for a blackbody is given by the well-known Planck formula. The power absorbed by the object per unit wavelength per unit area of the object surface is $\alpha(\lambda)\, M_\lambda(T)$; the remainder must either be transmitted through the object or be reflected. If the object is assumed opaque (no transmission), the amount reflected, the spectral reflected radiant exitance, is

$$[1 - \alpha(\lambda)]M_\lambda(T). \tag{2-16}$$

TABLE III RADIOMETRIC UNITS OF SPECTRAL QUANTITIES

Symbol and Defining Equation	Name	Description	Unit
$\Phi_\lambda = \partial\Phi/\partial\lambda$	Spectral radiant power	Radiant power per unit wavelength	watt cm^{-1} or watt μm^{-1}
$\Phi_\nu = \partial\Phi/\partial\nu$	Spectral radiant power	Radiant power per unit frequency interval	watt Hz^{-1} or watt sec
$\Phi_\sigma = \partial\Phi/\partial\sigma$	Spectral radiant power	Radiant power per unit wavenumber interval	watt cm
$W_\lambda = \partial W/\partial\lambda$	Spectral radiant power density	Radiant power density per unit wavelength interval	watt cm^{-2} cm^{-1}
$M_\lambda = \partial M/\partial\lambda$	Spectral radiant exitance	Radiant exitance per unit wavelength interval	watt cm^{-2} cm^{-1}
$E_\lambda = \partial E/\partial\lambda$	Spectral irradiance	Irradiance per unit wavelength interval	watt cm^{-2} cm^{-1}
$I_\lambda = \partial I/\partial\lambda$	Spectral radiant intensity	Radiant intensity per unit wavelength interval	watt sr^{-1} cm^{-1}
$L_\lambda = \partial L/\partial\lambda$	Spectral radiance	Radiance per unit wavelength interval	watt sr^{-1} cm^{-2} cm^{-1}

Because the object is in equilibrium with its surroundings, it must emit as much energy as it absorbs, $\alpha(\lambda)\, M_{\lambda}(T)$. This is the spectral emitted radiant exitance of this surface.

Emissivity $\epsilon(\lambda)$ is defined as the ratio of the energy emitted by the surface to the energy emitted by an equal area of a blackbody surface at the same temperature. Under the conditions assumed above, $\epsilon(\lambda)$ equals $\alpha(\lambda)$, that is, the emissivity equals the absorptance. This is known as Kirchhoff's law.

At the surface of an object where radiant energy at wavelength λ is incident upon the surface, a fraction $\alpha(\lambda)$ is absorbed, a fraction $\rho(\lambda)$ is reflected, and a fraction $\tau(\lambda)$ is transmitted [Refs. 14 and 15]. Because energy must be conserved,

$$\alpha(\lambda) + \rho(\lambda) + \tau(\lambda) = 1. \tag{2-17}$$

The reflection may occur at the surface by either specular or diffuse reflection, or it may return from within the material by scattering if the material is a translucent, inhomogeneous medium. The transmission may occur with or without scattering. The mechanism is immaterial provided $\rho(\lambda)$ and $\tau(\lambda)$ are the total fractions reflected and transmitted respectively in all possible directions; $\rho(\lambda)$ is called the spectral reflectance, and $\tau(\lambda)$ the spectral transmittance.

In summary, external to the object surface the incident, reflected, and emitted energy must be considered. Internal to the object surface, there is absorbed, transmitted, and scattered energy. As indicated earlier, the emitted flux density is a function of the temperature and the emissivity.

If $\alpha(\lambda)$ is found experimentally by measuring $\rho(\lambda)$ and $\tau(\lambda)$ and then using Eq. (2-17), $\rho(\lambda)$ and $\tau(\lambda)$ must be measured over all possible angles, which would be over the whole hemisphere for a plane surface. However, the experiment does not require that the incident energy come from all possible angles; but, since the value obtained for $\alpha(\lambda)$ may depend upon the angle of incidence, this angle must be specified whenever $\alpha(\lambda)$ is given. Fortunately, for most materials $\alpha(\lambda)$ is almost constant with angle of incidence.

As suggested by the comments above, one must keep in mind that if the energy is incident upon a sample from within some small solid angle and the reflected energy is measured within some other limited solid angle, nothing can be deduced concerning the total reflectance nor the absorptance except for certain surfaces whose characteristics are well known. For example, the reflecting properties of "good" mirrors are well known; they are specular reflectors. Experimentally, then, for surfaces in general, an optical system that can handle the reflected energy over at least an entire hemisphere must be used; the integrating sphere fulfills this requirement.

When $\epsilon(\lambda)$ is deduced from $\alpha(\lambda)$, the value applies only for conditions geometrically similar to those used in the determination of $\rho(\lambda)$ and $\tau(\lambda)$, from which $\alpha(\lambda)$ was found.

Generally, emissivity, transmittance, reflectance, and absorptance are wavelength dependent and have no meaning unless a wavelength is given. For heterogeneous incident energy, some sort of integration must be performed, either explicitly or implicitly. The total radiant power transmitted by an optical filter, for example, may be found by

$$\Phi = \int_0^\infty \Phi_\lambda \tau(\lambda)\, d\lambda, \tag{2-18}$$

where Φ_λ specifies the spectral composition of the incident power and $\tau(\lambda)$ the spectral transmittance of the filter. When a transmittance is given for a material without reference to wavelength, an effective transmittance, defined by the following equation, is usually meant:

$$\tau = \int_0^\infty \Phi_\lambda \tau(\lambda)\, d\lambda \Big/ \int_0^\infty \Phi_\lambda\, d\lambda. \tag{2-19}$$

We may define an integrated emissivity as

$$\epsilon = \frac{\int_0^\infty \epsilon(\lambda) M_{e\lambda}(T)\, d\lambda}{\sigma T^4}. \tag{2-20}$$

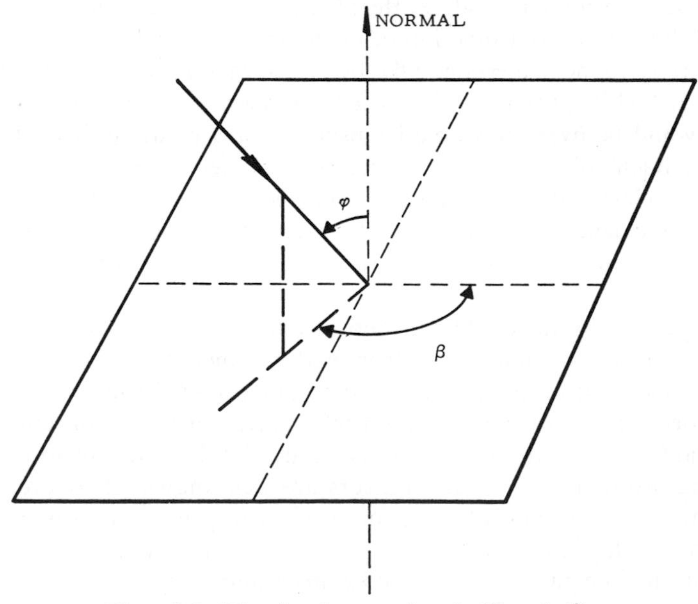

Figure 2.3 Directional properties of $\alpha(\lambda)$ and $\epsilon(\lambda)$.

For many surfaces $\epsilon(\lambda)$ remains nearly constant with wavelength; so, from the Stefan–Boltzmann law (discussed in the next chapter), we can state approximately $\epsilon = \epsilon(\lambda)$. Objects having a constant spectral emissivity less than unity are called graybodies to distinguish them from the ideal blackbodies.

We have already indicated that $\alpha(\lambda)$ is dependent not only upon wavelength but also upon the direction of the incident energy. The related $\epsilon(\lambda)$ was also said to be dependent on angle, both the angle with respect to the surface normal and the angle with respect to a reference direction in the surface, shown in Fig. 2.3. This can now be more explicitly stated as

$$\epsilon(\lambda,\beta,\varphi) = \alpha(\lambda,\beta,\varphi). \tag{2-21}$$

Extinction, Absorption, and Scattering Coefficients

Let us now consider what happens to a "collimated" beam of monochromatic radiant energy while the beam is passing through an isotropic medium. The power in the beam will certainly decrease because energy will be removed by two processes. First, there will be true absorption: radiant energy will be converted to thermal energy by photon absorption in the material, and molecules and atoms will accept quanta of energy and convert radiant energy to nonthermal molecular or atomic energy. Second, there will be scattering—any inhomogenity within the medium will cause some components of the beam to change, sometimes drastically, their direction of propagation. These components will consequently pass out of the beam while still in the form of radiant energy. A simple formula—known by several names: Bouguer's, Lambert's, and Beer's laws—to account for the change in power is

$$\Phi = \Phi_0 \exp(-kl). \tag{2-22}$$

The initial power in the beam is Φ_0, and Φ is the power remaining in the beam after passing through a thickness l of the medium. The constant k, a strong function of wavelength, is a characteristic of the material and is called the extinction coefficient. Since the extinction is caused by both an absorption process and a scattering process, the coefficient may be expressed as the sum of an absorption coefficient α_c and a scattering coefficient β_c.

$$k = \alpha_c + \beta_c. \tag{2-23}$$

In some mathematical developments, the imaginary part of the complex index of refraction is called either the absorption coefficient or the extinction coefficient. The extinction coefficient defined in this manner has a simple relationship to the coefficient defined by Bouguer's law. The complex index of refraction is discussed in Chapter 4.

We shall now assume a homogeneous medium, with $\beta_c = 0$, and discuss the absorption coefficient. The relationship,

$$\Phi = \Phi_0 \exp(-\alpha_c l), \tag{2-24}$$

was discovered early in the eighteenth century by Pierre Bouguer; hence the name Bouguer's law. Nearly a hundred years later A. V. Beer found the absorption coefficient to be proportional to the concentration of an absorbing material dissolved in a nonabsorbing solvent. In addition, if the absorbing material is a mixture, the absorption coefficient according to Beer's law is given by

$$\alpha_c = \epsilon_1 c_1 + \epsilon_2 c_2 + \epsilon_3 c_3 + \cdots. \tag{2-25}$$

The individual concentrations of the several components are $c_1, c_2, c_3, \ldots$; $\epsilon_1, \epsilon_2, \epsilon_3 \cdots$ are characteristic constants respectively for these components.

The law of absorption, Eq. (2-24), may be readily derived from the assumption that each infinitesimally thin layer of the absorbing material reduces the power by an amount $d\Phi$ that is proportional to two quantities, the power Φ of the incident radiant energy beam and the layer thickness dl:

$$d\Phi = -\Phi\alpha_c \, dl. \tag{2-26}$$

When this is integrated between limits,

$$\int_{\Phi_0}^{\Phi} (d\Phi/\Phi) = \alpha_c \int_0^l dl, \tag{2-27}$$

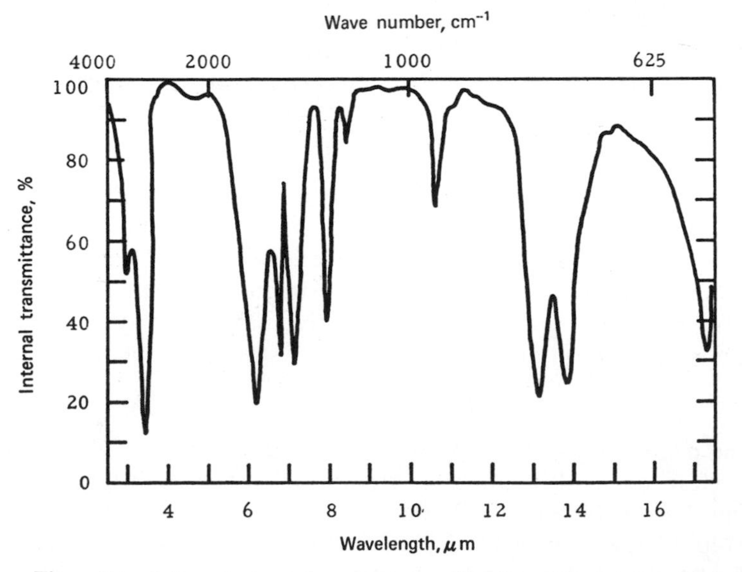

Figure 2.4 Infrared absorption spectrum of barium chloroacetate [Ref. 23].

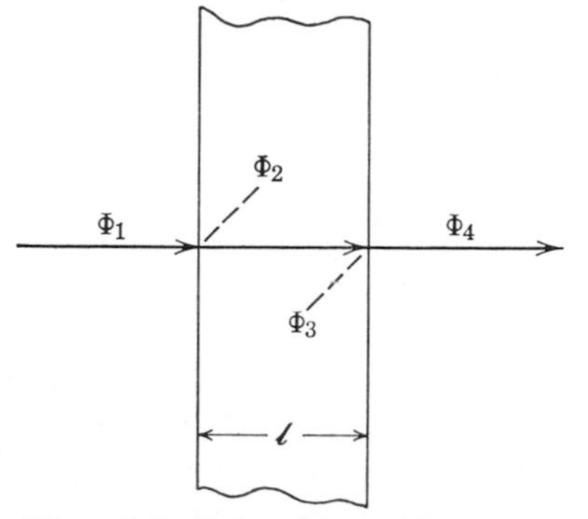

Figure 2.5 Internal transmittance.

it gives Eq. (2-24). The proportionality constant in Eq. (2-26) becomes the absorption coefficient. Beer's law follows from the assumption, which is not always valid, that each molecule of the absorbing material absorbs independently of every other molecule. Bouguer's and Beer's laws are quite generally found to be only approximations. Reasons for deviations from the simple laws become apparent as one studies more thoroughly molecular absorption and the problems of measuring absorption spectra.

If the frequency (or wavenumber, or wavelength) of the beam of radiant energy is varied continuously throughout a broad range, the ratio Φ/Φ_0 is found to vary. The plot of this function—plotted as a function of frequency, wavenumber, or wavelength—is called an absorption spectrum. An absorption spectrum is shown in Fig. 2.4 showing the transmittance Φ/Φ_0 through the sample of material as a function of wavelength and wavenumber. Equation (2-24) changed to show the wavelength dependence would be

$$\Phi(\lambda) = \Phi_0(\lambda) \exp[-\alpha_e(\lambda)l]. \tag{2-24a}$$

The exponential relationship expressed in Eq. (2-24) allows us to extend the discussion of transmittance undertaken earlier in this chapter. In Fig. 2.5 a beam of power Φ_1 is shown entering a material of thickness l. Because of losses at the interface, the beam power is reduced to Φ_2; and after transmission through the material, the power is further reduced to Φ_3. Finally, upon emergence from the material, the second interface brings the beam power down to Φ_4. The following definitions are often useful:

$$\text{External transmittance} = \Phi_4/\Phi_1,$$
$$\text{Internal transmittance} = \Phi_3/\Phi_2.$$

The path involved in the second definition fulfills the conditions of Eq. (2-24), and we can write:

$$\text{Internal transmittance} = \Phi_3/\Phi_2 = \exp(-\alpha_e l).$$

Internal transmittance for a unit path length is called transmissivity, so,

$$\text{Transmissivity} = \exp(-\alpha_c),$$

and

$$\text{Internal transmittance} = (\text{Transmissivity})^l.$$

Photometry

Thus far, we have dealt with the physical quantities related to radiant energy and the associated geometry. Now quantities that are not strictly physical will be discussed. Let us consider the radiant energy that reaches the retina of the eye as a stimulus and evaluate the responses that the human observer experiences. Since psychological stimuli and responses are involved, the quantities now to be defined can better be called psychophysical than physical. This remains fundamentally true even for such quantities as luminance, where standardized relationships have been worked out between radiometric and photometric units.

Light has been defined as the "visual aspect of radiant energy"; a more explicit definition is "The aspect of radiant energy of which the human observer is aware through the visual sensations which arise from the stimulation of the retina of the eye" [see Ref. 3, p. 220]. Light, therefore, cannot be separately described in terms of radiant energy or of visual sensation but is a combination of the two and is thus a psychophysical concept.

Brightness is defined as the attribute of sensation by which an observer is aware of *differences* of observed radiant energy [see Ref. 3, p. 42]. Brightness is reserved by this definition for a subjective sensory attribute. To illustrate: Let the field of view of each eye of an observer be completely separated from the field of view of the other eye. Let the "light" of one field have the same wavelength distribution but several thousand times the radiant power of the other field. Let each eye be fully adapted to the power in the field that it is viewing. In most experimental situations, the observer will see equally bright fields, that is, no difference in brightness will be sensed between the two fields. Our definition of brightness, therefore, is at odds with a current practice (no longer recommended by the Optical Society of America) of quantitatively basing brightness on the *fixed* level of radiant power under stated geometric and wavelength conditions.

Luminous Energy

When radiant energy at various wavelengths is evaluated experimentally, by using an eye adapted to relatively bright light, to determine its relative effectiveness in producing brightness, a curve similar to that in Fig. 2.6 is

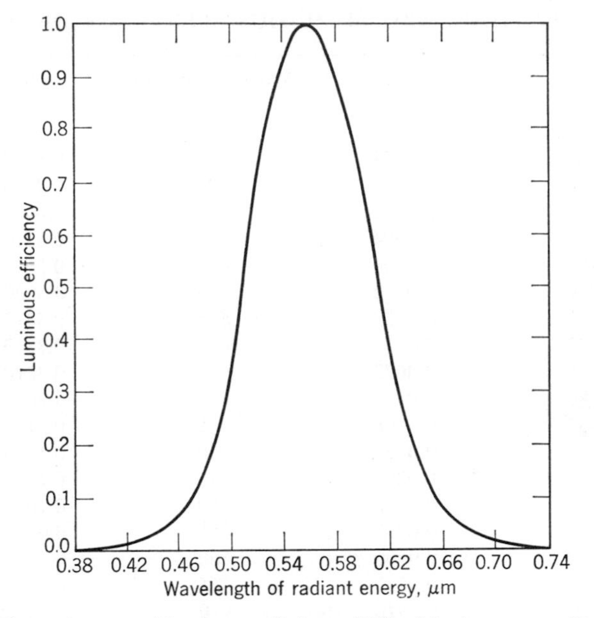

Figure 2.6 Photopic spectral luminous efficiency $V(\lambda)$ of the human eye [Ref. 24, p. 1–3].

obtained. The function $V(\lambda)$ represented by the curve, and tabulated in Table IV, is the relative spectral luminous efficiency; it is sometimes also called the relative spectral luminous efficacy, the luminous efficiency, or the relative luminosity. A standard curve of this function has been established by international agreement and may be considered the relative spectral sensitivity of the average normal, light-adapted human eye.

To convert radiant energy, distributed over the various wavelengths according to a spectral energy function U_λ, to luminous energy Q, we obviously have to use $V(\lambda)$ as a weighting function as indicated in the following expression,

$$Q_\lambda = K_m V(\lambda) U_\lambda, \tag{2-28}$$

or

$$Q = K_m \int_0^\infty V(\lambda) U_\lambda \, d\lambda, \tag{2-29}$$

where K_m is a constant that determines the size of the Q units.

Now that luminous energy has been defined, a photometric system of quantities parallel to the radiometric system based on radiant energy can be set up. When the system of units must be indicated explicitly, the subscript e should be used to indicate radiometric units and the subscript v to signify photometric units. Photometric Table V corresponds to radiometric Table II.

TABLE IV STANDARD LUMINOUS EFFICIENCY DATA

Wavelength, nm	$V(\lambda)$	Wavelength, nm	$V(\lambda)$
380	0.0000	580	0.8700
385	0.0001	585	0.8163
390	0.0001	590	0.7570
395	0.0002	595	0.6949
400	0.0004	600	0.6310
405	0.0006	605	0.5668
410	0.0012	610	0.5030
415	0.0022	615	0.4412
420	0.0040	620	0.3810
425	0.0073	625	0.3210
430	0.0116	630	0.2650
435	0.0168	635	0.2170
440	0.0230	640	0.1750
445	0.0298	645	0.1382
450	0.0380	650	0.1070
455	0.0480	655	0.0816
460	0.0600	660	0.0610
465	0.0739	665	0.0446
470	0.0910	670	0.0320
475	0.1126	675	0.0232
480	0.1390	680	0.0170
485	0.1693	685	0.0119
490	0.2080	690	0.0082
495	0.2586	695	0.0057
500	0.3230	700	0.0041
505	0.4073	705	0.0029
510	0.5030	710	0.0021
515	0.6082	715	0.0015
520	0.7100	720	0.0010
525	0.7932	725	0.0007
530	0.8620	730	0.0005
535	0.9149	735	0.0004
540	0.9540	740	0.0003
545	0.9803	745	0.0002
550	0.9950	750	0.0001
555	1.0002	755	0.0001
560	0.9950	760	0.0001
565	0.9786	765	0.0000
570	0.9520	770	0.0000
575	0.9154	775	0.0000

TABLE V PHOTOMETRIC UNITS: PREFERRED SYMBOLS, NAMES, AND UNITS

Symbol and Defining Equation	Name	Description	Unit
Q	Luminous energy		talbot
$q = dQ/dv$	Luminous density		talbot cm^{-3}
$\Phi_v = dQ/dt$	Luminous flux (power)	Rate of transfer of luminous energy	lm
$F = \partial\Phi_v/\partial s$	Luminous flux density	Luminous flux incident upon or leaving a surface divided by the area of that surface	lm cm^{-2}
$M_v = \partial\Phi_v/\partial s$	Luminous exitance	Luminous flux leaving an infinitesimal element of area divided by that area	lm cm^{-2}
$E_v = \partial\Phi_v/\partial s$	Illuminance	Luminous flux per unit area incident upon a surface	lm cm^{-2}
$I_v = \partial\Phi_v/\partial\omega$	Luminous intensity	Luminous flux leaving a point source per unit solid angle	lm $sr^{-1} \equiv$ cd
$L_v = \dfrac{\partial^2\Phi_v}{(\partial s\ \partial\omega \cos\theta)}$	Luminance	Luminous flux per unit solid angle per unit of projected area from a source	lm sr^{-1} cm^{-2}

A quantity corresponding to power, but in the photometric system, can be obtained by taking the derivative of Eq. (2-28) with respect to time:

$$dQ_\lambda/dt = K_m V(\lambda)\, dU_\lambda/dt = \Phi_{v\lambda} = K_m V(\lambda)\Phi_{e\lambda}, \tag{2-30a}$$

or

$$\Phi_v = \int_0^\infty \Phi_{v\lambda}\, d\lambda = K_m \int_0^\infty V(\lambda)\Phi_{e\lambda}\, d\lambda. \tag{2-30b}$$

The symbol Φ_v represents luminous flux, the rate of transfer of luminous energy which corresponds to radiant power, and has the lumen as its unit.

If Eq. (2-30a) is differentiated with respect to area,

$$d\Phi_v/ds = K_m V(\lambda)\, d\Phi_{e\lambda}/ds, \tag{2-31a}$$

and if the left-hand side is replaced by the symbol F_λ, then

$$F_\lambda = K_m V(\lambda) W_\lambda. \tag{2-31b}$$

Radiant power per unit area W_λ appearing on the right of this equation is radiant flux density, irradiance or radiant exitance depending upon the direction of energy flow. The units for the corresponding quantity F_λ on the left would be luminous flux per unit area for luminous flux density, illuminance, or luminous exitance.

The other terms in Table V correspond so precisely to the radiometric units, which have been explained earlier, that no further explanation is needed here. Alternate units to some of those in Table V are given in Table VI.

TABLE VI CONVERSION TABLE OF PHOTOMETRIC UNITS

Some common units of illuminance

One	Equals		
Lux	*1 lm/m^2	10^{-4} phot	9.3×10^{-2} ft candle
Phot	1 lm/cm^2	10^4 lux	9.3×10^2 ft candles
Foot candle	1 lm/ft^2	10.76 lux	10.76×10^{-4} phot

Some common units of luminance

One	Equals		
Stilb	1 cd/cm^2	π lamberts	2.919×10^3 ft lamberts
Nit	*1 cd/m^2	$\pi \times 10^{-4}$ lambert	0.2919 ft lambert
Lambert	$1/\pi$ cd/cm^2	0.318 stilb	9.29×10^2 ft lamberts
Foot lambert	$1/\pi$ cd/ft^2	3.42×10^{-4} stilb	1.1×10^{-3} lambert

* Recommended unit.

Standard Source of Luminous Energy

The International Standard Source of luminous energy is a square centimeter of a blackbody operating at the temperature of solidification of melted platinum. The luminous intensity I_v (luminous flux per unit solid angle) of such a source is arbitrarily assigned the value of 60 cd. The candela is sometimes referred to as the "new candle," but it should not be confused with the old source unit of candle, which has a slightly different magnitude and is now obsolete.

By definition, one candela is a luminous intensity of one lumen per steradian. From the relationship between the lumen and the unit of Q in Eq. (2-29), this definition establishes the value of K_m. This can be seen more clearly by writing an equation for luminous exitance corresponding to Eq. (2-31b):

$$M_{v\lambda} = K_m V(\lambda)\, d\Phi_{e\lambda}/ds, \tag{2-32}$$

or

$$M_v = K_m \int_0^\infty V(\lambda) M_{e\lambda}\, d\lambda. \tag{2-33}$$

Now let us reconsider the standard source of luminous energy and review some of the photometric quantities that were involved in the preceding discussion. The luminous intensity normal to the surface is

$$I_v = 60 \text{ cd} = 60 \text{ lm/sr}.$$

Since the source area is 1 cm^2 the luminance is

$$L_v = 60 \text{ lm cm}^{-2} \text{ sr}^{-1}.$$

We discuss a problem in Chapter 3 in which it is shown that the luminous exitance of a blackbody or other diffuse radiator is π times the luminance. The luminous exitance of our standard source, then, is

$$M_v = 60\pi = K_m \int_0^\infty V(\lambda) M_{e\lambda}(T)\, d\lambda. \tag{2-34}$$

The temperature T is the temperature of solidifying platinum, approximately 2042 K. The necessary quantities in Eq. (2-34) are known, so that K_m, the luminous efficacy at the wavelength of maximum spectral luminous efficacy, can be found. Both the Optical Society of America and the Illuminating Engineering Society use 680 lm/watt for this constant, but recent experimental results [Refs. 16 and 17] indicate 685 or 686 lm/watt to be a more accurate value.

After K_m is established from standard conditions, Eq. (2-33) becomes the general means for transforming watts to lumens. Note particularly that no simple ratio exists between these two units. The product $V(\lambda)\, K_m$ is known as

TABLE VIIa APPROXIMATE LUMINANCE VALUES OF VARIOUS SOURCES*

Source	Luminance, cd/m^2
Atomic fission bomb; 0.1 msec after firing, 90-ft diameter ball	2×10^{12}
Blackbody; 6500 K	3×10^9
Sun; at surface	2.25×10^9
Sun; observed at zenith from the earth's surface	1.6×10^9
High intensity carbon arc; 13.6-mm rotating—positive carbon	0.75×10^9 to 1.50×10^9
Photoflash lamps	1.6×10^8 to 4.0×10^8
Blackbody; 4000 K	2.5×10^8
High intensity mercury short arc; type SAH1000A, 30 atoms	2.4×10^8
Xenon short arc; 900-watt, direct current	1.8×10^8
Zirconium concentrated arc; 300-watt size	4.5×10^7
Tungsten filament incandescent lamp; 1200-watt projection, 31.5 lm/watt	3.3×10^7
Tungsten filament; 750-watt, 26 lm/watt	2.4×10^7
Tungsten filament; gas filled, 20 lm/watt	1.2×10^7
Sun; observed from the earth's surface, sun near horizon	6.0×10^6
Blackbody; 2042 K	6.0×10^5
Inside-frosted bulb; 60-watt	1.2×10^5
Acetylene flame; Mees burner	1.05×10^5
Welsback mantle; bright spot	6.2×10^4
Sodium arc lamp; 10,000-lm size	5.5×10^4
Low pressure mercury arc; 50-in. rectifier tube	2.0×10^4
T-12 bulb-fluorescent lamp; 1500-mA extra high loading	1.7×10^4
T-12 bulb-fluorescent lamp; 800-mA high loading	1.13×10^4
Clear sky; average brightness	8.0×10^3
T-17 bulb-fluorescent; 420-mA low loading	4.3×10^3
Illuminating gas flame; fish-tail burner	4.0×10^3
Moon; observed from the earth's surface, bright spot	2.5×10^3
Sky; overcast	2.0×10^3
Clear glass neon tube; 15 mm, 60 mA	1.6×10^3
Clear glass blue tube; 15 mm 60 mA	8.0×10^2
Self-luminous paint	0.0 to 0.17

* Data obtained from reference 24, Fig. A-12.

luminous efficacy, the ratio of any photometric unit to the corresponding radiometric unit at a given wavelength. For convenience, a number of the blackbody function tables [Refs. 18 and 19] include tabulated values of both L and $L(\lambda)$ for blackbodies at various temperatures. Tables VIIa and VIIb list approximate or typical luminances for several sources and backgrounds.

Because a *diffuse* surface having a luminous exitance of 1 lm/cm^2 or 1 lm/ft^2 has a luminance of 1 L or 1 ft-L respectively, the term lambert is

TABLE VIIb TABLE OF APPROXIMATE LUMINANCES OF BACKGROUNDS*

Source	Luminance, cd/m^2
Horizon sky	
Overcast, no moon	3.4×10^{-5}
Clear, no moon	3.4×10^{-4}
Overcast, moon	3.4×10^{-3}
Clear, moonlight	3.4×10^{-2}
Deep twilight	3.4×10^{-1}
Twilight	3.4
Very dark day	34
Overcast day	3.4×10^{2}
Clear day	3.4×10^{3}
Clouds, sun-lighted	3.4×10^{4}
Daylight fog	
Dull	3.4×10^{2} to 10×10^{2}
Typical	10×10^{2} to 34×10^{2}
Bright	3.4×10^{3} to 17×10^{3}
Ground	
On sunny day	3.4×10^{2}
On overcast day	34 to 100
Snow, full sunlight	17×10^{3}

* Data obtained from reference 24, pp. 2–27.

sometimes used when lumen per centimeter^2 is meant. Inasmuch as this equivalence does not, in general, hold for other than diffuse surfaces, this practice should be avoided.

The infrared spectrum and the ultraviolet spectrum are generally accepted as having no effect on sight; hence, there should be no common ground for these spectra and photometric units. However, in a few instances, the separation is not quite complete. For instance, a light source may be rated in photometric units because of its intended use; yet, it may also be an excellent radiator in either or both the infrared and the ultraviolet. In fact, if the general characteristics of the source are known, the photometric rating may tell something about performance in the nonvisible spectra. The nature of the visible band boundaries also leaves room for application of photometric units in the infrared and ultraviolet spectra. Visible wavelengths are generally said to extend from 0.4 to 0.7 μm; however, the curve of Fig. 2.6 may, in fact, be extended to near 0.3 μm in the ultraviolet [Refs. 20 and 21] and to near 1 μm in the infrared [Ref. 22]. The limitation on seeing ultraviolet is largely caused by ultraviolet absorption of the lens of the eye before this radiant energy reaches the retina. Without the lens the eye could respond to

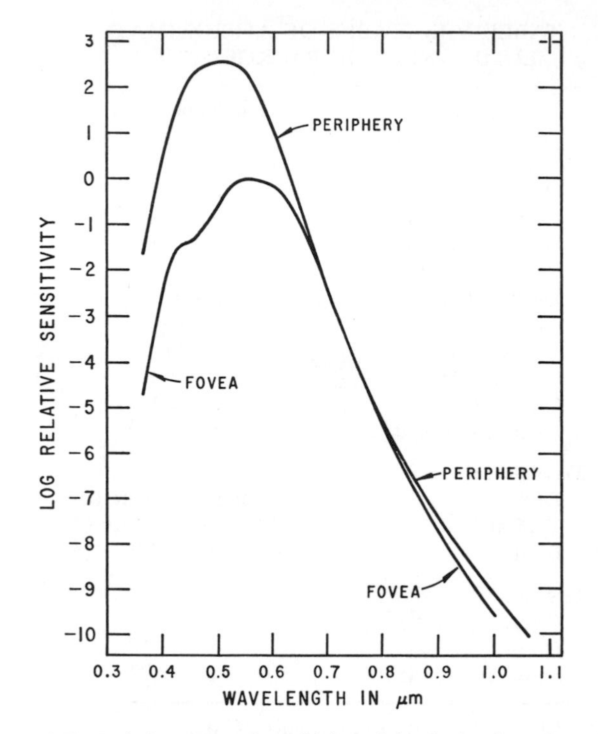

Figure 2.7 Relative spectral sensitivity of the dark adapted eye [Ref. 22].

wavelengths down to 0.320 μm. Figure 2.7 shows the relative sensitivity of the dark-adapted eye extended into the infrared. The fovea is the most sensitive area of the eye, near the center of the retina, and the periphery is near the edge of the retina. Thus, $V(\lambda)$ may be obtained from Fig. 2.7 and have small but meaningful values in both the infrared and ultraviolet.

REFERENCES

1. A. Walther, "Radiometry and Coherence," *J. Opt. Soc. Am.*, **58,** 1256 (1968).
2. SUN (Commission for Symbols, Units and Nomenclature of the International Union of Pure and Applied Physics), "Symbols, Units, and Nomenclature in Physics," *Phys. Today*, **15,** 20 (June, 1962).
3. OSA Committee on Colorimetry, *Science of Color*, Optical Society of America, 2100 Pennsylvania Ave., Washington, D.C. 20037, 1953.
4. "USA Standard Nomenclature and Definitions for Illuminating Engineering," Standard Z7.1-1967 of American Standards Association, sponsored by Illuminating Engineering Society, 345 East 47th Street, New York, N.Y. 10017.
5. Joint Committee on Nomenclature in Applied Spectroscopy, "Suggested Nomenclature in Applied Spectroscopy," Rept No. 6, *Anal. Chem.*, **24,** 1349 (1952).

6. "Definition of Terms and Symbols Relating to Molecular Spectroscopy," *1969 Book of Standards*, American Society of Testing Materials, Part 30, No. E131-68, 1916 Race Street, Philadelphia 19103.
7. "Definitions of Terms and Symbols Relating to Emission Spectroscopy," *1969 Book of Standards*, American Society of Testing Materials, Part 30, No. E135-67, 1916 Race Street, Philadelphia 19103.
8. Editor's Page, *J. Opt. Soc. Am.*, **57**, 854 (1967).
9. J. R. Meyer-Arendt, "Radiometry and Photometry: Units and Conversion Factors," *Appl. Optics*, **7**, 2081 (1968).
10. Ely E. Bell, "Radiometric Quantities, Symbols and Units," *Proc. IRE*, **47**, 1432 (1959).
11. A. V. H. Masket, *Tables of Solid Angles*, TID-14975, University of North Carolina, (Available from Office of Technical Services, Department of Commerce, Washington. D.C., $5.)
12. Fritz Kasten, "Note on the Unit Solid Angle," *J. Opt. Soc. Am.*, **54**, 845 (1964).
13. F. E. Nicodemus, "Radiance," *Am. J. Phys.*, **31**, 368 (1963).
14. E. L. DeLa Perrelle, T. S. Moss, and H. Herbert, "The Measurement of Absorptivity and Reflectivity," *Infrared Phys.* **3**, 35 (1963).
15. "Selection of Geometric Conditions for Measurement of Reflectance and Transmittance," *1969 Book of Standards*, American Society of Testing Materials, Part 30, No. 179-66, 1916 Race Street, Philadelphia 19103.
16. J. S. Preston, "A Radiometric Method of Perpetuating the Unit of Light," *Proc. Roy. Soc.*, **A272**, 133 (1963).
17. E. J. Gilham, "Further Work on a Radiometric Method of Perpetuating the Unit of Light," *Proc. Roy. Soc.*, **A278**, 137 (1964).
18. M. Pivovonsky and M. R. Nagel, *Tables of Blackbody Radiation Functions.* Macmillan, New York, 1961.
19. D. Hahn et al., *Seven Place Tables of the Planck Function for the Visible Spectrum.* Friedr, Uierveg & Sohn, Braunschweig (Academic Press, New York), 1964.
20. N. I. Pinegin, "Absolute Photopic Sensitivity of the Eye in the Ultraviolet and in the Visible Spectrum," *Nature*, **154**, 770 (1944).
21. N. I. Pinegin, "Absolute Scotopic Sensitivity of the Eye in the Ultraviolet and in the Visible Spectrum," *Nature*, **155**, 20 (1945).
22. D. E. Griffin, R. Hubbard, and G. Wald, "The Sensitivity of the Human Eye to Infrared Radiation," *J. Opt. Soc. Am.*, **37**, 546 (1947).
23. A. V. R. Warrier and P. S. Narayanan, "Infrared Spectra of Crystalline Chloroacetates of Cu, Ca, Sr, Ba, and Pb," *Spectrochim. Acta*, **23A**, 1061, 1064 (1967).
24. *IES Lighting Handbook*, Fourth Edition. Illuminating Engineering Society, 345 E. 47th St., New York, 10017.
25. F. E. Nicodemus, "Optical Resource Letter on Radiometry," *J. Opt. Soc. Am.*, **59**, 243 (1969).

3

Blackbodies

The blackbody radiator is an idealized source of radiant energy having defined radiating properties. It is perfectly diffuse, radiates at all wavelengths, and at any given temperature its spectral radiant exitance, at all wavelengths, is the maximum possible for any actual, thermal source at the same temperature. The characteristics of many actual radiators approach those of blackbodies; in fact, this radiator concept is useful as an approximation even when there is appreciable divergence between the ideal and the actual.

As suggested by the indicated color, a blackbody is a body (or a surface) that absorbs all incident radiant energy. To be in radiative equilibrium with its surroundings, it must emit as much as it absorbs. The expression that describes emission in terms of frequency (or wavelength) and temperature is known as Planck's Radiation law. One expression of the law is

$$M_\nu(T) = c_1\nu^3[\exp(c_2\nu/T) - 1]^{-1}, \tag{3-1}$$

where $M_\nu(T)$ is spectral radiant emitted exitance, and c_1 and c_2 are constants known as the first and second radiation constants. Equation (3-1) is firmly established by experimental observations. It can also be derived by several different methods [Refs. 1, 2, and 3]. An abbreviated derivation is given below.

Planck's Radiation Law

Let us assume a completely enclosed cavity in thermal equilibrium at a temperature T. Even with a vacuum within the cavity, there will be radiant energy present because photons are radiated from the cavity walls. A "photon gas" of uniform density is assumed to fill the cavity, with the direction of photon motion being strictly random. Let us consider an imaginary surface of unit area somewhere within the cavity. The number of photons passing through this area in each direction per second within the frequency range ν to $\nu + d\nu$ can be found by making a statistical analysis of a photon gas, the gas being assumed to conform to the Bose–Einstein distribution. The detailed derivation will not be given here, but the result is as follows:

$$\mathscr{M}_\nu(T)\,d\nu = (2\pi\nu^2/c^2)[\exp(h\nu/kT) - 1]^{-1}\,d\nu. \tag{3-2}$$

The energy flow is obtained by multiplying the number of photons by the energy per photon as given by Eq. (1-9). The result is

$$M_\nu(T)\,d\nu = (2\pi h\nu^3/c^2)[\exp(h\nu/kT) - 1]^{-1}\,d\nu. \tag{3-3}$$

In Eq. (3-2), k is the Boltzmann constant equal to 1.38045×10^{-23} J deg^{-1}, and the other symbols are defined in the preceding chapters as they occur. The quantity $M_\nu(T)$ is also the rate of energy flow per unit frequency range into a unit area of the cavity surface. If the walls are perfectly absorbing, and in equilibrium, they are also emitting energy at the same rate, so the spectral radiant exitance of a blackbody surface is $M_\nu(T)$. In terms of wavelength this quantity is

$$M_\lambda(T) = (2\pi hc^2/\lambda^5)[\exp(hc/k\lambda T) - 1]^{-1} \tag{3-4}$$

because

$$\nu = c/\lambda \tag{3-5}$$

and

$$d\nu = -(c/\lambda^2)\,d\lambda. \tag{3-6}$$

In terms of the number of photons emitted, corresponding to Eq. (3-4), the spectral exitance is

$$\mathscr{M}_\lambda(T) = (2\pi c/\lambda^4)[\exp(hc/k\lambda T) - 1]^{-1}. \tag{3-7}$$

$\mathscr{M}_\lambda(T)$ is the spectral radiant emitted exitance from a blackbody surface in photons sec^{-1} cm^{-3}. The centimeter is taken as the unit wavelength zone of the spectrum. Figure 3.1 is a plot of $M_\lambda(T)$ and $\mathscr{M}_\lambda(T)$.

It is not necessary that the blackbody cavity walls have an absorptance of one to maintain equality between energy flow into and away from the walls. If the irradiance at the walls is quantitatively equal to $M_\nu(T)$, it is required only that the sum of emitted exitance and reflected exitance be $M_\nu(T)$, since the exitance at the walls includes both. We assume of course that the transmittance of the walls is zero.

The quantity $2\pi hc^2$ from Eq. (3-4) is commonly defined as the first radiation constant; but any of the quantities $2\pi h/c^2$, $2\pi/c^2$, or $2\pi c$ from Equations (3-3), (3-2), or (3-7), respectively, might equally have been called the first radiation constant. When h is given the units joule sec, c the units cm sec^{-1},

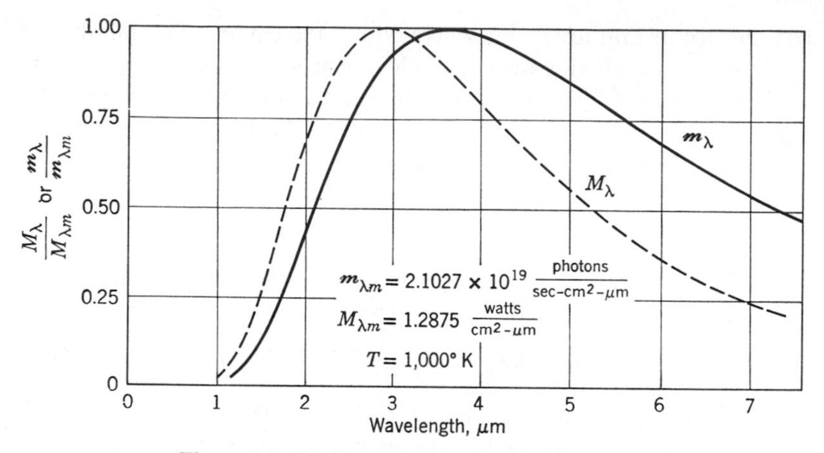

Figure 3.1 Radiant exitance of a blackbody at 1000 K.

and λ the units cm, then, at a given temperature, M_λ (radiant emitted exitance as a function of wavelength) has the units watt cm^{-2} cm^{-1}, and the unit wavelength region is 1-cm wide. Sometimes M_λ is given in watt cm^{-2} μm^{-1} where μm is a micrometer (or micron). This is an odd unit because in the formula, Eq. (3-4), the factor $1/\lambda$ would have to be used four times with the unit cm^{-1} and a fifth time with the unit μm^{-1}.

Although there are a number of ways of writing the Planck radiation formula and of defining the radiation constant, the second radiation constant c_2 is uniformly defined as hc/k. Whenever Planck formulas or tables of blackbody radiation functions are encountered, the user must note the units involved and the definition of the first radiation constant used in computing the tables. As already indicated, the Planck formula may be written as a function of wavelength, frequency, or wavenumber. Also, it may be written for polarized or unpolarized radiant energy; for radiant exitance, radiance, or radiant flux density; for energy or photons.

When M_λ (or W_λ or H_λ) is in units of watt cm^{-2} cm^{-1}, other selected Planck formulas can be obtained as follows:

$$M_\lambda{}^\mu = M_\lambda \times 10^{-4} \text{ watt cm}^{-2}\, \mu\text{m}^{-1}, \tag{3-8}$$

$$L_\lambda{}^\mu = M_\lambda{}^\mu/\pi \text{ watt cm}^{-2} \text{ sr}^{-1}\, \mu\text{m}^{-1}, \tag{3-9}$$

$$u_\lambda = M_\lambda(4/c) \text{ j cm}^{-3} \text{ cm}^{-1}, \tag{3-10}$$

$$L_\nu{}^\mu = L_\lambda{}^\mu(\nu^3\lambda^5/c^4) \text{ watt cm}^{-2} \text{ sr}^{-1} \text{ Hz}^{-1}. \tag{3-11}$$

These formulas are for unpolarized radiant energy. The radiance of a polarized beam of radiant energy, from Eq. (3-9), is

$$L_{p\lambda}{}^\mu = L_\lambda{}^\mu/2 \text{ watt cm}^{-2} \text{ sr}^{-1}\, \mu\text{m}^{-1}. \tag{3-12}$$

Extensions of Planck's Law

From Planck's law of blackbody radiation, several other interesting and useful relationships may be derived.

Rayleigh–Jeans Law. The exponential from Planck's Radiation Law may be written as

$$\exp(hc/k\lambda T) = 1 + (hc/k\lambda T) + \frac{1}{2!}(hc/k\lambda T)^2 + \cdots. \tag{3-13}$$

Then, if $hc/k\lambda T \ll 1$

$$[\exp(hc/k\lambda T) - 1] \approx hc/k\lambda T \tag{3-14}$$

by neglecting terms of the second power and higher. Then,

$$M_\lambda(T) \approx 2\pi ckT/\lambda^4. \tag{3-15}$$

This is known as the Rayleigh–Jeans law and is useful at high temperatures and long wavelengths. It is correct to within 1% if $\lambda T > 72$ cm deg. At a given wavelength, within this useful range of λT, the radiant exitance is linear with temperature. The relative error involved in using the Rayleigh–Jeans expression is

$$\begin{aligned}\mathscr{E}_r &= [(M_\lambda)_r - (M_\lambda)_p]/(M_\lambda)_p \\ &= \left[\frac{(2\pi ckT/\lambda^4)}{(2\pi hc^2/\lambda^5)[\exp(hc/k\lambda T) - 1]^{-1}}\right] - 1\end{aligned} \tag{3-16}$$

$$\mathscr{E}_r = (\tfrac{1}{2})(hc/k\lambda T) + (\tfrac{1}{6})(hc/k\lambda T)^2 + \cdots. \tag{3-17}$$

Wien Law. If the Planck formula, Eq. (3-4), is written in the form

$$M_\lambda(T) = (2\pi hc^2/\lambda^5)\exp(-hc/k\lambda T)[1 - \exp(-hc/k\lambda T)]^{-1}, \tag{3-18}$$

the binomial can be expanded to obtain

$$M_\lambda(T) = (2\pi hc^2/\lambda^5)\exp(-hc/k\lambda T)[1 + \exp(-hc/k\lambda T) + \cdots]. \tag{3-19}$$

If $hc/k\lambda T \gg 1$, we may neglect all terms of the series except the first. Then the radiant exitance is given approximately by

$$M_\lambda(T) \approx (2\pi hc^2/\lambda^5)\exp(-hc/k\lambda T), \tag{3-20}$$

which is somewhat easier to work with mathematically and, over much of the wavelength range, differs but little from the exact expression. It is correct to within 1% if $\lambda T < 0.31$ cm deg. The relative error involved is

$$\mathscr{E}_w = [(M_\lambda)_w - (M_\lambda)_p]/(M_\lambda)_p = -\exp(-hc/k\lambda T). \tag{3-21}$$

By including higher power terms in Equations (3-13) and (3-19), more accurate approximations of the Planck formula are, of course, obtained at the expense of increased complication. Figure 3.2 is a plot of $\mathscr{E}_r$ and $\mathscr{E}_w$ as functions of λT for the approximations given and for approximations with one additional term. Obviously, there is an intermediate region where neither the Rayleigh–Jeans equation nor the Wien equation is useful as an approximation to the Planck formula. Erminy [Ref. 4] has given several equations of simple mathematical form that are closer approximations in this intermediate region.

Stefan–Boltzmann Law. To find the total radiant exitance from a blackbody surface, one can integrate $M_\nu(T)$ over all frequencies as follows:

$$M(T) = \int_0^\infty M_\nu(T)\,d\nu = \int_0^\infty (2\pi h\nu^3/c^2)[\exp(h\nu/kT) - 1]^{-1}\,d\nu. \quad (3\text{-}22)$$

Let

$$x = h\nu/kT; \qquad \text{then} \quad dx = (h/kT)\,d\nu \quad (3\text{-}23)$$

and

$$M(T) = (2\pi k^4T^4/c^2h^3)\int_0^\infty x^3(\exp x - 1)^{-1}\,dx. \quad (3\text{-}24)$$

The integrand can be transformed as follows by expanding the binomial:

$$x^3/(\exp x - 1) = x^3[\exp(-x)][1 - \exp(-x)]^{-1}$$
$$= x^3\sum_{j=1}^{\infty}\exp(-jx). \quad (3\text{-}25)$$

Then, the integral is

$$M(T) = (2\pi k^4T^4/c^2h^3)\int_0^\infty x^3\sum_{j=1}^{\infty}\exp(-jx)\,dx. \quad (3\text{-}26)$$

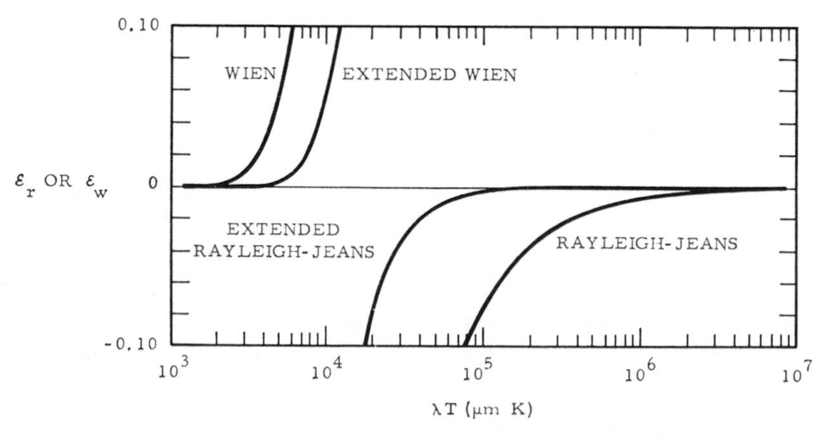

Figure 3.2 Errors in the approximations of the Planck formula [Ref. 4].

The series can be integrated term by term. The integral of the jth term is [Ref. 5, integral 305]

$$\int x^3[\exp(-jx)]\,dx = \left\{\frac{-1}{j^4}\left[\frac{(jx)^3}{\exp jx} + \frac{3(jx)^2}{\exp jx} + \frac{6(jx)}{\exp jx} + \frac{6}{\exp jx}\right]\right\}. \tag{3-27}$$

When the limits of integration are substituted, all four terms in the brackets become zero as x approaches infinity; and the first three terms are zero when x is zero. The definite integral then is obtained by summing $6/j^4$ over all terms. The sum of the series is [Ref. 6]

$$6\sum_{j=1}^{\infty}(1/j^4) = \pi^4/15. \tag{3-28}$$

By making the indicated substitutions in Eq. (3-26),

$$M(T) = (2\pi^5k^4/15c^2h^3)T^4 = \sigma T^4, \tag{3-29}$$

where σ is known as the Stefan–Boltzmann constant and has the value 5.6697×10^{-12} watt cm^{-2} deg^{-4}. This useful expression, Eq. (3-29), is the Stefan–Boltzmann law, which gives the total rate of emission of electromagnetic energy into a 2π solid angle from an element of area on a blackbody radiator surface.

Wien Displacement Law. To find the wavelength at which the emission has its highest value, Planck's law is differentiated (at a constant temperature) with respect to wavelength and then the derivative of $M_\lambda(T)$ is set equal to zero. The solution of this equation gives λ_m as follows:

$$\frac{dM}{d\lambda} = \frac{c_1}{\lambda^6}\cdot\frac{1}{[\exp(c_2/\lambda T) - 1]}\cdot\left\{\frac{(c_2/\lambda T)\exp(c_2/\lambda T)}{[\exp(c_2/\lambda T) - 1]} - 5\right\} = 0 \tag{3-30}$$

where $c_1 = 2\pi hc^2$ and $c_2 = hc/k$. Setting the first two factors each equal to zero leads to the nonmaximum zero slope solutions, where the emission is zero. When the third factor is set equal to zero the resulting equation reduces to

$$\exp(-\beta) + (\beta/5) - 1 = 0, \tag{3-31}$$

where

$$\beta = hc/k\lambda_m T.$$

The evaluation of this equation gives $\beta = 4.9651142317\cdots$, so

$$(hc/k\lambda_m T) = 4.9651142317\cdots \tag{3-32}$$

and

$$\lambda_m T = 0.28978 \text{ cm deg}. \tag{3-33}$$

This is known as the Wien Displacement law.

For real bodies, $\lambda_m T$ does not remain constant but is a function of both the temperature and wavelength [Ref. 7]. It is also of interest to note that an error will result from the assumption that a blackbody having $\lambda_m =$ 555 nm—the wavelength at which the average eye is most sensitive—produces the greatest visual response. For example,

$$T_0 = 0.28978/(555 \times 10^{-7}) = 5221 \text{ K}, \tag{3-34}$$

but a blackbody at 6600 K has the greatest luminous efficiency [Ref. 8].

A substitution for λ_m from Eq. (3-33) into Eq. (3-4) gives the maximum value of $M_\lambda(T)$:

$$M(T)_m = 1.2875 \times 10^{-11} T^5 \text{ watt cm}^{-2} \text{ cm}^{-1}. \tag{3-35}$$

Similarly, from Eq. (3-7) the wavelength for which the photon emission is maximum is obtained, and in turn the condition for maximum photon emission:

$$\lambda_m' T = 0.36689 \text{ cm deg}. \tag{3-36}$$

Again, by substitution,

$$\mathscr{M}(T)_m = 2.1027 \times 10^{11} T^4 \text{ photon sec}^{-1} \text{ cm}^{-2} \text{ cm}^{-1}. \tag{3-37}$$

A normalized blackbody formula can be developed by writing $\lambda = n\lambda_m$, where n varies from 0 to ∞, and substituting into Eq. (3-4). The form can be simplified by setting $c_1 = 2\pi hc^2$:

$$M_n(T) = c_1 n^{-5} \lambda_m^{-5} [\exp (hc/kn\lambda_m T) - 1]^{-1}, \tag{3-38}$$

then if $\lambda_m T = 0.2898$,

$$M_n(T) = (0.2898)^{-5} c_1 n^{-5} T^5 [\exp (4.96511/n) - 1]^{-1}. \tag{3-39}$$

Finally if Eq. (3-39) is divided by Eq. (3-35), the following results:

$$\begin{aligned} M_n &= [M_n(T)/M(T)_m] \\ &= 3.8010 \times 10^{13} c_1 n^{-5} [\exp (4.96511/n) - 1]^{-1}. \end{aligned} \tag{3-40}$$

If λ_m is treated as a unit wavelength and $M(T)_m$ as a unit exitance, a normalized blackbody curve shown in Fig. 3.3, plotted using logarithmic scales, results. This, therefore, represents all possible blackbody curves (Ref. 9).

Other blackbody equations. If $x_1 = (hc/k\lambda_1 T)$ is substituted for 0 as the lower limit in Eq. (3-26), and the indicated integration is performed, the total power radiated at all wavelengths less than λ_1 for a given T is obtained. Figures 3.4 and 3.5 show how the total blackbody power radiated at wavelengths less than a specified wavelength varies with λT and λ. It is interesting to note, for instance, that one-fourth of the blackbody power is radiated at wavelengths less than λ_m.

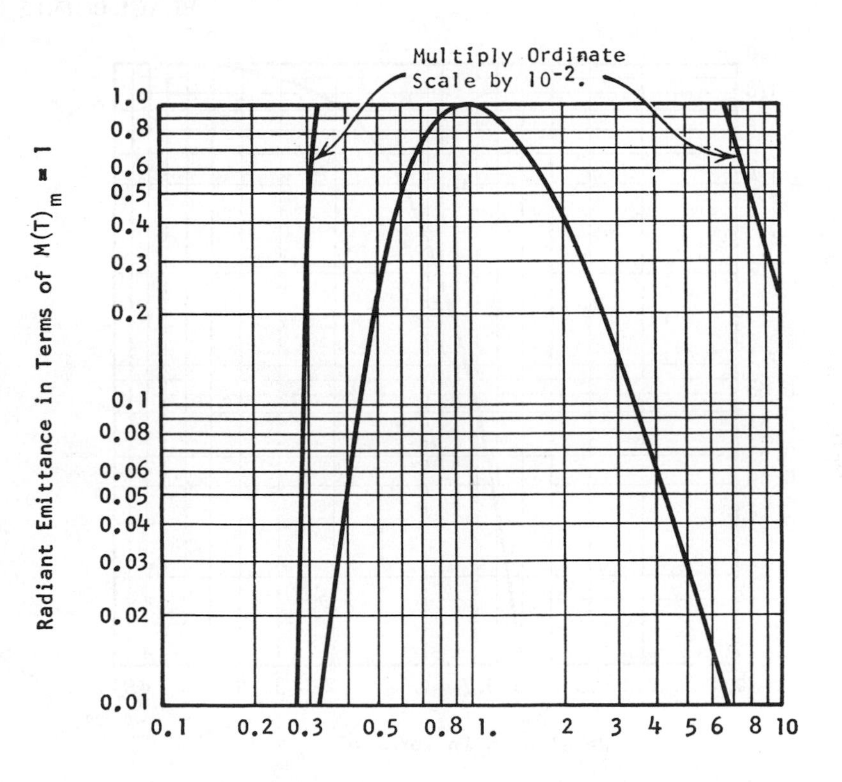

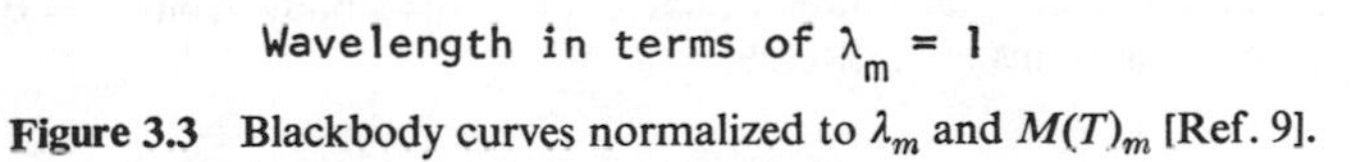

Figure 3.3 Blackbody curves normalized to λ_m and $M(T)_m$ [Ref. 9].

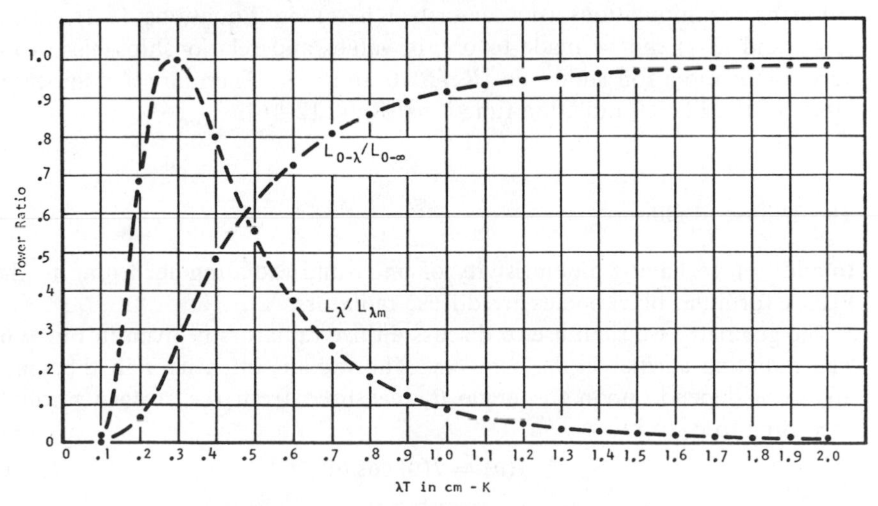

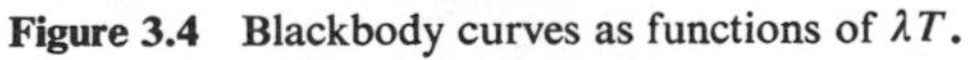

Figure 3.4 Blackbody curves as functions of λT.

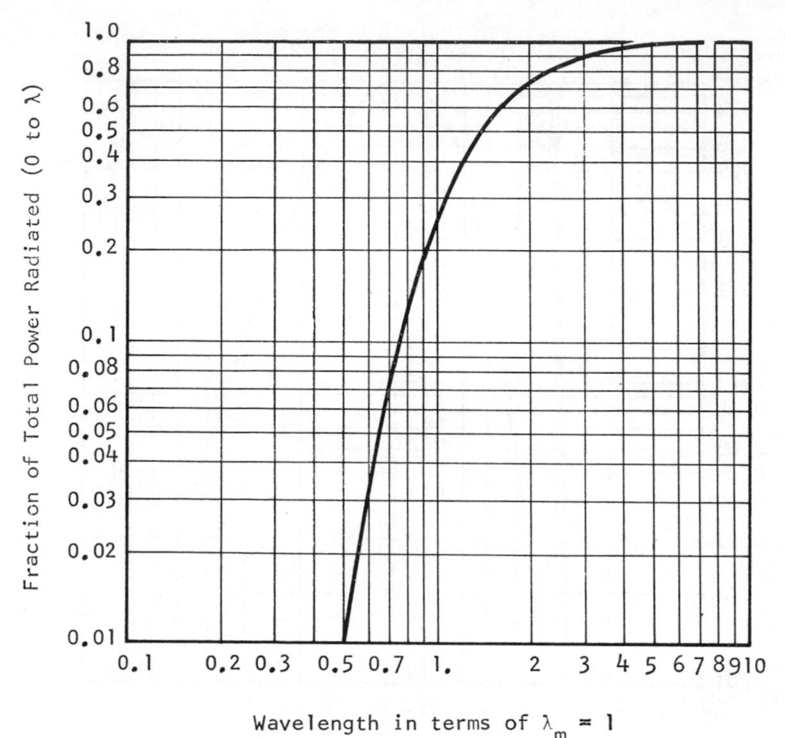

Figure 3.5 Fraction of total power radiated by a blackbody from $\lambda = 0$ up to any wavelength, given in terms of λ_m [Ref. 9].

Further manipulations (not described here) on Equations (3-2), (3-3), (3-4), and (3-7) can be made to obtain values and relationships basic to a very useful radiation slide rule [Refs. 10 and 11]. There are a number of tables of the Planck radiation functions [Refs. 12–16].

The Diffuse Radiator

In addition to having an emissivity of unity and radiating according to the Planck formula, blackbodies are diffuse radiators.

The geometry we shall use to discuss diffuse radiators is given in Fig. 3.6. The radiating surface is the area δa. The radiant intensity I in the space about δa depends upon the angle θ, measured from the surface normal, according to:

$$I(\theta) = I(0) \cos \theta, \tag{3-41}$$

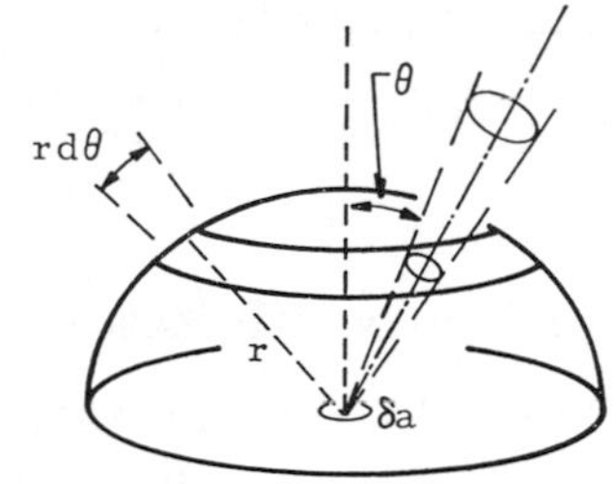

Figure 3.6 Geometry about a diffuse radiator.

where $I(0)$ is the radiant intensity in the direction of the normal. The intensity is said to follow the Lambert cosine law, so the surface is sometimes called a Lambertian surface.

To find the radiance in the direction of θ, we consider an annulus on the surface of a sphere, the sphere being centered on the surface and the annulus being concentric with the normal to the surface. The area da of the annulus is

$$da = (2\pi r \sin \theta) r \, d\theta. \tag{3-42}$$

This area is everywhere normal to a radius of the sphere. The solid angle $d\omega$ subtended by this area is

$$\begin{aligned} d\omega &= (2\pi r^2 \sin \theta \, d\theta)/r^2 \\ &= 2\pi \sin \theta \, d\theta. \end{aligned} \tag{3-43}$$

The radiant flux (power) passing out through this solid angle, since the direction to the area is always the angle θ, is

$$d\Phi = I(\theta) \, d\omega. \tag{3-44}$$

Then, by using the definition of radiance, one can obtain

$$\begin{aligned} L_e(\theta) &= d\Phi/(\delta a \cos \theta) \\ &= [I(0) \cos \theta] \, d\omega/(d\omega \delta a \cos \theta), \end{aligned} \tag{3-45}$$

$$L_e(\theta) = I(0)/\delta a. \tag{3-46}$$

This result shows that radiance is the same in all directions about a diffuse radiator.

The radiant emitted exitance can be found by first integrating the radiant intensity over a hemisphere about the surface. By substituting in Eq. (3-44),

$$d\Phi = [I(0) \cos \theta](2\pi \sin \theta \, d\theta). \tag{3-47}$$

Then by integrating over a range of θ from 0 to $\pi/2$,

$$\begin{aligned} \Phi &= 2\pi I(0) \int_0^{\pi/2} \cos \theta \sin \theta \, d\theta \\ &= \pi I(0). \end{aligned} \tag{3-48}$$

Then the radiant emitted exitance is

$$M = \Phi/\delta a = \pi I(0)/\delta a$$
$$= \pi L_e. \tag{3-49}$$

Cavity-Type Sources of Blackbody Radiant Energy

In order to construct a near-perfect blackbody (absorptance approaching one), using available materials whose absorptances are less than one, a way to compensate for this deficiency is sought. One method is to arrange a path for the incoming energy so that many surfaces are encountered before the energy finds its way back as "reflected" energy. A simple approach is to cut a small hole into an otherwise enclosed cavity and regard the hole as the simulated blackbody. Since absorptance is quantitatively equal to the emissivity, it is not surprising that among the various practical sources of radiant energy the temperature-controlled cavity-type sources approach the radiating characteristics of blackbodies most closely. A cavity-type source is made by forming a small opening in an enclosed, isothermal cavity—thus exposing a field of radiant energy flux that has almost the same properties as a field produced within a completely enclosed, isothermal (blackbody) cavity. The opening is the source, at the temperature of the cavity, and the radiant energy passing through the opening constitutes a transmitted radiant

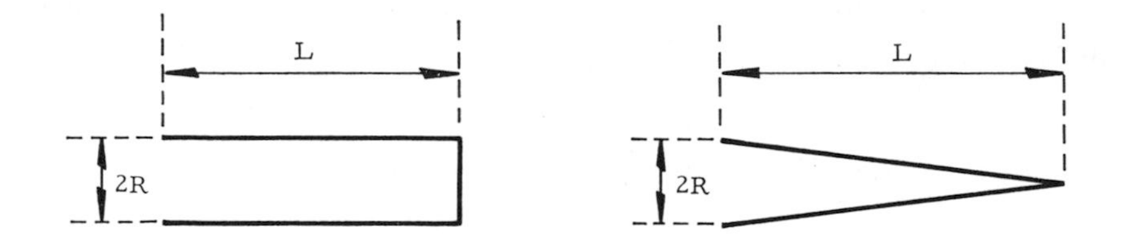

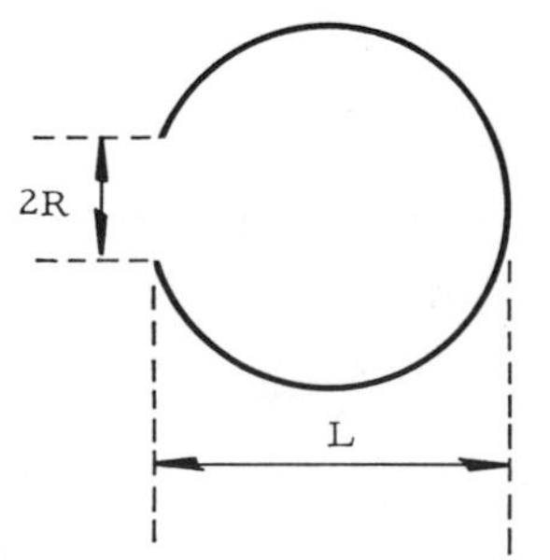

Figure 3.7 Three geometries of cavity-type sources of radiant energy.

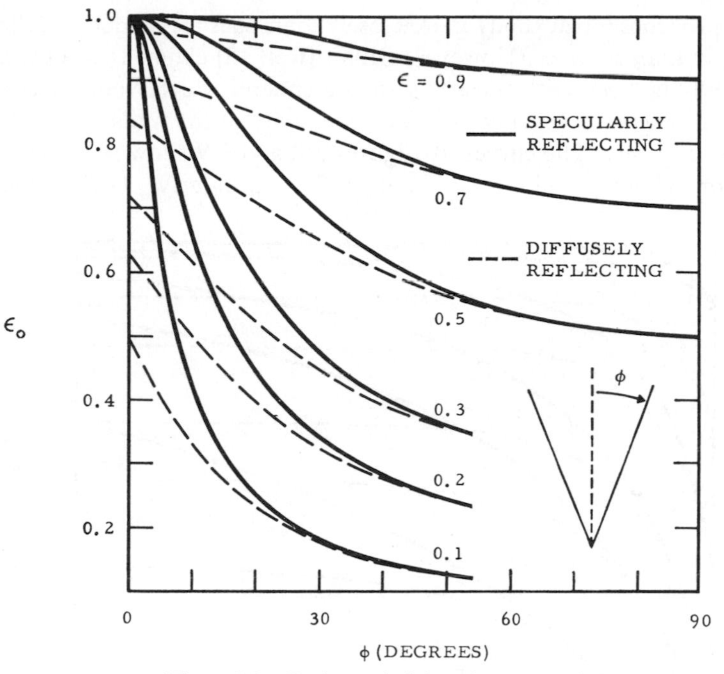

Figure 3.8 Cavity emissivity of cones [Ref. 24].

exitance of a hypothetical surface covering the opening. The power leaving the opening actually upsets the thermal equilibrium within the cavity; thus the radiating properties depart somewhat from those of a true blackbody. Pertinent questions, therefore, are: How closely can these cavity-type "blackbodies" be made to approach the ideal? What shape and size are most suitable for a given size opening? What is the value of the cavity emissivity ϵ_0? How is ϵ_0 related to ϵ, the emissivity of the walls? What is the directional distribution of the radiated energy, of ϵ_0? Should the walls be made to reflect diffusely, or specularly (like a mirror)?

In Fig. 3.7, three common shapes—circular cones, circular cylinders, and spheres—are shown in cross section, the section containing the axis or radius. A significant parameter in the design of these shapes is the aspect ratio $\rho = L/2R$. In general, the larger the ratio, the higher the cavity emissivity. In cavities of any shape, a variation in temperature over the walls causes a corresponding uncertainty in ϵ_0 [Ref. 17]. Among the shapes given, the cone is easiest to construct under the constraints of large aspect ratio and uniform temperature, and the cone turns out to be the most commonly found among commercially available sources. However, one finds—after investigating the relationship between cavity emissivity and shape as indicated in Figures 3.8, 3.9, and 3.10, especially for diffusely reflecting walls—that a sphere approaches a blackbody more closely than does a cylinder, a cylinder more nearly than a cone. (However, Quinn [Ref. 18] claims that a cylinder has a slightly higher cavity emissivity than a sphere, which would make the cylinder more nearly a blackbody.) The emissivity of the walls should be as high as practicable. The curves of Figures 3.8 and 3.9 show that the walls at least for cylinders and cones, should reflect specularly. Unfortunately,

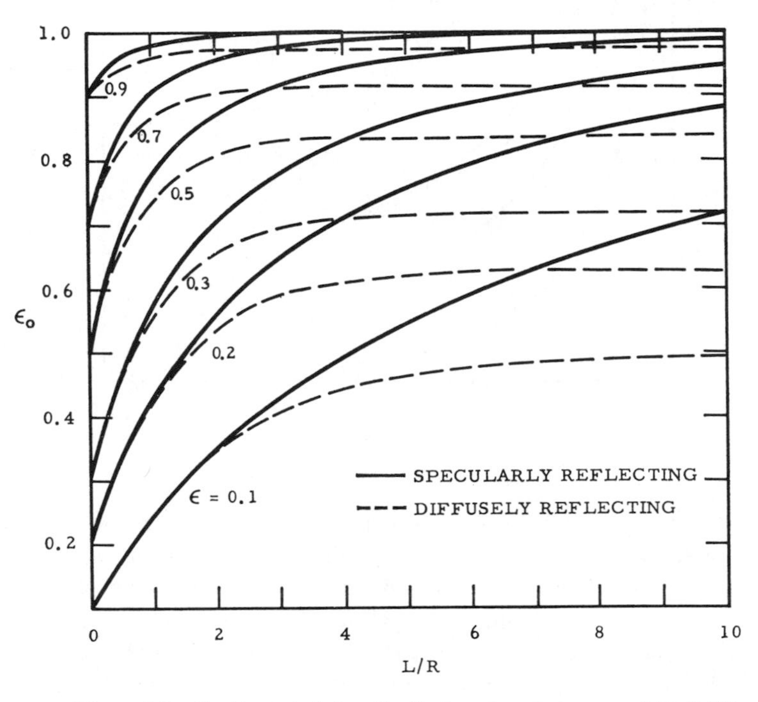

Figure 3.9 Cavity emissivity of cylinders closed at one end [Ref. 24].

high emissivity and specular reflectance are rarely achieved on the same surface because surfaces reflecting specularly tend to have low emissivities, and surfaces having high emissivities tend to reflect a large proportion of the reflected energy diffusely.

The emissivity ϵ_0 is defined as the ratio of the total power radiated by the opening to the power radiated by a blackbody having the same area as the opening and the same temperature as the cavity. Generally, the directional distribution of the radiant power about the cavity opening is not uniform; that is, the cavity is not a diffuse radiator. However, as ϵ_0 approaches unity, the cavity source also more nearly assumes the directional characteristics of a diffuse radiator. It has been shown [Ref. 19] that if one is able, by a suitable optical system, to observe a back region, near the intersection of the

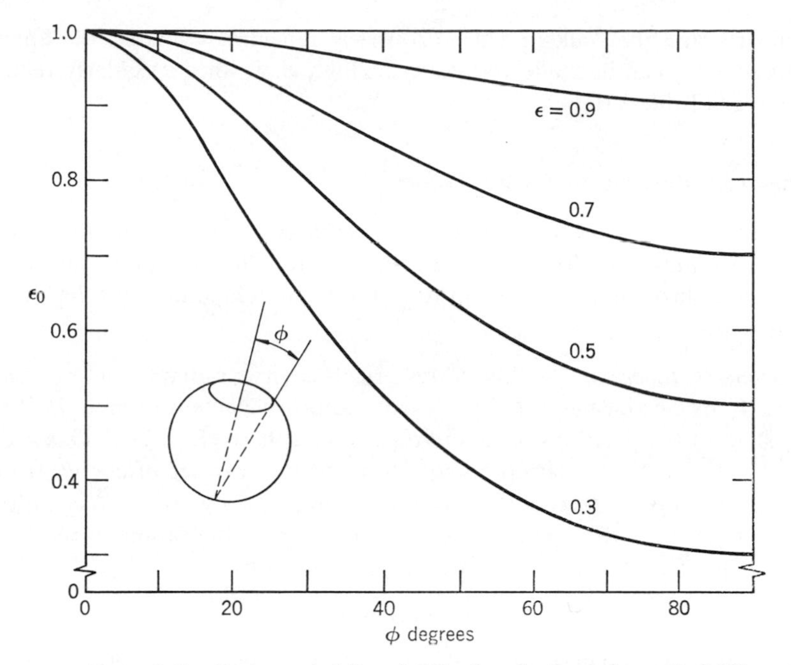

Figure 3.10 Cavity emissivity of diffusely reflecting spheres [Ref. 30].

TABLE VIII EFFECTIVE EMISSIVITIES FOR CYLINDERS CLOSED AT ONE END HAVING SPECULAR AND DIFFUSE REFLECTANCE*

$\rho = L/2R$	$\epsilon = 0.5$	
	Diffuse	Specular
2	0.9460–0.9540	0.9422
3	0.9768–0.9793	0.9677
4	0.9880–0.9887	0.9797

$\rho = L/2R$	$\epsilon = 0.75$	
	Diffuse	Specular
2	0.9815–0.9842	0.9885
4	0.9956–0.9961	0.9966

* For the diffuse cases, the apparent emissivity applied only when observing deep into the cavity (see Ref. 19). For the specular case the cavity emissivity applied to the total radiated power leaving the opening.

end disk and the wall, in a deep diffusely reflecting cylinder, the apparent emissivity ϵ_a can be made essentially as high as ϵ_0 for a specularly reflecting cylinder (Table VIII).

Theorems Relating to Cavity Theory

The theorems discussed below will be useful later for deriving formulas for spherical and cylindrical cavities (Ref. 20). The radiation surfaces are assumed throughout to be diffuse. Uniform temperature walls are also assumed.

Sumpner's theorem. Within a spherical cavity, radiant energy received directly by one element of area ds from another element of area dS is independent of the positions of ds and dS on the sphere [Ref. 21]. This is easily demonstrated by setting up an equation for the transfer of energy from dS to ds. An important corollary is that the power radiated from dS is uniformly distributed over the sphere. Because this is true, the power Φ received by an extended area s from another area S is given by

$$\Phi = W(sS/A_s), \tag{3-50}$$

in which W is radiant emitted exitance of the walls and A_s is the area of the sphere. Similarly, if a beam of radiant energy enters the spherical cavity by a small opening and is then diffusely reflected, the incident energy is uniformly distributed over the spherical walls after the first reflection.

Bartlett's theorem. In Fig. 3.11, the radiant power received by a disk C from an annulus of a cylinder between a and b is the power that disk C would receive from a disk at position a minus what it would receive from a disk at b. The assumed disks at a and b have the same emissivity, temperature, and radius as the cylinder. The radius of the disk C is assumed not larger than the radius of the cylinder. To prove this theorem, consider two surfaces S and s, having the same temperature and emissivity, bounded by a common closed curve c. A remote element of area da is oriented so that no point of

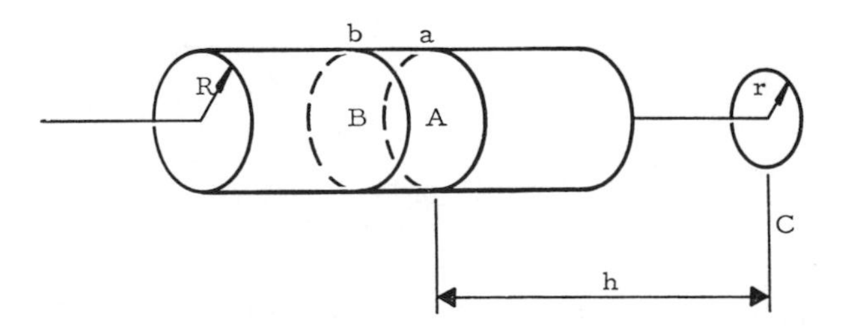

Figure 3.11 Radiant energy received by disk C from cylindrical annulus between a and b.

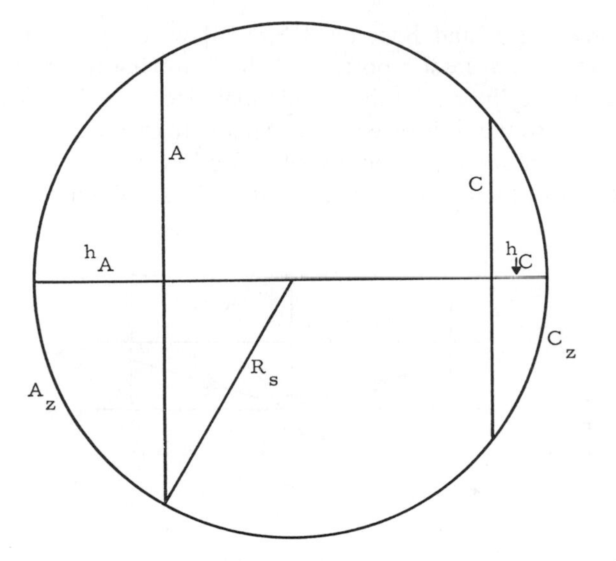

Figure 3.12 Radiant energy from disk A to disk C (also shown in Fig. 3.11) is radiant energy from spherical zone A_z to spherical zone C_z.

either S or s is outside a cone having its vertex on da and a generatrix that traces the curve c. If S and s were interchanged, there would be no change in radiant power received by da. This radiant power is

$$d\Phi = L_e \int_S (dS\ da)/h^2 = L_e \int_s (ds\ da)/h^2, \tag{3-51}$$

where h is distance between da and ds or between da and dS. Let the imaginary disk A (Fig. 3.11) be the surface s, and the annulus plus the disk B be S. Therefore, the radiant power received by disk C from the annulus is the power that would be received by disk A minus what would be received by disk B, which is Bartlett's theorem [Ref. 22].

Walsh's theorem. Walsh (Ref. 23) determined the power received by disk C from disk A (Fig. 3.11). The circles bounding disks A and C determine a sphere (Fig. 3.12). The radiant power received by C from A is the same as the power that would be received by C from the zone of the sphere A_z. Similarly, the radiant power received by C from A_z is the same as the power that would be received from A_z by C_z, which is determined by using Sumpner's theorem as follows: Writing Φ from Eq. (3-50) and using the geometry of the sphere, one obtains

$$\Phi = (\pi^2 N/2)\{(h^2 + r^2 + R^2) - [(h^2 + r^2 + R^2)^2 - 4r^2R^2]^{1/2}\} \tag{3-52}$$

where h is distance between the disks and R and r are their radii.

Lin's theorem. Lin and Sparrow [Ref. 24] were apparently the first to publish a proof of at least a portion of the following theorem: When a ray passes down a cylindrical tube of circular cross section with specularly reflecting walls so that it is reflected repeatedly from the walls, both the axial distance between successive points of reflection and the angular rotation about the axis of these points are constant. To see the nature of the proof,

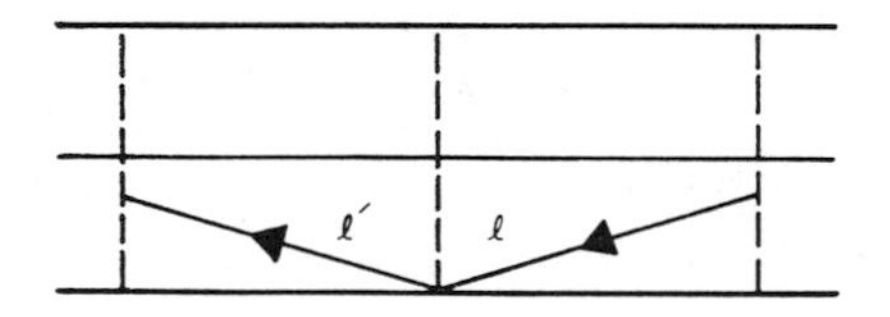

(a)

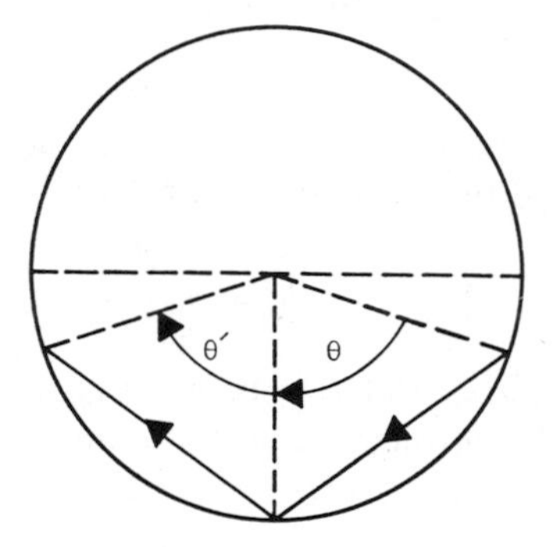

(b)

Figure 3.13 Symmetry of a ray reflected from a cylindrical wall.

consider a ray at any point of reflection and two projection planes passing through this point, one a meridional plane and the other perpendicular to it. (A meridional plane contains the cylinder axis.) Project both the incident and reflected rays between points of reflection onto the meridional plane and onto the perpendicular plane. The two projections are shown in Figures 3.13a and 3.13b respectively. From cylindrical geometry and the equality of incidence and reflection angles, it can be shown that the distances l and l' and the angles θ and θ' are equal.

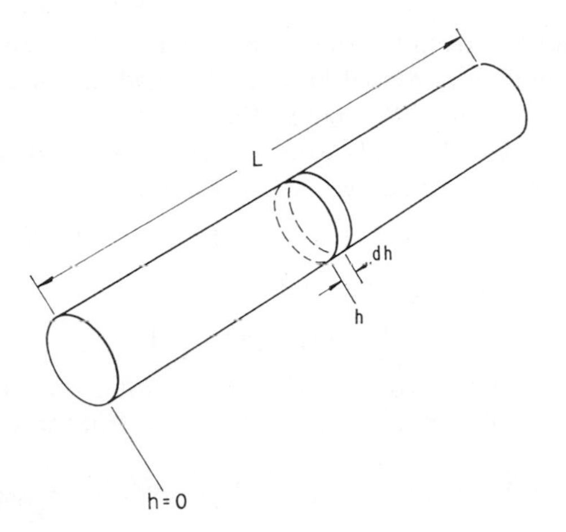

Figure 3.14 Geometry and notation for analyzing a radiating cylinder.

Cylinders Reflecting Specularly

Jain [Ref. 25] and Krishnan [Ref. 26] have described a technique for finding the radiant power leaving a cylindrical cavity whose walls radiate diffusely and reflect specularly. The notation and geometry are shown in Fig. 3.14. The power $\Phi(h)$ radiated from an imaginary disk at h and leaving the end at $h = 0$ directly without being reflected is known from Bartlett's and Walsh's theorems. It is found by using Eq. (3-50) and Eq. (3-52) with $r = R = D/2$ and $M = \pi L_e = \sigma T^4$:

$$\Phi(h) = \sigma T^4 \epsilon \xi(h), \tag{3-53}$$

where

$$\xi(h) = (\pi/4)[2h^2 + D^2 - 2h(h^2 + D^2)^{1/2}]. \tag{3-54}$$

The emissivity of the walls is assumed to be independent of temperature and of wavelength; $(1 - \epsilon)$ is reflectance and is assumed independent of angle of incidence; and h is distance along the axis measured from the open end that is being observed. Assume further that the cylinder is open at both ends. Then by writing the differential of $\Phi(h)$, one obtains the power from an annular ring dh at h that leaves the cylinder at $h = 0$ directly:

$$(d\Phi)_0 = -\sigma T^4 \epsilon f(h)\, dh, \tag{3-55$_0$}$$

where

$$f(h) = -d\xi/dh. \tag{3-56}$$

We observe for use later that

$$\int_0^L f(h)\, dh = \xi(0) - \xi(L). \tag{3-57}$$

The power radiated from the ring dh at h that first encounters the wall on any other elemental ring located between the end and $h/2$ will be reflected only once before exiting. We know this from Lin's theorem because a second encounter with the walls would have to occur at least a distance h from dh. The power reflected once and exiting is

$$(d\Phi)_1 = -\sigma T^4\epsilon(1 - \epsilon)[f(h/2) - f(h)]\, dh. \qquad (3\text{-}55_1)$$

Similarly, the power exiting after r reflections is

$$(d\Phi)_r = -\sigma T^4\epsilon(1 - \epsilon)^r\{f[h/(r + 1)] - f(h/r)\}\, dh. \qquad (3\text{-}55_r)$$

The total power radiated from the ring and exiting the end is obtained by summing a set of equations ($3\text{-}55_r$), where r goes from 0 to ∞:

$$d\Phi = \sum_{r=0}^{\infty}(d\Phi)_r = -\sigma T^4\epsilon^2\sum_{r=0}^{\infty}(1 - \epsilon)^r f[h/(r + 1)]\, dh. \qquad (3\text{-}58)$$

Then, by integrating over h from zero to the length of the cavity L, one obtains the total power radiated by the tube and leaving the open end.

$$\Phi_L = \sigma T^4\epsilon^2 \sum_{r=0}^{\infty}(1 - \epsilon)^r \int_0^L f[h/(r + 1)]\, dh. \qquad (3\text{-}59)$$

Let $h' = h/(r + 1)$; $\rho = L/D$; and $A = \pi D^2/4$, the area of the open end as a hypothetical radiating surface. Then

$$\Phi_L = \sigma T^4\epsilon^4 \sum_{r=0}^{\infty}(1 - \epsilon)^r(r + 1)\int_0^{L/(r+1)} f(h')\, dh', \qquad (3\text{-}60)$$

$$\Phi_L = (\sigma T^4 A)\left(-2\rho^2\epsilon^2 \sum_{r=0}^{\infty} \frac{(1 - \epsilon)^r}{(r + 1)}\left\{1 - \left[1 + \left(\frac{r + 1}{\rho}\right)^2\right]^{1/2}\right\}\right). \qquad (3\text{-}61)$$

By this grouping of terms, the cavity emissivity ϵ_0 can be seen by definition to be the quantity in the large parentheses of Eq. (3-61). Values for ϵ_0 defined

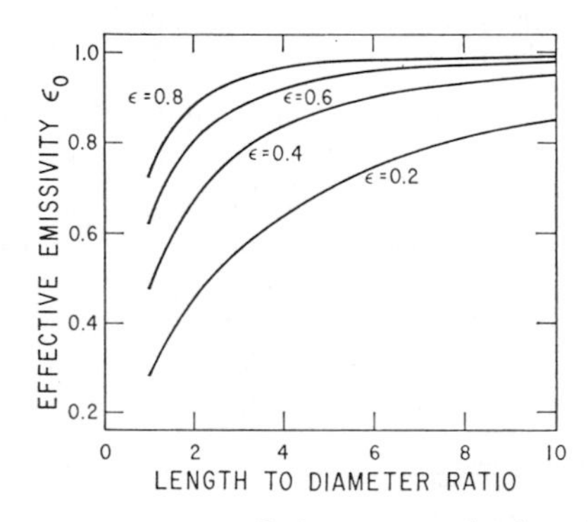

Figure 3.15 Cavity emissivity of a cylinder open at both ends computed from Jain's formula [Ref. 27].

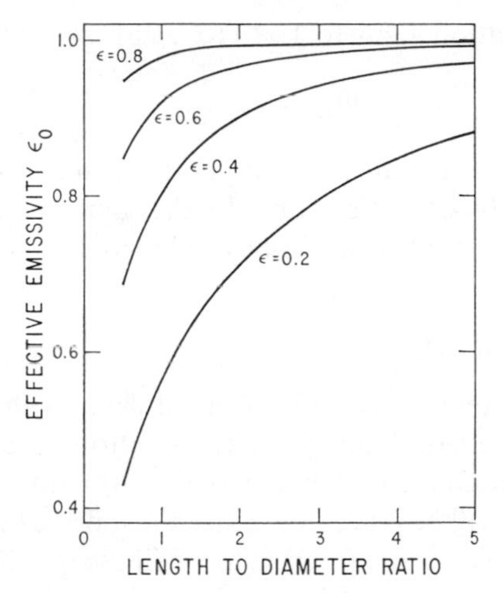

Figure 3.16 Cavity emissivity of a cylinder closed at one end computed from Jain's formula [Ref. 27].

in this way are shown plotted in Figures 3.15 and 3.16 [Ref. 27]. Jain has shown that $(1 - \epsilon_0)$ for a tube closed at the far end is $(1 - \epsilon)$ times the $(1 - \epsilon_0)$ for an open end tube of twice the length. The disk that closes the tube is assumed to have a surface with the same radiating and reflecting properties as the side walls. The cavity emissivity ϵ_{0c} for the closed tube is given by

$$\epsilon_{0c} = \epsilon - 8\rho^2\epsilon^2 \sum_{r=0}^{\infty} \frac{(1-\epsilon)^{r+1}}{(r+1)} \left\{ 1 - \left[1 + \left(\frac{r+1}{2\rho} \right)^2 \right]^{1/2} \right\}. \tag{3-62}$$

The similar curves of Fig. 3.9 were first calculated by Lin and Sparrow, who derived Eq. (3-62) by an entirely different procedure.

A conclusion reached both by Jain and Krishnan, and supported by the trend in the curves of Figures 3.9 and 3.16, is that a specularly reflecting cylinder approaches a blackbody as closely as one wishes if the length is extended sufficiently. We can illustrate what length is sufficient by considering the number of terms used by the computer to sum the series of Eq. (3-61). The computer was instructed to terminate the series at a term that became 10^{-5} times the accumulated sum. For the "worst case"—an open tube with $\epsilon = 0.1$ and $\rho = 10$—the computer used 97 terms. The number of terms increases very slowly with ρ. When L becomes such a length that $L/98$ [$98 = 97 + 1$, see $(r + 1)$ in Eq. (3-60)] is considered unbounded [i.e., $(L/98) \gg D$ or $\rho \gg 98$] the upper limit, $L/(r + 1)$, in Eq. (3-60) can be written ∞. Then the integration gives

$$\Phi_L = \sigma T^4 A \epsilon^2 \sum_{r=0}^{\infty} (1 - \epsilon)^r (r + 1). \tag{3-63}$$

The series in this equation sums to $1/\epsilon^2$ (Ref. 28), so that

$$\Phi_L = \sigma T^4 A. \tag{3-64}$$

Thus, provided $\epsilon \geq 0.1$ and $\rho \gg 98$, Φ_L is independent of ϵ, and the tube radiates as a blackbody. A closed tube, by the same criterion, would accept $2\rho \gg 98$ as being adequate for blackbody behavior.

Spheres Reflecting Diffusely

To review the theory that applies to spherical cavities that reflect diffusely, suppose unit radiant energy enters the cavity through a small opening of area s. At the first encounter with the walls a fraction ρ (ρ being the reflectance of the walls) will be reflected. Since the reflected energy is uniformly distributed over the sphere, the fraction s/S will escape through the opening if S is the area of the sphere, including the opening. The energy escaping after one reflection in the cavity is $\rho - (\rho s/S)$. The amount escaping after two reflections is $(\rho^2 s/S)[1 - (s/S)]$; after three reflections, $(\rho^3 s/S)[1 - (s/S)]^2$. The total amount escaping after all reflections is found by summing the series:

$$\sum = (\rho s/S) + (\rho^2 s/S)[1 - (s/S)] + (\rho^3 s/S)[1 - (s/S)]^2 + \cdots + (\rho^{n+1} s/S)[1 - (s/S)]^n, \tag{3-65}$$

$$\sum = (\rho s/S) \sum_{n=0}^{\infty} \rho^n [1 - (s/S)]^n. \tag{3-66}$$

The series is a geometric series with a common ratio less than one; its sum is known:

$$\sum = (\rho s/S)\{1 - \rho[1 - (s/S)]\}^{-1}. \tag{3-67}$$

The fraction $(1 - \sum)$ of the radiant energy entering the opening is absorbed by the opening. The cavity emissivity is then given by $\epsilon_0 = (1 - \sum)$, and it can be written in the following form by using $\rho = (1 - \epsilon)$:

$$\epsilon_0 = 1/\{[1 - (s/S)] + (s/\epsilon S)\}. \tag{3-68}$$

This theory, when first derived by Gouffé [Ref. 29], was intended to be an approximate theory for any of the three shapes of cavities. The theory is exact for diffusely reflecting spheres. The curves of Fig. 3.10 discussed earlier are based on an entirely different approach to the spherical cavity problem by Sparrow and Jonsson [Ref. 30]. However, Fecteau [Ref. 31] has shown that the three methods of Gouffé, Sparrow and Jonsson, and DeVos [Ref. 32] are equivalent for diffusely reflecting spheres.

Cones

Theories for diffusely reflecting and specularly reflecting cones, which will not be given in detail here, have been developed by Sparrow and Jonsson [Ref. 33] and by Lin and Sparrow [see Ref. 24] respectively. The curves of Fig. 3.8 are from Lin and Sparrow.

REFERENCES

1. F. K. Richtmeyer, E. H. Kennard, and T. Lauritsen, *Introduction to Modern Physics*, Fifth Edition. Chapter 4. McGraw-Hill Book Company, Inc., New York, 1955.
2. A. Sommerfeld, *Lectures in Theoretical Physics*, Vol. V, Thermodynamics and Statistical Mechanics, pp. 135–152. Academic Press, Inc., New York, 1956.
3. F. W. Constant, *Theoretical Physics: Thermodynamics, Electromagnetism, Waves, and Particles*, pp. 108–111. Addison-Wesley Publishing Company, Inc., Reading, Mass., 1958.
4. D. E. Erminy, "Some Approximations to the Planck Function in the Intermediate Region with Applications in Optical Pyrometry," *Appl. Optics*, **6**, 107 (1967).
5. R. S. Burrington, *Handbook of Mathematical Tables and Formulas*. McGraw-Hill Book Company, Inc., New York, 1965.
6. K. Knopp, *Theory and Application of Infinite Series*, p. 237. Blackie and Son, London, 1951.
7. B. S. Sadykov, "Position of the Maximum in the Emission Spectrum of Metals and its Effect on the Measurement of Temperature," *Zh. Priklad Spektrosk.* (*USSR*) **8**, 511 (1968).
8. M. M. Gurvich, "Spectral Distribution of Radiant Energy," *Sov. Phys.—Usp.*, **5**, 908 (1963).
9. F. Benford, "Laws and Corollaries of the Blackbody," *J. Opt. Soc. Am.*, **29**, 92 (1939).
10. M. W. Makowski, "A Slide Rule for Radiation Calculations," *Rev. Sci. Inst.*, **20**, 876 (1949).
11. A radiation slide rule, designated GEN-15C, is available from the General Electric Company, 1 River Road, Schenectady, N.Y.
12. M. Pivovonsky and M. R. Nagel, *Tables of Blackbody Radiation Functions*. Macmillan, New York, 1961.
13. M. Czerny and A. Walther, *Tables of the Planck Radiation Law*. Springer-Verlag, Berlin, 1961.
14. A. N. Lowan and G. Blanch, "Tables of Planck's Radiation and Photon Functions," *J. Opt. Soc. Am.*, **30**, 70 (1940).
15. S. A. Twomey, "Table of the Planck Function for Terrestrial Temperatures," *Infrared Phys.* **3**, 9 (1963).
16. S. A. Golden, "Spectral and Integrated Blackbody Radiation Functions," Research Report 60-23, Rocketdyne Division, North American Aviation, Inc., Canoga Park, Calif. (1960).
17. P. F. O'Brien and B. J. Heckert, "Effective Emissivity of a Blackbody Cavity at Nonuniform Temperature," *Illum. Eng.*, **60**, 187 (1965).
18. T. J. Quinn, "The Calculation of the Emissivity of Cylindrical Cavities Giving Near Blackbody Radiation," *Brit. J. Appl. Phys.* **18**, 1105 (1967).
19. E. M. Sparrow, L. U. Alberts, and E. R. G. Eckert, "Thermal Radiation Characteristics of Cylindrical Enclosures," *J. Heat Transfer*, **84C**, 73 (1962).

20. C. S. Williams, "Discussion of the Theories of Cavity-Type Sources of Radiant Energy," *J. Opt. Soc. Am.*, **51,** 564 (1961).
21. W. E. Sumpner, "Diffusion of Light," *Proc. Phys. Soc.* (London), **12,** 10 (1892).
22. A. C. Bartlett, "Radiation from a Cylindrical Wall," *Phil. Mag.*, **40,** 111 (1920).
23. W. T. Walsh, "Radiation from a Perfectly Diffusing Circular Disc," *Proc. Phys. Soc.* (London), **32,** 59 (1919–1920).
24. S. H. Lin and E. M. Sparrow, "Radiant Interchange Among Curved Specularly Reflecting Surfaces—Application to Cylindrical and Conical Cavities," *J. Heat Transfer*, **87C,** 229 (1965).
25. S. C. Jain, "Radiation from Inside a Long Cylinder as Approximation to that from a Blackbody," *Indian J. Pure Appl. Phys.*, **1,** 7 (1963).
26. K. S. Krishnan, "Effect of Specular Reflexions on the Radiation Flux from a Heated Tube," *Nature*, **187,** 135 (1960).
27. C. S. Williams, "Specularly vs Diffusely Reflecting Walls for Cavity-Type Sources of Radiant Energy," *J. Opt. Soc. Am.*, **59,** 249 (1969).
28. K. S. Krishnan and R. Sundaram, "The Distribution of Temperature along Electrically Heated Tubes and Coils," *Proc. Ray. Soc.*, **A256,** 302 (1960).
29. A. Gouffé, "Corrections d'ouverture des corps-noir artificials compte tenu des diffusions multiples internes," *Rev. Opt.*, **24,** Nos. 1–3 (1945).
30. E. M. Sparrow and V. K. Jonsson, "Absorption and Emission Characteristics of Diffuse Spherical Enclosures," *J. Heat Transfer*, **84C,** 188 (1962).
31. M. T. Fecteau, "The Emissivity of the Diffuse Spherical Cavity," *Appl. Optics*, **7,** 1363 (1968).
32. J. C. De Vos, "Evaluation of the Quality of a Blackbody," *Physica*, **20,** 669 (1954).
33. E. M. Sparrow and V. K. Jonsson, "Radiant Emission Characteristics of Diffuse Conical Cavities," *J. Opt. Soc. Am.*, **53,** 816 (1963).

4

Properties of Waves and Materials

In a broad sense "optics" is the science of interaction between electromagnetic waves and matter. Examples of common optical processes, which are often accepted without a concern for the underlying phenomena, are: reflection at metallic surfaces (because free electrons of the metals so readily absorb and reradiate the incident radiant energy), focusing of light rays by a lens (because a wave changes direction of propagation with a change in velocity), and chromatic aberration of a lens (because of dispersion near wavelength regions of resonance absorption). An electromagnetic wave behaves differently in various materials; properties, to account for the differences, are assigned to a material by comparing the wave behavior in the material to the behavior in free space. The underlying wave propagation phenomena, which can be treated either as material properties or as transmission characteristics, constitute the subject matter of this chapter: Hence the title.

Two perspectives are commonly used that are distinguished by the scale of the region considered. In a macroscopic perspective, the region is very large compared with the dimensions of atoms and molecules. We use this perspective in the study of wave phenomena, that is, the propagation of electromagnetic disturbances in various media. The media are, in general, considered to be continuous. The second perspective considers a much smaller region in which we might deal with encounters between the wave, or photons, and the individual atoms or electrons. The present chapter generally assumes the first perspective; later chapters, particularly the ones on absorption and emission spectra and detectors of radiant energy, adopt the second, the microscopic perspective.

Optical Materials

What are the essential characteristics of a material for use in optical instruments? What materials have these properties? Suitable materials have a number of required characteristics and are, in fact, relatively few. Preliminary to studying the interaction of electromagnetic waves and materials, these characteristics are described in the following paragraphs.

(1) Optical materials, to be useful in practical instrumentation, must have mechanical properties adaptable to those of the holders and supporting structures with which they are associated. This is because surface shape and position of optical elements must typically be held within a fraction of a micrometer throughout all changes in environmental conditions. Differential temperature effects (for instance, relative movement and distortion caused by temperature change or gradients), and sagging because of gravity, require particular attention.

(2) Before an optical element is mounted, it must be given the proper dimensions, shape, and polish in the optic shop. To facilitate this "figuring," the material should be easy to work. Sapphire is so hard, for example, and the surface pits so readily during grinding, that a specified, optically polished surface is difficult to achieve. By contrast, arsenic trisulfide glass, which is sometimes used as an infrared-transmitting material, is soft, but working tends to make it waxy and, therefore, mechanically plastic.

(3) Finally, the optical properties themselves—transmittance, reflectance, refractive index, optical homogeneity, isotropy, and the wavelength variation of these properties—are significant. These properties become manifest during the study of electromagnetic wave phenomena. Wave phenomena, as we shall discuss them, have been studied for a long time but continue to be timely—even for one seeking the most up-to-date optical information. In pursuing these phenomena, we shall postulate Maxwell's equations, and since they are accepted as classical, we feel free immediately to outline theories of various phenomena based upon them.

The Field Vectors and Material Parameters

In the theory of electromagnetism, Maxwell's equations have become so basic and so firmly established that the study of wave phenomena is started with them. They are:

$$\nabla \times \mathbf{E} = -\partial \mathbf{B}/\partial t, \tag{4-1}$$

$$\nabla \times \mathbf{H} = \mathbf{j} + \partial \mathbf{D}/\partial t. \tag{4-2}$$

Two additional equations, which are often included with Maxwell's equations, may be derived from the above by assuming that electric charge is conserved [Ref. 1, Chapter 1]:

$$\nabla \cdot \mathbf{B} = 0, \tag{4-3}$$

$$\nabla \cdot \mathbf{D} = q_d. \tag{4-4}$$

In these equations **E**, **D**, **H**, and **B** are the field vectors (electric field strength, electric induction, magnetic field strength, and magnetic induction, respectively), **j** is current density, and q_d is electric charge density. We assume the equations to be valid, that is, that they form a self-consistent set whose predictions concerning static and slowly varying (nonrelativistic) fields are in agreement with experimental data.

The four field vectors can also be related by using three material parameters: ϵ, μ, and σ, which are called permittivity, permeability, and conductivity, respectively. These quantities, which characterize the medium, are point functions constant in time. They may be scalars or, for anisotropic materials, tensors. Their values in a given medium depend upon the frequency components constituting the wave disturbance. The relationships between the field vectors established by the material parameters may be linear or nonlinear, the latter occurring particularly in connection with lasers and their high-intensity, coherent beams. In the present text we shall assume linearity. If, for the present, we also assume isotropy, the relationships are:

$$\mathbf{D} = \epsilon\mathbf{E}, \tag{4-5}$$

$$\mathbf{B} = \mu\mathbf{H}, \tag{4-6}$$

$$\mathbf{j} = \sigma\mathbf{E}. \tag{4-7}$$

For anisotropic materials, the field vectors are related by sets of tensor equations. The only set that is of practical interest in optics is the one relating electric induction and the electric field strength in crystals and other anisotropic materials.

$$D_j = \epsilon_{jk}E_k. \tag{4-8}$$

In this equation j and k are each successively x, y, and z.

Which of the field effects are of practical importance depends upon the class of materials—for example, insulators (or dielectrics), crystals, metals (or conductors), semiconductors, and magnetic materials. Although there are no materials belonging perfectly to any one of these classes, these distinctions are often useful. How Maxwell's equations are applied, therefore, depends on the nature of the material and the corresponding nature and relative values of ϵ, μ, and σ. With one set of values for these constants, the field equations apply to the whole volume of one medium; with different values, to the whole volume of a second. The theory can be extended to give an account of what happens at the interface between the two media.

A glimpse of the nature of **E** and **B** may be seen by examining the Lorentz equation of force **F** on an electric charge q moving with a velocity **v** in a magnetic and electric field (Ref. 2, p. 460):

$$\mathbf{F} = q\mathbf{E} + q(\mathbf{v} \times \mathbf{B}). \tag{4-9}$$

The vectors **E** and **B** may be operationally defined in terms of force, charge, and velocity by using this equation. The units for **E**, in the MKS system of units that we are using, are newton coulomb^{-1} or the equivalent volt meter^{-1}. The magnetic induction is in newton ampere^{-1} meter^{-1} or the equivalent weber meter^{-2}. In a vacuum the conductivity is zero, and ϵ and μ are designated ϵ_0 and μ_0 and have the values $\epsilon_0 = 4.85 \times 10^{-12}$ farad meter^{-1} and $\mu_0 = 4\pi \times 10^7$ henry meter^{-1}. In materials, $\epsilon = \epsilon_0\epsilon_r$ and $\mu = \mu_0\mu_r$, where ϵ_r and μ_r are relative values characterizing the material. **H** has the units ampere meter^{-1}, and **D** the units coulomb meter^{-2}.

Dielectrics

If we assume that a solid material has no free charges (i.e., no electrons are free to move throughout the body of material so that σ is everywhere zero) the material is called a dielectric. If σ is zero, the current density **j** also is zero. Any accumulated excess charge q will be a static charge and will neither contribute to nor respond to changes in field. The four "Maxwell's equations," assuming isotropy and homogeneity, are

$$\nabla \times \mathbf{E} = -\mu(\partial\mathbf{H}/\partial t), \tag{4-10}$$

$$\nabla \times \mathbf{H} = \epsilon(\partial\mathbf{E}/\partial t), \tag{4-11}$$

$$\nabla \cdot \mathbf{H} = 0, \tag{4-12}$$

$$\nabla \cdot \mathbf{E} = 0. \tag{4-13}$$

From these, two vector wave equations can be derived [Ref. 2, p. 519; Ref. 3, p. 327]:

$$\nabla^2\mathbf{H} = \epsilon\mu(\partial^2\mathbf{H}/\partial t^2), \tag{4-14}$$

$$\nabla^2\mathbf{E} = \epsilon\mu(\partial^2\mathbf{E}/\partial t^2). \tag{4-15}$$

In a Cartesian coordinate system, six scalar wave equations are represented by Equations (4-14) and (4-15) and are of the form

$$\nabla^2 U = \epsilon\mu(\partial^2 U/\partial t^2), \tag{4-16}$$

where U is any one of the field components E_x, E_y, E_z, H_x, H_y, and H_z. A completely general solution of Eq. (4-16) would be of the following form, which is actually more general than is necessary for the type of material assumed:

$$U(\mathbf{r},t) = U_1[\mathscr{k}\zeta(\mathbf{r}) + vt] + U_2[\mathscr{k}\zeta(\mathbf{r}) - vt]. \tag{4-17}$$

Here U_1 and U_2 are arbitrary functions of the indicated arguments; $\mathbf{r}$ is a position vector from the origin to a point (x,y,z); $\mathscr{k}$ is a propagation constant, the value of which will be determined later; $\zeta(\mathbf{r})$ is a real scalar quantity, a function of position; v is a velocity. The arguments show that any spatial waveform represented by the function existing at a time t_1 is displaced a distance $v(t_2 - t_1)$ at a later time t_2; the displacement is in the negative direction for U_1, the positive direction for U_2. Whether or not the waveform is preserved—that is, propagated without distortion—depends upon the function $\zeta(\mathbf{r})$. As will be seen later for the general case, only the sinusoidal components of a waveform can be transmitted undistorted through dielectric materials, and a variation of velocity with wavelength will alter the amplitudes and phases of the components. In the simple waves, usually sinusoidal plane waves, that we consider where $\mathscr{k}\zeta(\mathbf{r})$ has the form $(\mathscr{k} \cdot \mathbf{r})$, the waveform is preserved. We use spherical waves in the following chapter.

The equation representing the most general type of periodic waveform is of the form

$$\hat{U} = \hat{U}_0(\mathbf{r}) \exp i[\omega t \pm \mathscr{k}\zeta(\mathbf{r})]. \tag{4-18}$$

We shall usually deal with periodic waveforms in which $\omega = 2\pi\nu$, ν being the frequency. $\hat{U}_0(\mathbf{r})$ is a complex number the real part of which represents a scalar field quantity. Before considering special cases of this general equation, some of its properties should be noted. Materials for which the general form is required are inhomogeneous. Surfaces where the real part of $\hat{U}_0(\mathbf{r})$,

$$U_0^r(\mathbf{r}) = \text{constant}, \tag{4-19}$$

are surfaces of constant amplitude, and surfaces where

$$\zeta(\mathbf{r}) = \text{constant}, \tag{4-20}$$

are surfaces of constant phase. The two surfaces in general do not coincide as they do in the simpler waves. Also, the phase surfaces and the amplitude surfaces may not be propagating in the same direction, and neither direction is necessarily the direction of energy flow.

Let us assume a particular solution of Eq. (4-16) of the form

$$\begin{aligned} \hat{U} &= U_0 \exp - (i\mathscr{k} \cdot \mathbf{r} - i\omega t) \\ &= \hat{U}_0 \exp i\omega t, \end{aligned} \tag{4-21}$$

where

$$\hat{U}_0 = U_0 \exp(-i\mathscr{k} \cdot \mathbf{r}). \tag{4-22}$$

The propagation vector $\boldsymbol{k}$ has a magnitude k and the direction of wave propagation.

$$\boldsymbol{k} = \mathbf{I}_x k_x + \mathbf{I}_y k_y + \mathbf{I}_z k_z,$$

$$k^2 = k_x^2 + k_y^2 + k_z^2. \tag{4-23}$$

$\mathbf{I}_x$, $\mathbf{I}_y$, $\mathbf{I}_z$ are unit vectors parallel to the coordinate axes. A substitution of Eq. (4-21) into Eq. (4-16) gives

$$\nabla^2 \hat{U}_0 + \epsilon\mu\omega^2 \hat{U}_0 = 0. \tag{4-24}$$

This equation is known as the Helmholtz equation. A wave function that satisfies the wave equation must be both a space and a time function. Whenever the time function can be separated out as in Eq. (4-21), the remaining space function need satisfy only the Helmholtz equation. A separable time function is usually omitted (suppressed) in wave theory problems.

Suppose a transformation of coordinates has been made so that $\boldsymbol{k}$ is parallel to the y-axis. Then

$$\boldsymbol{k} \cdot \mathbf{r} = k(\mathbf{I}_y \cdot \mathbf{r}). \tag{4-25}$$

The wave surfaces are now planes perpendicular to the y-axis. The wave is a plane wave; and if $\mathbf{r}$ is taken along the y-axis, $\boldsymbol{k} \cdot \mathbf{r} = ky$, and

$$\hat{U} = U_0 \exp - i(ky - \omega t), \tag{4-26}$$

the real part of which is

$$U^{\imath} = U_0 \cos(\omega t - ky). \tag{4-27}$$

This is equivalent to Eq. (1-14) of Chapter 1. Earlier in Chapter 1, the sinuosidal function, Eq. (1-4), was discussed to develop the concepts of a traveling wave, including definitions of the period $\mathscr{T}$ and the wavelength λ. By pursuing a similar development with the argument of Eq. (4-27), we arrive at the equality

$$\omega\mathscr{T} = k\lambda. \tag{4-28}$$

From the definitions, $\omega = 2\pi\nu$ and $\mathscr{T} = 1/\nu$, we find

$$k = 2\pi/\lambda. \tag{4-29}$$

If Eq. (4-27) is substituted into Maxwell's equations, **E** and **H** can be shown to have no component parallel to the direction of propagation; the wave is transverse (Ref. 3, p. 327). Also, **E** and **H** are mutually perpendicular and are in phase in nonconducting, isotropic, and homogeneous materials.

By a separate argument **E** and **H** can be related to a vector **S** called *Poynting's vector* which gives the magnitude and direction of energy flow, the magnitude being energy per unit time per unit area (radiant flux density) crossing a fictitious surface perpendicular to the direction of flow [see Ref. 2, p. 518]. The relationship is

$$\mathbf{S} = \mathbf{E} \times \mathbf{H}. \tag{4-30}$$

The phase given by the argument of the cosine, Eq. (4-27), is propagating with a velocity that can be determined by setting the phase equal to a constant (representing a surface of fixed phase) and differentiating with respect to time, as follows:

$$\omega t - k y = C,$$

$$v = dy/dt = \omega/k = \lambda \nu. \tag{4-31}$$

Substitution of Eq. (4-27) into Eq. (4-16) gives

$$k^2 = \epsilon \mu \omega^2,$$

$$\epsilon \mu = k^2/\omega^2 = 1/\lambda^2 \nu^2 = 1/v^2. \tag{4-32}$$

Thus, the magnitude of the phase velocity is given by

$$v = 1/\sqrt{\epsilon \mu}. \tag{4-33}$$

In a vacuum, this phase speed is

$$c = 1/\sqrt{\epsilon_0 \mu_0}, \tag{4-34}$$

and in dielectrics

$$v = c/\sqrt{\epsilon_r \mu_r}. \tag{4-35}$$

The ratio,

$$c/v = \sqrt{\epsilon_r \mu_r} = n, \tag{4-36}$$

is called the *index of refraction* of the dielectric material through which the wave is propagating. Generally, in practical optical materials, μ_r is so nearly unity that we may say $n = \sqrt{\epsilon_r}$.

The phase velocity is also called the *wave velocity* because, for a wave represented by Eq. (4-27), the amplitude surfaces are propagating with the phase velocity. In a general wave, the phase velocity can actually exceed the speed of light, but the reader is reminded that this is not the velocity of energy flow and that no signal can be transmitted with a velocity greater than the velocity of light.

The implication in Eq. (4-36) that the index of refraction is constant for all wavelengths is actually not generally true. It is constant only in free space, Eq. (4-34), and tends to be constant in materials at long wavelengths; certain dielectrics, however, exhibit constant n over a considerable range of wavelengths.

The variation of n (refractive index) with wavelength is called dispersion. At wavelengths much greater than the linear dimensions of an atom, wave theory is quantitatively satisfactory. However, at optical frequencies—near infrared, visible, and ultraviolet—the theory must be augmented by microscopic considerations.

We can define another kind of velocity, group velocity, by considering a beam consisting of two special waves. The waves have equal amplitudes and the same direction of linear polarization. If the frequency of each has a difference $\frac{1}{2}\delta\nu$ from a mean frequency ν_0, one above and one below, the two phases are $[(\omega_0 - \frac{1}{2}\,\delta\omega)t - (k_0 - \frac{1}{2}\,\delta k)y]$ and $[(\omega_0 + \frac{1}{2}\,\delta\omega)t - (k_0 + \frac{1}{2}\,\delta k)y]$. The equation representing the resultant of the two beams is, in the form of Eq. (4-18),

$$\begin{aligned}\hat{U} &= U_0\{\exp\,[(i/2)(\delta\omega t - \delta k y)] + \exp\,[-(i/2)(\delta\omega t - \delta k y)]\} \\ &\quad \times \exp\,[i(\omega_0 t - k_0 y)] \\ &= 2U_0 \cos\,[\tfrac{1}{2}(\delta\omega t - \delta k y)] \exp\,[i(\omega_0 t - k_0 y)]. \qquad (4\text{-}37)\end{aligned}$$

This represents a wave whose phase function is $(\omega_0 t - k_0 y)$ and whose amplitude is varying as $\cos\,[\frac{1}{2}(\delta\omega t - \delta k y)]$ (Fig. 4.1). The phase velocity of the wave is, according to Eq. (4-32),

$$v_p = \omega_0/k_0. \qquad (4\text{-}38)$$

But the envelope of the wave, determined by the cosine factor, has a different velocity called the group, or envelope, velocity, and its magnitude is

$$v_g = \delta\omega/\delta k. \qquad (4\text{-}39)$$

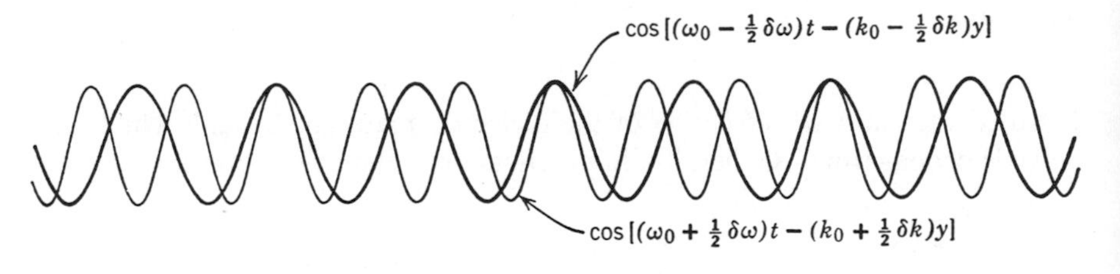

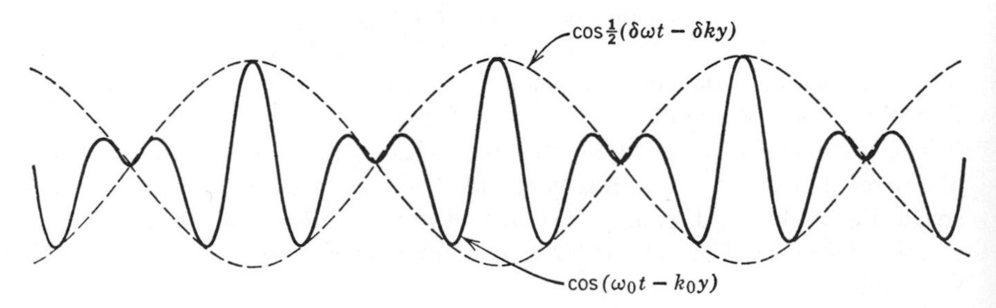

Figure 4.1 Group velocity illustrated by dispersion with waves of slightly different wavelengths.

If the velocity given by Eq. (4-33) were constant with frequency, the phase velocity and the group velocity would be the same since the two waves would have a constant phase difference. However, if there is dispersion, the two waves will propagate with slightly different velocities; the group, or envelope, moves with a velocity different from that of the waves themselves. The relative velocity of the envelope with respect to one of the waves shows that one wave is continuously gaining on the other. The group velocity is the velocity of energy flow and will not exceed the velocity of light.

As a beam of radiant energy passes through the interface between two different media, the frequency ν is invariant. Thus, the wavelength λ must change, inasmuch as the product of frequency and wavelength is the magnitude of the velocity:

$$\begin{gathered}\nu\lambda_0 = c,\ \nu\lambda_1 = v_1,\\ \lambda_0/\lambda_1 = c/v_1,\ \lambda_1 = \lambda_0(v_1/c),\\ \lambda_1 = \lambda_0/n,\ k = 2\pi/\lambda = 2\pi n/\lambda_0.\end{gathered} \tag{4-40}$$

The Eikonal

The function $\zeta(\mathbf{r})$, a term appearing in Eq. (4-17), is called the *eikonal.* A study of this function, which is not intended here to be a rigorous development [Refs. 4 and 5], will ease the transition from physical to geometrical optics and justify the concept of rays. We start with a wave expression in the form of Eq. (4-18):

$$\hat{U}(\mathbf{r},t) = U(\mathbf{r}) \exp\,[i\omega t - ik\zeta(\mathbf{r})]. \tag{4-41}$$

Define $a(\mathbf{r})$ as a real function such that $U(\mathbf{r}) = C \exp a(\mathbf{r})$, C being a scalar constant that we assume here to be unity. Also, we suppress the time function and substitute the following into the Helmholtz equation, Eq. (4-24), using $k^2n^2 = (2\pi n/\lambda_0)^2$ for $\epsilon\mu\omega^2$:

$$\hat{U}(\mathbf{r}) = \exp\,[a(\mathbf{r}) - ik\zeta(\mathbf{r})]. \tag{4-42}$$

The substitution gives

$$\nabla^2 a(\mathbf{r}) + [\nabla a(\mathbf{r})]^2 - k^2[\nabla\zeta(\mathbf{r})]^2 + k^2n^2 - ik\{\nabla^2\zeta(\mathbf{r}) + 2[\nabla a(\mathbf{r})\cdot\nabla\zeta(\mathbf{r})]\} = 0. \tag{4-43}$$

For this equation to be true, the real and imaginary components must each be separately equal to zero. Hence,

$$[\nabla\zeta(\mathbf{r})]^2 - n^2 = (1/k)^2\{\nabla^2 a(\mathbf{r}) + [\nabla a(\mathbf{r})]^2\}. \tag{4-44}$$

Whenever k is sufficiently large, that is, when $\lambda_0 \to 0$, the right side of this equation approaches zero. Then:

$$[\nabla\zeta(\mathbf{r})]^2 = n^2. \tag{4-45}$$

To appreciate the significance of this result, we return to Eq. (4-41) and note that $\zeta(\mathbf{r})$ is the function, of considerable generality, that determines how the phase of the wave varies with displacement in any direction. The simple result of Eq. (4-45) that the gradient of $\zeta(\mathbf{r})$ is a constant, if the index of refraction is constant, indicates that the maximum space rate of change of the phase takes place in the direction of propagation as set up in the plane wave expression of Chapter 1. Thus, the relatively simple concept of the plane wave can be carried into the general wave, provided the wavelength is very small compared with the geometry of the optical system. Furthermore, if time is fixed and C represents a constant,

$$\zeta(\mathbf{r}) = C \tag{4-46}$$

represents a surface in space. By giving C a sequence of constant values, we obtain a family of such surfaces, no two of which can intersect. The function, $\nabla\zeta(\mathbf{r})$, because it is the gradient of a scalar, is normal to the surfaces at each point and indicates the direction of the greatest rate of change in $\zeta(\mathbf{r})$ with position.

In homogeneous, isotropic and nonconducting media, as time is allowed to progress, the surfaces associated with given phases of the wave move in the direction of the gradient. These normals to the phase surfaces are called rays and are used to identify the optical path of waves through optical systems. The surfaces are wave fronts, and a ray is generally defined as any line perpendicular to the successive wave fronts.

Reflection and Refraction at an Interface

Let us now examine the phenomena that take place as a plane wave encounters a plane interface between two dielectric materials. The values of ϵ, μ, and the index of refraction will generally differ in the two materials. The index of refraction does vary with frequency, but at a given wavelength it is constant in each material. It is known from experience that some of the radiant energy from one medium enters the second medium with a change in direction; it is also known that a portion of the energy is reflected back into the first medium.

To analyze what takes place at the interface, we choose a normal to this surface as the z-axis and the plane of the interface as the x–y-plane (see Fig. 4.2). Then we write an equation for a plane wave approaching the surface, using for a phase function the expression

$$(\omega t - \mathbf{k} \cdot \mathbf{r}). \tag{4-47}$$

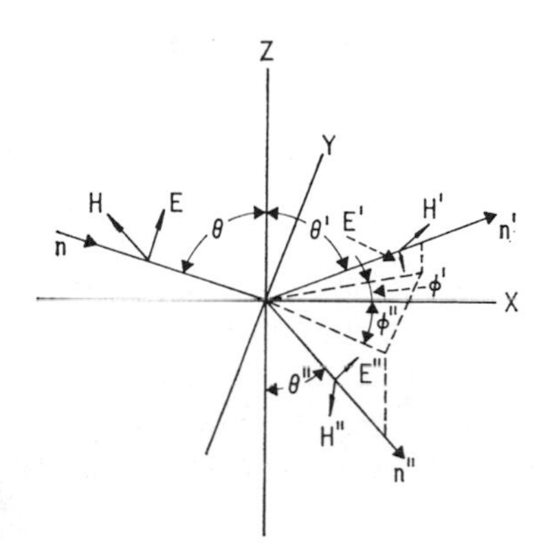

Figure 4.2
Analysis of wave behavior at a dielectric–dielectric interface.

Let **n** be a unit vector in the direction of k (the direction of propagation) so that we can develop $k \cdot \mathbf{r}$ as follows:

$$k \cdot \mathbf{r} = k(\mathbf{n} \cdot \mathbf{r}) = (2\pi\nu/\lambda\nu)(\mathbf{n} \cdot \mathbf{r}) = (\omega/v)(\mathbf{n} \cdot \mathbf{r}). \quad (4\text{-}48)$$

Then the equation for the wave is

$$\mathbf{E} = \mathbf{E}_0 \cos \omega[t - (\mathbf{n} \cdot \mathbf{r}/v)]. \quad (4\text{-}49)$$

An equation for the reflected wave is

$$\mathbf{E}' = \mathbf{E}_0' \cos \{\omega'[t - (\mathbf{n}' \cdot \mathbf{r}/v)] - \delta'\}, \quad (4\text{-}50)$$

and an equation for the wave that enters the second medium is

$$\mathbf{E}'' = \mathbf{E}_0'' \cos \{\omega''[t - (\mathbf{n}'' \cdot \mathbf{r}/v'')] - \delta''\}. \quad (4\text{-}51)$$

There are three corresponding equations for the magnetic wave. Across the surface, certain components of the field vectors must be continuous, resulting in the following six equations [see Ref. 3, p. 343]:

$$\begin{aligned} E_x + E_x' &= E_x'', & H_x + H_x' &= H_x'', \\ E_y + E_y' &= E_y'', & H_y + H_y' &= H_y'', \\ D_z + D_z' &= D_z'', & B_z + B_z' &= B_z''. \end{aligned} \quad (4\text{-}52)$$

By choosing the x–z-plane as the plane containing an incident ray and the surface normal, and calling it the plane of incidence, the components of the three unit vectors can be written as indicated in the following equations:

$$\begin{aligned} \mathbf{n} &= \mathbf{I}_x \sin\theta - \mathbf{I}_z \cos\theta, \\ \mathbf{n}' &= \mathbf{I}_x \sin\theta' \cos\theta' + \mathbf{I}_y \sin\theta' \cos\varphi' + \mathbf{I}_z \cos\theta', \\ \mathbf{n}'' &= \mathbf{I}_x \sin\theta'' \cos\theta'' + \mathbf{I}_y \sin\theta'' \cos\varphi'' - \mathbf{I}_z \cos\theta''. \end{aligned} \quad (4\text{-}53)$$

Six equations of the following form result from the six equalities in Eq. (4-52):

$$E_{0x} \cos \{\omega[t - (\mathbf{n} \cdot \mathbf{r}/v)]\} + E_{0x}' \cos \{\omega'[t - (\mathbf{n}' \cdot \mathbf{r}/v)] - \delta'\}$$
$$= E_{0x}'' \cos \{\omega''[t - (\mathbf{n}'' \cdot \mathbf{r}/v'')] - \delta''\}. \quad (4\text{-}54)$$

These six equations, and the requirement that they be true at all times and for all $\mathbf{r}$ (on the interface), place certain restrictions on the parameters in the phase functions of the three waves. Because the cosine functions must change together, the frequency must be invariant across the interface. At $\mathbf{r} = 0$, equality of the phase functions gives

$$\omega t = \omega' t - \delta' = \omega'' t - \delta'', \quad (4\text{-}55)$$

so the phase terms δ' and δ'' must be equal. Because $\omega = \omega' = \omega''$ and $\delta' = \delta'' = 0$, equality of the phases in Eq. (4-54) at $t = 0$ gives

$$(\mathbf{n} \cdot \mathbf{r})/v = \mathbf{n}' \cdot \mathbf{r}/v = \mathbf{n}'' \cdot \mathbf{r}/v''. \quad (4\text{-}56)$$

Since $\mathbf{n}$ has no y component, neither $\mathbf{n}'$ nor $\mathbf{n}''$ can have a y component, so φ' and φ'' of Fig. 4.2 must each be zero, which means that the incident ray, the reflected ray, and the refracted ray are all in the plane of incidence. By substituting the components as given in Eq. (4-53) into the first two parts of Eq. (4-56), we get

$$\frac{x \sin \theta}{v} - \frac{z \cos \theta}{v} = \frac{x \sin \theta'}{v} + \frac{z \sin \theta'}{v}. \quad (4\text{-}57)$$

Then at $z = 0$, which defines the interface,

$$\sin \theta = \sin \theta'; \qquad \theta = \theta'. \quad (4\text{-}58)$$

Stated briefly, *the angle of incidence equals the angle of reflection.* Similarly,

$$\frac{x \sin \theta}{v} = \frac{x \sin \theta''}{v''}. \quad (4\text{-}59)$$

Because $v = c/n$ from Eq. (4-36),

$$n \sin \theta = n'' \sin \theta'', \quad (4\text{-}60)$$

which is known as *Snell's law*, the fundamental law of refraction at an interface (Fig. 4.3).

The magnitudes of the reflected and refracted waves, as well as their directions, are of interest to us. To find these, resolve each electric field vector into two components, one, E_i, in the plane of incidence and the other, E_s, perpendicular to the plane of incidence, which would make it

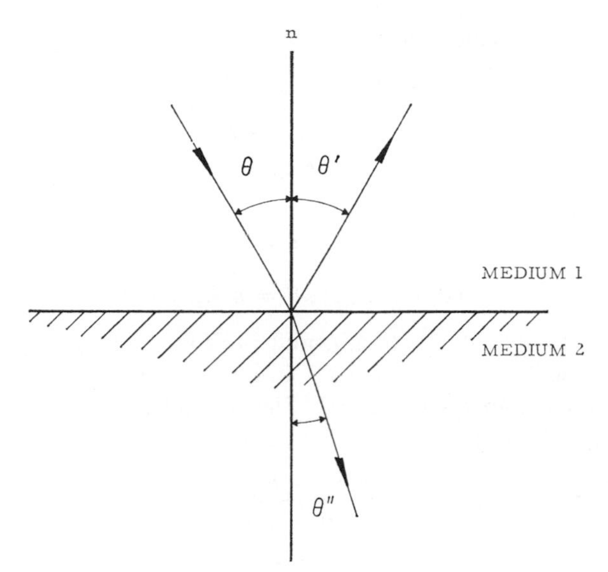

Figure 4.3 Refraction and reflection at a dielectric–dielectric interface.

parallel to the interface. As a result of our choice of coordinates, the components of these three waves are related at the interface by the first of Equations (4-52):

$$(E_i - E_i') \cos \theta = E_i'' \cos \theta'', \tag{4-61}$$

$$E_s + E_s' = E_s''. \tag{4-62}$$

(A note is in order to explain the negative sign in Eq. (4-61). From earlier discussion, we know that the direction of $\mathbf{E}$ (or $\mathbf{E}'$ or $\mathbf{E}''$), and so also the direction of E_i (or E_i' or E_i''), is perpendicular to the direction of propagation (direction of propagation shown by arrow) in each wave of Fig. 4.3. Whatever convention we adopt for a positive E_i direction (say upward), we must also adopt for E_i' so that when θ is expanded to the limit of $\pi/2$—depicting an unreflected wave parallel to the interface—the positive convention is not reversed at the now meaningless position of the normal. If, then, this positive convention is maintained for E_i and E_i', we note that their x components indicated by Eq. (4-61) reverse sign; therefore the negative sign is required because of the implied transformation from the i, s to the x, z coordinate system.)

The magnetic field vector is perpendicular to the electric field vector and the two magnitudes are related by Eq. (1-7), but with ϵ and μ substituted for ϵ_0 and μ_0. The components of H are also related by Equations (4-52):

$$(H_i - H_i') \cos \theta = H_i'' \cos \theta'', \tag{4-63}$$

$$H_s + H_s' = H_s''. \tag{4-64}$$

When μ is 1, one finds from Eq. (1-7) and Eq. (4-36) that $H_s = nE_i$ and $H_i = nE_s$, so

$$n(E_s - E_s') \cos \theta = n'' E_s'' \cos \theta'', \tag{4-65}$$

$$n(E_i + E_i') \cos \theta = n'' E_i'' \cos \theta''. \tag{4-66}$$

Equations (4-61), (4-62), (4-65), and (4-66) are independent and, therefore, determine E_i', E_s', E_i'', and E_s'' in terms of n, n'', and θ. Application of Snell's law eliminates n and n''. The relationships obtained, known as Fresnel's equations, are

$$E_i' = E_i \frac{\tan (\theta - \theta'')}{\tan (\theta + \theta'')}, \tag{4-67}$$

$$E_s' = E_s \frac{\sin (\theta - \theta'')}{\sin (\theta + \theta'')}, \tag{4-68}$$

$$E_i'' = E_i \frac{2 \sin \theta'' \cos \theta}{\sin (\theta + \theta'') \cos (\theta - \theta'')}, \tag{4-69}$$

$$E_s'' = E_s \frac{2 \sin \theta'' \cos \theta}{\sin (\theta + \theta'')}. \tag{4-70}$$

Since $\tan (\theta - \theta'') = -\tan (\theta'' - \theta)$ and $\sin (\theta - \theta'') = -\sin (\theta'' - \theta)$, there may or may not be a phase change in the reflected wave depending upon the relative sizes of θ and θ''. According to Snell's law, which relates the several θ and n, when the transmitted wave goes from a more dense to a less dense medium, $n'' < n$, there will be a change in phase in the reflected wave. There is also a phase change in E_i' when $\pi/2 < (\theta + \theta'') < \pi$ because the tangent is negative.

At the particular angle of incidence where $(\theta + \theta'') = \pi/2$, the denominator of Eq. (4-67) becomes unbounded and the component of the reflected wave in the plane of incidence is zero. In a general beam of quasi-monochromatic, unpolarized light incident at this angle, the reflected light will be completely polarized with the electric vector perpendicular to the plane of incidence. Solving for this angle by using Snell's law, we obtain the polarizing angle, which is known as Brewster's angle (Fig. 4.4):

$$n \sin \theta = n'' \sin (\pi/2 - \theta) = n'' \cos \theta,$$

$$\theta'' = \text{arc tan } (n''/n). \tag{4-71}$$

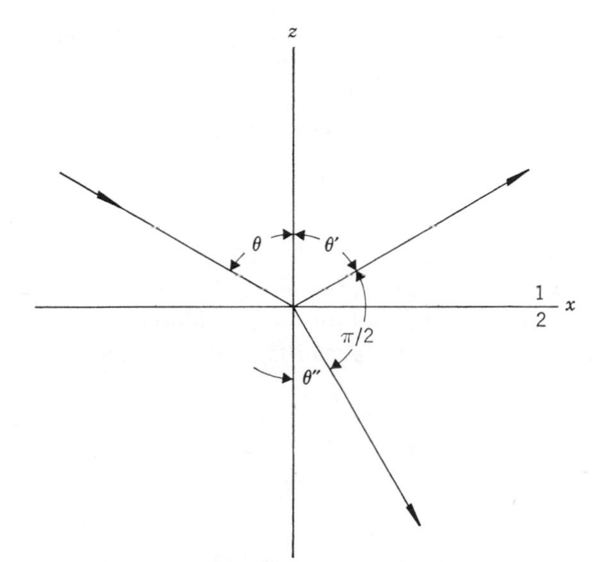

Figure 4.4 An illustration of the polarizing angle.

Whenever a beam is passing from a more dense into a less dense medium, $n > n''$, as from glass into air, the following expression from Snell's law is unity for some value of θ:

$$\sin\theta'' = (n/n'')\sin\theta. \tag{4-72}$$

This particular angle is called the critical angle θ_c. When a beam is incident at an angle greater than θ_c, none of the incident radiant energy will be transmitted through the interface, but all will be totally reflected. This phenomenon is called total internal reflection.

Reflectance and Transmittance at a Dielectric Interface

Whenever we are trying to achieve maximum overall transmittance through a sequence of optical media (e.g., several lenses of an optical system) significant power losses result from reflections at the interfaces. These are referred to as Fresnel reflection losses. The reflectance at the interface depends upon the polarization of the incident beam. By Equations (4-67) and (4-69) for polarization parallel to the plane of incidence, the reflectance ρ_i and transmittance τ_i are

$$\rho_i = \left(\frac{E_i'}{E_i}\right)^2 = \frac{\tan^2(\theta - \theta'')}{\tan^2(\theta + \theta'')}, \tag{4-73}$$

$$\tau_i = \left(\frac{E_i''}{E_i}\right)^2 = \frac{4\sin^2\theta''\cos^2\theta}{\sin^2(\theta + \theta'')\cos^2(\theta - \theta'')}. \tag{4-74}$$

By Equations (4-68) and (4-70) for the other plane of polarization, we have (Fig. 4.5):

$$\rho_s = \frac{\sin^2(\theta - \theta'')}{\sin^2(\theta + \theta'')},$$
$$\tau_s = \frac{4\sin^2\theta''\cos^2\theta}{\sin^2(\theta + \theta'')}. \tag{4-75}$$

The reflectance at normal incidence cannot be obtained from these equations because the expressions become indeterminate at $\theta = 0$ and $\theta'' = 0$. However, by simultaneous solution of Equations (4-61), (4-62), (4-65), and (4-66), the reflectance and transmittance can be found:

$$\rho = \left(\frac{n_{12} - 1}{n_{12} + 1}\right)^2,$$
$$\tau = \frac{4n_{12}}{(n_{12} + 1)^2}. \tag{4-76}$$

Here we have used $n_{12} = n/n''$, the relative index, which applies at the interface. Figure 4.6 shows the Fresnel reflection loss (normal incidence) as a function of n_{12} for an optical flat, taking account of the reflectance at both surfaces and the multiple reflections between surfaces within the flat. It is interesting to note that the Fresnel reflection loss at a single surface is only about 4% when the relative index is 1.5 but grows to 25% when the relative index increases to 3.

Reflection losses can be reduced within a particular wavelength region by properly coating the surface. The coating material must have a refractive

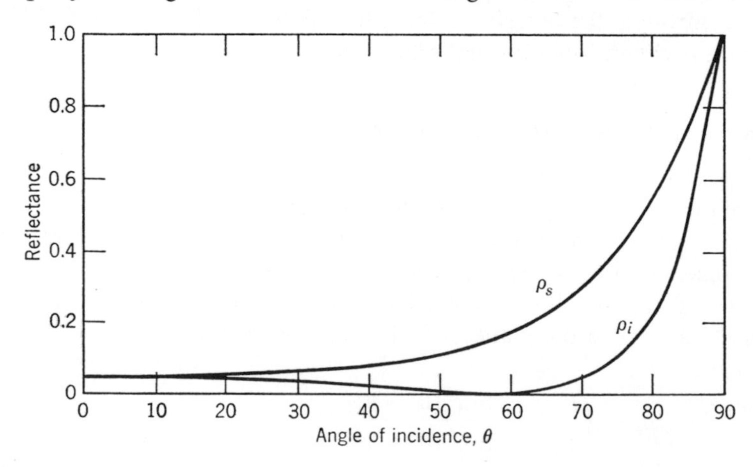

Figure 4.5 Variation of reflectance with angle of incidence for components of the electric field in the plane of incidence (ρ_i) and perpendicular to the plane of incidence (ρ_s), $(n''/n) = 1.5$.

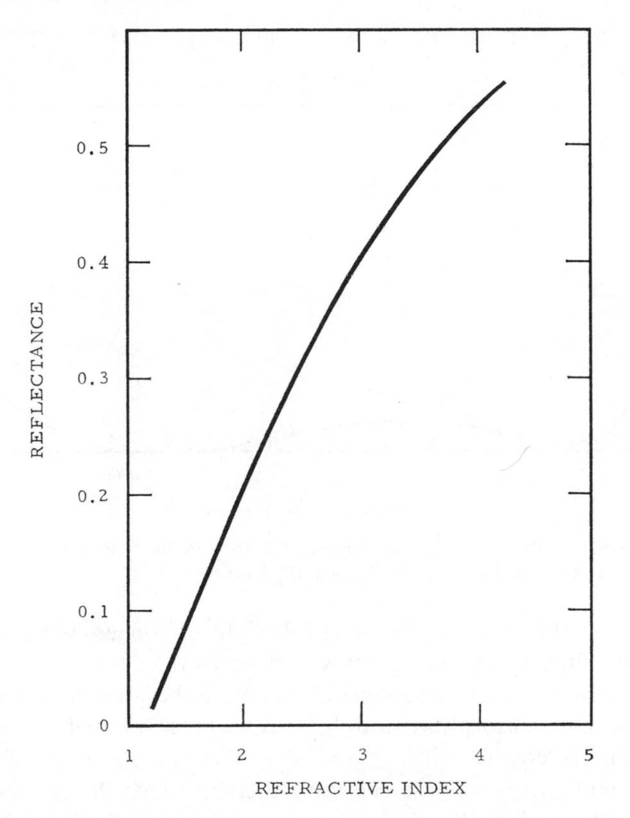

Figure 4.6 Fresnel reflection losses in a dielectric plate as function of relative refractive index. Losses at both surfaces and multiple reflections were included. (normal incidence)

index related to the refractive index of the two media, and the thickness of the film coating depends upon the wavelength at the particular wavelength region chosen. Figure 4.7 shows the reflectance of a glass optical element, for the air–glass interface, before and after application of an antireflection coating. Commercial nonglare, or "invisible," glass has an antireflection coating designed to reduce reflection in a broad band centered near 555 nm, at which wavelength the eye is most sensitive.

Absorbing Materials

Contrary to our assumption in the preceding discussion, no material is perfectly nonabsorbing at any frequency, and every material shows strong absorption at some frequencies. Among the various materials, metals have the highest absorption, it being so strong that no measurable amount of

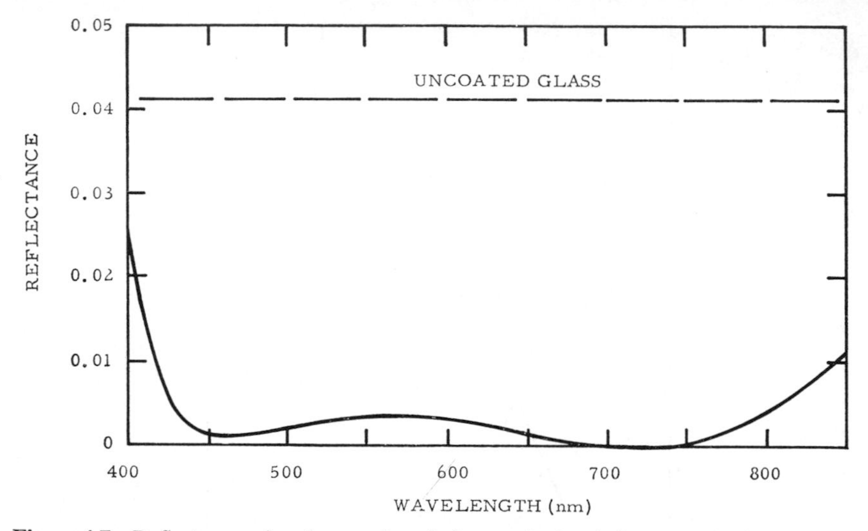

Figure 4.7 Reflectance of a glass surface before and after being coated with a quarter–half–quarter wave coating of $MgF_2 + ZnO_2 + CsF_3$ [Ref. 6].

radiant energy, at the wavelengths of visible light or longer, is transmitted through metallic films more than a few wavelengths thick.

We are interested in metals as optical materials because they are good reflectors. The high absorption and high reflectance of metals are both related to electrical conductivity; classically, the free electrons of metals absorb the radiant energy. These electrons then reradiate the energy at the same frequency to explain the phenomenon of metallic reflection. Because reflectance is related in this way to conductivity, the usual reduction of conductivity with increasing temperature accounts for a reflectance decrease with an increase in temperature. However, one must be careful in an experiment to distinguish this cause of reflectance decrease from the effects of surface corrosion, which occurs rapidly at elevated temperatures.

When radiant energy is reflected from a metallic surface at other than normal incidence, it is partially polarized. Also, the two components of the wave change phase by different amounts at the surface, so that the reflected beam is elliptically polarized.

In absorbing materials, $\sigma \neq 0$, and in the presence of an electric field, $\mathbf{j} \neq 0$. The applicable forms of Maxwell's equations are Eq. (4-1) to Eq. (4-4). Since the material is a conductor, no steady-state charge can accumulate; so $q = 0$. Again, we can obtain vector wave equations which contain six scalar wave equations, one for each of the field components, corresponding to Eq. (4-16). The general scalar equation is

$$\nabla^2 U - \epsilon\mu(\partial^2 U/\partial t^2) - \sigma\mu(\partial U/\partial t) = 0. \qquad (4\text{-}78)$$

By substituting $\hat{U} = \hat{U}_0 \exp(-i\omega t)$ into this equation we obtain

$$\nabla^2 \hat{U}_0 - [\epsilon\mu\omega^2 - i\sigma\mu\omega]\hat{U}_0 = 0. \tag{4-79}$$

By defining a complex propagation constant $\hat{k}$ and a complex permittivity $\hat{\epsilon}$,

$$\hat{k}^2 = \mu\omega^2\hat{\epsilon} = \mu\omega^2[\epsilon - (i\sigma/\omega)], \tag{4-80}$$

we can put Eq. (4-78) into the form of the Helmholtz equation:

$$\nabla^2 \hat{U}_0 - \hat{k}^2 \hat{U}_0 = 0. \tag{4-81}$$

Also, let us define a complex phase velocity and a complex index of refraction:

$$\hat{v} = c/\sqrt{\mu\hat{\epsilon}},$$

$$\hat{n} = c/\hat{v} = (c/\omega)\hat{k}; \tag{4-82}$$

$$\hat{n} = n(1 - i\mathscr{K}). \tag{4-83}$$

We call $\mathscr{K}$ the attenuation index; n and $\mathscr{K}$ are real. Solving for the real quantities in terms of the material constants and frequency gives

$$n^2 = (\tfrac{1}{2})c^2\{[\epsilon^2\mu^2 - (\sigma^2\mu^2/\omega^2)]^{1/2} + \epsilon\mu\}, \tag{4-84}$$

$$n^2\mathscr{K}^2 = (\tfrac{1}{2})c^2\{[\epsilon^2\mu^2 - (\sigma^2\mu^2/\omega^2)]^{1/2} - \epsilon\mu\}. \tag{4-85}$$

From Equations (4-82) and (4-83),

$$\hat{k} = (\omega n/c)(1 - i\mathscr{K}). \tag{4-86}$$

Let us choose our coordinate system so that $\mathbf{n} \cdot \mathbf{r} = y$ in the phase terms of equations like Eq. (4-49) so that a solution of Eq. (4-81) is $\hat{U} = U_0 \exp(i\omega t - i\hat{k}y)$. By substituting for $\hat{k}$ as indicated in Eq. (4-86), we get, using $k = \omega n/c$,

$$\hat{U} = U_0 \exp[-(\omega n\mathscr{K}/c)y] \exp(i\omega t - iky). \tag{4-87}$$

The radiant flux density is proportional to $(\hat{U} \cdot \hat{U}^*)$. At some $y_0 = 0$ let the radiant flux density be

$$W_0 = U_0^2. \tag{4-88}$$

The constant of proportionality has been assumed unity. At some $y = y_1$ the flux density is

$$W_1 = U_0^2 \exp[-(2\omega n\mathscr{K}/c)y_1]. \tag{4-89}$$

If y_1 is made a unit distance, the ratio W_1/W_0 becomes the transmissivity, so

$$\tau = W_1/W_0 = \exp[-(2\omega n\mathscr{K}/c)]. \tag{4-90}$$

When this equation is compared with Eq. (2-24), one sees that the absorption coefficient is $2\omega n\mathscr{K}/c$. If $\sigma/\omega \gg \epsilon$ in Eq. (4-85), the attenuation index is found to be:

$$\mathscr{K} = (c/n)(\sigma\mu/2\omega)^{1/2}. \tag{4-91}$$

If y_s is the distance into a metallic surface that reduces the incident flux density to $1/e$ of its value at $y = 0$, then:

$$(2\omega n \mathscr{K} y_s/c) = 1$$

$$y_s = 1/(2\omega\sigma\mu)^{1/2}. \tag{4-92}$$

When frequency and conductivity are both large, the skin depth y_s is quite small.

The study of waves in metals, because metals have high conductivity, is seen to be a study of surface phenomena, which, though highly interesting and important, cannot be given in detail in the present book [Ref. 7]. Because the material parameters do not allow the simplifying approximations valid for dielectrics, even the angle of reflection—as well as the propagation constant, velocity, index of refraction, and permittivity—is complex. We note that Equations (4-24) and (4-81), with $\epsilon\mu\omega^2$ of the dielectric equation represented by the more general $\hat{k}^2$ in metals, are of the same form. By a method similar to that used for dielectrics, one could derive equations for metals corresponding to Equations (4-73) and (4-74). A complex Snell's law would be obtained:

$$n \sin \theta = \hat{n}'' \sin \theta'', \tag{4-93}$$

where $\hat{n}''$ and the angle θ'' are both complex.

The reflectance at normal incidence can be obtained as with dielectrics by using Equations (4-61), (4-62), (4-65), and (4-66) but with the stipulation that $\hat{n}''$ is now complex. The ratio of the amplitudes of the reflected and incident waves turns out to be, for metals,

$$(E'/E) = (\hat{n} - 1)/(\hat{n} + 1), \tag{4-94}$$

where $\hat{n}$ is now equal to $n/\hat{n}''$. In complex notation, the energy is proportional to amplitude times its complex conjugate, so, from Equations (4-83) and (4-94),

$$\rho = \left(\frac{E'}{E}\right)\left(\frac{E'^*}{E^*}\right) = \frac{(n-1)^2 + n^2\mathscr{K}^2}{(n+1)^2 + n^2\mathscr{K}^2}. \tag{4-95}$$

If $n^2\mathscr{K}^2 \gg (n+1)^2$, the reflectance is nearly unity. At oblique incidence, the calculation becomes quite complicated and will not be undertaken here. Figure 4.8 shows the reflectance for each component of an incident wave as a function of the angle of incidence. The computations were made for $n = 1.50$ and $\mathscr{K} = 100$, which would be typical for a partially transparent, strongly absorbing material. The phase changes of the two components also differ; when linearly polarized light is incident, it becomes elliptically polarized after reflection, provided the direction of polarization is not parallel to either E_i or E_s. At an angle of incidence near 60 deg, ρ_i is a

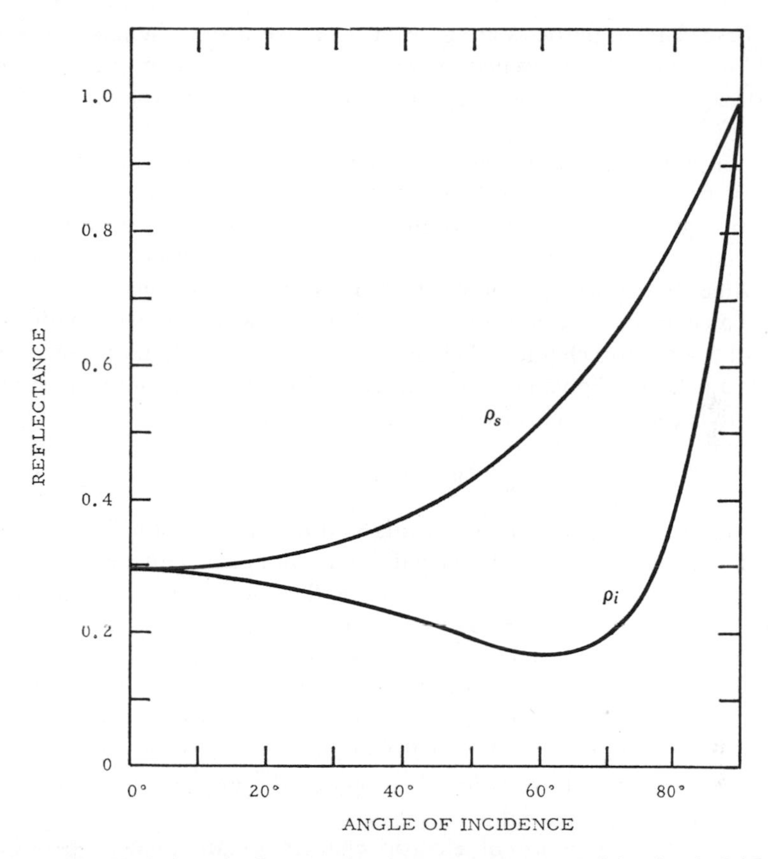

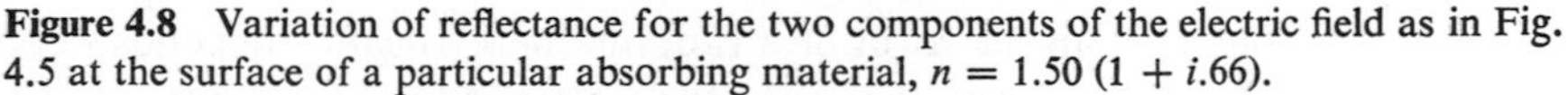

Figure 4.8 Variation of reflectance for the two components of the electric field as in Fig. 4.5 at the surface of a particular absorbing material, $n = 1.50\ (1 + i.66)$.

minimum but does not become zero as it would at the surface of a non-absorbing material. At this angle the reflected beam is strongly polarized perpendicular to the plane of incidence.

Dispersion

A closer look at the refractive index of dielectrics is required for an explanation of dispersion [Ref. 8, pp. 90–98]. Our perspective is brought to a microscopic scale so that we can examine the response of electric charges to an applied, external electromagnetic field. We shall use the classical model in which electrons, or in some cases ions, are held bound to their atoms, or molecules, by atomic or interatomic potential wells. Though bound, these charged particles are allowed to move within the limits of the wells. This model allows oscillation of electric charges with energy-dissipating, damping forces provided by some kind of thermodynamic mechanism.

In a transparent, dielectric (low conductivity) material, undisturbed by any electric or magnetic field, the distribution of both positive and negative charges in a given macroscopic volume is uniform and the quantities of negative and positive charges are equal. However, on a scale of the order of a few times the dimensions of an atom or molecule, the negative charges may have a centroid, a "center of gravity," that does not coincide with the centroid of the positive charges. This is said to be a polar atom or molecule, which has an electric dipole moment $\mathbf{p}$ directed from the centroid of positive charge to the centroid of negative charge:

$$\mathbf{p} = \mathbf{r}q. \tag{4-96}$$

Here $\mathbf{r}$ is the separation of the centroids and q is the amount of positive charge. The dipoles, because of thermal agitation, normally have a random orientation so that on a macroscopic scale there is no net electric field produced within the material. However, in the presence of an applied, external field, the dipoles tend to align themselves in the direction of the field. Even if the molecules are not polar molecules, the applied field causes a small average separation of positive and negative charges, each molecule becoming an induced dipole. The redistribution of charges in this manner produces an internal field that locally opposes the applied field, and the material is said to be polarized.

The polarization response of electric charges to an applied, periodic electromagnetic field—in the form of a radiant energy beam—depends upon the nature of the elastic forces that hold the electrons to their atoms, or ions to their positions in the lattice structure, and upon the thermal effects within the material. The completely general explanation of the response, and the consequent characteristics of the material, is a complicated problem of quantum mechanics. Here we can give only a classical explanation which, though much simplified, gives plausibility to experimental observations and can be verified by a rigorous theoretical study.

According to the Lorentz equation we should consider both electric and magnetic forces. However, since the velocities and displacements are small, and since the magnetic field associated with an electromagnetic wave, for most problems of general optical interest, is much smaller in magnitude than the electric field, we ignore the magnetic forces. Let us assume a sinusoidal applied field:

$$\mathbf{E} = \mathbf{E}_0 \exp(-i\omega t). \tag{4-97}$$

Within the dielectric, this produces an effective field $\mathbf{E}'$ which, in turn, produces an induced dipole moment proportional to the effective field:

$$\mathbf{p} = \alpha \mathbf{E}'. \tag{4-98}$$

The proportionality constant α is called the mean polarizability. If there are N dipoles per unit volume, the polarization $\mathbf{P}$ of the material is

$$\mathbf{P} = N\mathbf{p} = Nq\mathbf{r} = N\alpha\mathbf{E}'. \tag{4-99}$$

In gases the charges are separated so far that the effective field is essentially the applied field, and $\mathbf{D} = \epsilon_0\mathbf{E}$. In solids, particles are so close together that they are influenced not only by the applied field but also by the field produced by neighboring dipoles. The electric field $\mathbf{E}$ in a polarized dielectric is the resultant of the external field and the field produced by the polarization, a field proportional to $\mathbf{P}$; but $\mathbf{E}$ is not identical to the effective field $\mathbf{E}'$ acting on an individual dipole because in computing $\mathbf{E}'$ the contribution of the dipole itself must be excluded. The effective field $\mathbf{E}'$ differs from $\mathbf{E}$ sometimes appreciably but is always proportional to $\mathbf{E}$. However, there are already constants involved in the equations—N and α of Equations (4-98) and (4-99) and later N_i; these constants have to be determined by experiment. In this simplified analysis, we make the constant of proportionality unity and use $\mathbf{E}$ for $\mathbf{E}'$ in the equations. The vectors $\mathbf{D}$ and $\mathbf{P}$ are each proportional to $\mathbf{E}$, so we may define a permittivity ϵ by the equation:

$$\mathbf{D} = \epsilon\mathbf{E} = \epsilon_0\mathbf{E} - \mathbf{P}. \tag{4-100}$$

If the material is assumed to be isotropic, then the vector notation may be dropped. The permittivity is related to the index of refraction by Eq. (4-36) or Equations (4-82) and (4-83), with $\mu = 1$:

$$\epsilon = \epsilon_0 - P/E = \epsilon_0 - N\alpha, \tag{4-101}$$

$$\epsilon_r = 1 - (N\alpha/\epsilon_0), \tag{4-102}$$

$$\hat{\epsilon} = \hat{n}^2 = n^2(1 - i\mathscr{K})^2. \tag{4-103}$$

In the following discussion we refer only to the motion of electrons, although other charged particles are often involved in polarization phenomena. In the model we are using, electrons are accelerated by the electric force qE. The equation of motion for an electron is

$$m(\partial^2 r/\partial t^2) + g(\partial r/\partial t) + hr = qE_0 \exp(-i\omega t), \tag{4-104}$$

where m is the electronic mass, g is a damping factor, and hr is an elastic restoring force. If both sides of the equation are multiplied by Nq/m and Eq.

(4-99) is substituted into Eq. (4-104), an equation for polarization results,

$$\partial^2 P/\partial t^2 + (g/m)(\partial P/\partial t) + (h/m)P = (Nq^2E_0/m) \exp(-i\omega t). \quad (4\text{-}105)$$

Any transient polarization will soon die out because of the dissipative forces. The steady-state solution is found by assuming for a solution:

$$P = P_0 \exp(-i\omega t), \quad (4\text{-}106)$$

and by substituting this into Eq. (4-105):

$$[-\omega^2 - (i\omega g/m) + (h/m)]P_0 = (Nq^2E_0/m). \quad (4\text{-}107)$$

A natural frequency of oscillation,

$$\omega_0 = [(h/m) - (g^2/4m^2)]^{1/2} \approx (h/m)^{1/2}, \quad (4\text{-}108)$$

can be found by studying undamped oscillatory motion. Then, by letting $\tau = m/g$ be a damping time constant, Eq. (4-107) can be written as

$$P_0 = (Nq^2E_0/m)[(\omega_0^2 - \omega^2) - i\omega/\tau]^{-1} \quad (4\text{-}109)$$

$$P_0 = \frac{Nq^2E_0}{m} \cdot \frac{\exp(i\varphi)}{[(\omega_0^2 - \omega^2)^2 + (\omega^2/\tau^2)]^{1/2}} \quad (4\text{-}110)$$

where

$$\varphi = \arctan[(\omega/\tau)/(\omega_0^2 - \omega^2)]. \quad (4\text{-}111)$$

A comparison of Eq. (4-109) with Eq. (4-99), shows that the polarizability is

$$\alpha = (q^2/m)[(\omega_0^2 - \omega^2) - i\omega/\tau]^{-1}. \quad (4\text{-}112)$$

So α is, in general, complex. The relationships between α and n in Equations (4-102) and (4-103) show that the index of refraction is related to the applied frequency.

In gases and in many dielectric solids where the damping constant is small and ω is far from ω_0, the damping term may be neglected. Then φ is less than $\pi/2$ when $\omega < \omega_0$ and greater than $\pi/2$ when $\omega > \omega_0$. When damping is low, causing $\tan\varphi$ to be small, the applied field and the polarization are almost in phase when $\omega < \omega_0$, and the two are almost opposite in phase when $\omega > \omega_0$. For low damping

$$P_0 = Nq^2E_0/[m(\omega_0^2 - \omega^2)]. \quad (4\text{-}113)$$

If the electrons in the material have a number of different resonant frequencies ω_{i0}, then there is an equation like Eq. (4-113) for each of these frequencies; the total polarization is obtained by taking a sum. If the

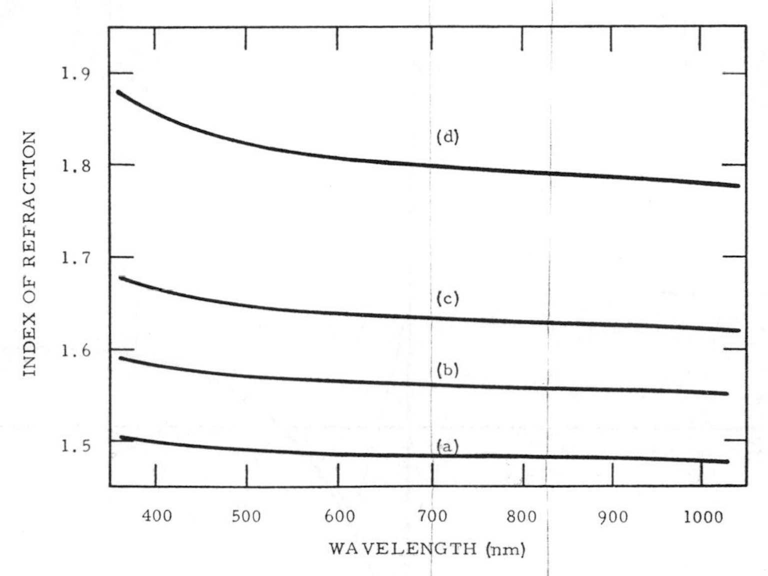

Figure 4.9 Dispersion of several optical glasses: (a) FK 50-486 815, (b) Bak 3-565 559, (c) BaF 12-639 452, (d) LaSF 8-807 316 [Ref. 9].

oscillations are all electronic, the equations for polarizability and polarization are

$$\alpha_i = q^2/[m(\omega_{i0}^2 - \omega^2)], \tag{4-114}$$

$$P_i = N_i \alpha_i E_0 = \frac{q^2 E_0}{m} \cdot \frac{N_i}{(\omega_{i0}^2 - \omega^2)}, \tag{4-115}$$

$$P = \frac{q^2 E_0}{m} \sum_i N_i (\omega_{i0}^2 - \omega^2)^{-1}. \tag{4-116}$$

The index of refraction can be obtained by using the relations in Equations (4-36) and (4-102):

$$n^2 = \epsilon_r = 1 - \frac{q^2}{\epsilon_0 m} \sum \frac{N_i}{(\omega_{i0}^2 - \omega^2)}. \tag{4-117}$$

It may be noted that, in general, α, ϵ, and n^2 are all complex, and that because φ is nonzero, the polarization is not instantaneously proportional to the applied field. We have defined permittivity for propagating electromagnetic fields that vary sinusoidally, and this permittivity turns out to be a function of frequency. In ranges where, according to Eq. (4-117), the index of refraction increases gradually with ω (decreases with increasing wavelength) and where $\omega < \omega_0$, the variation is called normal dispersion. Figure 4.9 shows the normal dispersion for several optical materials.

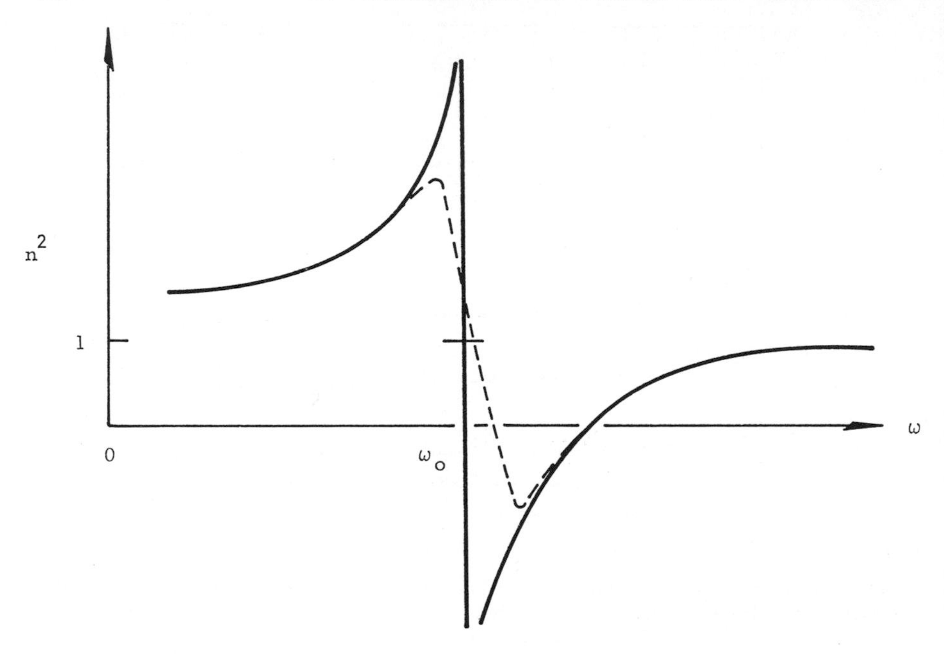

Figure 4.10 Dispersion near an absorption band showing region of anomalous dispersion (dashed curve).

A plot typical of a complete analysis is shown in Fig. 4.10. Transparent materials usable as transmitting optical elements are applied in regions of normal dispersion. A study of resonance absorption, material properties, and atomic and molecular structure would be particularly concerned, however, with the region of absorption. The anomalous dispersion in a region of absorption is identified by the dotted portion of the curve in Fig. 4.10.

REFERENCES

1. J. A. Stratton, *Electromagnetic Theory*. McGraw-Hill Book Company, Inc., New York, 1941.
2. L. Page, *Introduction to Theoretical Physics*. D. Van Nostrand Company, Inc., New York, 1952.
3. G. Joos, *Theoretical Physics*. Blackie & Son Limited, London, 1951.
4. M. Kline and I. W. Kay, *Electromagnetic Theory and Geometrical Optics*, Introduction. Interscience Publishers, New York, 1965.
5. E. L. O'Neill, *Introduction to Statistical Optics*, p. 47. Addison-Wesley Publishing Company, Inc., Reading, Mass., 1963.

6. J. T. Cox and G. Hass, "Triple-Layer Coatings on Glass for the Visible and Near Infrared," *J. Opt. Soc. Am.*, **52,** 965 (1962).
7. A. V. Sokolov, *Optical Properties of Metals*, English translation by S. Chomet. American Elsevier Publishing Company, Inc., New York, 1967.
8. M. Born and E. Wolf, *Principles of Optics*, Third Edition. Pergamon Press, Oxford, 1965.
9. *Catalogue*, Schott Optical Glass, Inc., York Ave., Duryea, Pa. 18642.

5

Diffraction and Interference

Our Point of View

Historically, the interpretation of diffraction and interference phenomena helped to establish the wave theory of light; however, when we undertake a logical study of light, we find the order reversed. Once we accept the concepts that radiant energy consists of waves, that the electric and magnetic field strengths are the oscillating quantities, that the wave transports energy, and that the wave interacts—by reflection, refraction, and absorption—with materials at an interface, we are led to expect that radiant energy exhibits both diffraction and interference effects. Certain wave manifestations cannot be clearly categorized as exclusively diffraction or exclusively interference; in fact, the reader will discover that our treatment of diffraction also involves interference. Nevertheless, the topics in this chapter are separated so that those in the first part are conventional diffraction problems and those in the latter part are interference problems. "Diffraction" is commonly used with several different meanings, but here it will apply only to the phenomena observed when a beam passes by the edge of an opaque screen (Fig. 5.1). Usually the edge forms a closed figure such as a rectangular or circular opening, an aperture, in the screen.

Besides the strictly optical concern with diffraction effects, interest has developed in radio waves having wavelengths of the same order as a dimension of the aperture. This has focused on a sort of quasi-optical wavelength region where optical diffraction theory becomes important wherever wave propagation is involved in chemistry, physics, mathematics, and engineering. These broadened applications have been sufficient to generate a recent renewed interest in diffraction theory. Because inconsistencies have been uncovered in the classical scalar theory, because polarization information cannot be ignored in radio wave problems, and because discrepancies have emerged between theory and experiment at short waves, new theories are being investigated. Unfortunately, space allows us only to give references to a number of significant papers in this new field.

Exact solutions of diffraction problems are extremely intricate; in fact, the most difficult calculations in optics are probably diffraction problems. To learn precisely how the electromagnetic wave interacts with the obstructing screen, one would have to solve the vector wave equation with appropriate boundary conditions applied at the surface of the screen. It is known from the preceding chapter that the boundary conditions must include the properties of the screen material, the angle of incidence at all points of the surface including the edge, and the relationships between field vectors at the surface.

Generally, neither rigorous nor exact diffraction solutions are necessary in optics. An approach using scalar theory and certain approximations can explain the commonly observed diffraction effects. Vector treatment would still be required to find conditions at the surface, but this precision contributes significantly only out to distances of a few wavelengths. Ambiguities of this order in dimensions of diffracting apertures are usually unimportant at optical wavelengths because openings and obstacles are relatively quite large.

We, therefore, confine our discussion to a classical scalar diffraction theory which, though simpler than vector theory, is still quite involved. One immediate loss is all information concerning polarization effects; but,

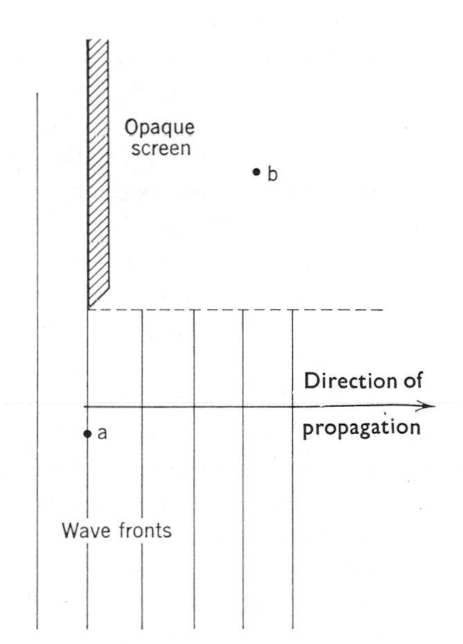

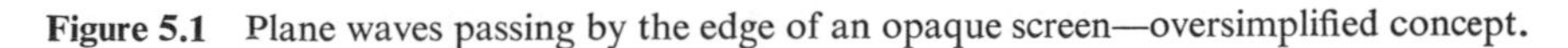
Figure 5.1 Plane waves passing by the edge of an opaque screen—oversimplified concept.

on the whole, the theory has been highly successful in the analysis of diffraction problems and gives excellent agreement with experiment in most practical cases in optics [Ref. 1]. The development of the theory may be found in several standard references [Ref. 2; Ref. 3, Chapter 8; Ref. 4; Ref. 5, Chapter 9; and Ref. 6]. Discussions of more rigorous theories may be found in the additional references [Ref. 3, Chapter 12 and Refs. 7–14].

Partially Obstructed Waves

The positions, at one instant, of several wavefronts, a few of which have just passed by an opaque screen, are depicted in Fig. 5.1. As shown, the waves are cut off sharply by the obstructing edge. The electric vector would be constant along a line parallel to a wavefront and would, according to the diagram, abruptly become zero where the screen had chopped off the wave. However, in an analogous water system, we have never seen a wave whose crest suddenly disappeared in this way; furthermore, we can show that a water wave would never display this abrupt boundary, for example, after having passed through a gate in a fence. A light wave will not behave in this way either, which can be demonstrated by performing a simple experiment. If the plane waves illustrated in the figure are experimentally passed through a circular aperture, each wavefront does not actually continue as a "circular disk." The distribution of the energy in a residual wavefront in a plane parallel to the "disks" changes significantly and rapidly. The actual distribution depends upon several things, including wavelength, size and shape of the aperture, distance behind the screen [Ref. 15], and, in the precise analysis, the material of the screen. Diffraction problems typically involve the study of this distribution for various conditions.

Although the exact analysis of energy distribution in the wave after passage through the circular aperture would be quite involved, a review of the continuity requirements expressed in Eq. (4-52) makes evident the impossibility of the discontinuity represented by the broken line in Fig. 5.1. In fact, some of the energy in the vicinity of the region around point a in the aperture will be directed to the region around point b behind the screen.

Late in the seventeenth century, Christian Huygens made an intellectual guess that a light wave behaves in the way we have suggested. His explanation is now known as the Huygen's principle, and it is well established for radiant energy waves in the infrared, visible, and ultraviolet wavelength regions. He said, briefly, that each point on a wavefront may be considered as instantaneously and continuously the origin of a new spherical wavefront moving outward from the point. The secondary wavelets from all points along the wavefront overlap, and the superposition of all of them accounts for the forward motion of the original wavefront and, as in the example of Fig. 5.1, the appearance of radiant energy behind the screen.

In our discussions of both diffraction and interference, we assume a perfectly coherent beam without concerning ourselves with the practical question of what source size would be necessary for coherence.

Spherical Wave Equation

Radiant intensity I in the space about a "point" source in watts per steradian is defined in Chapter 2. The equivalent radiant flux density can be calculated if the flux is distributed uniformly throughout a solid angle Ω. Let A be the area intercepted by the solid angle on a hypothetical sphere of radius r. Then the flux density on the sphere is

$$W = I\Omega/A = I/r^2 \tag{5-1}$$

since $\Omega = A/r^2$. The relationship in Eq. (5-1) is known as the inverse square law: the flux density varies inversely as the square of the distance from a point source. Because the radiant flux density in a wave is proportional to the square of the amplitude in the wave, it follows from the inverse square law that the amplitude in a spherical wave will vary inversely as the first power of r. Therefore, an expression that satisfies the wave equation, Eq. (4-16), and whose real part represents an expanding spherical wave, is

$$\hat{U}_0 = (U_0/r) \exp i(\omega t - \not{k}r), \tag{5-2}$$

where U_0 is the amplitude at unit distance from the source.

The Diffraction Formula

In Fig. 5.2 radiant energy originating near a point P reaches a second point P' by passing through an opening in a screen. The aperture is small compared with both the distance from the source and the distance to the point of observation P'. No point on the edge of the aperture is far from a straight line joining P to P'. Spherical wavefronts move outward from P so that one point on each wavefront passes through a point Q in the plane of the screen within the aperture. Let the medium everywhere (except the screen) have a refractive index of approximately one. An expression for this wave is

$$\hat{U}(Q) = (U_0/r) \exp i(\omega t - \not{k}r). \tag{5-3}$$

Here, r is the distance PQ.

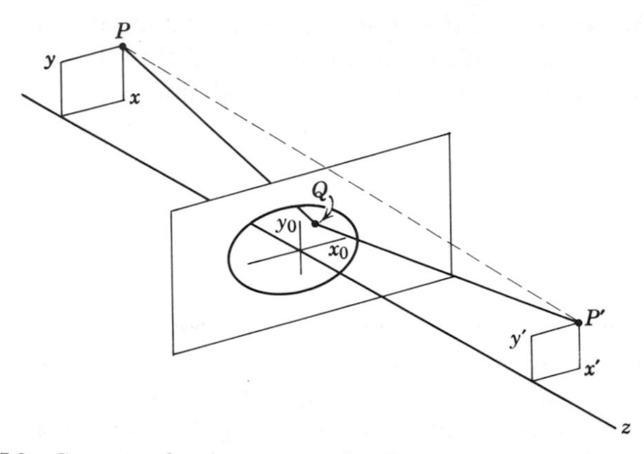

Figure 5.2 Geometry for the passage of radiant energy through an aperture.

According to Huygens' principle, every point (such as Q) on each wavefront is the source of a new spherical wavelet traveling toward P'. An equation for one of these wavelets originating at an element of area $d\sigma$ at Q is

$$d[\hat{U}(P')] = (i/\lambda)(\hat{U}(Q)/s) \exp(-i\,ks)\, d\sigma. \tag{5-4}$$

Here, s is the distance QP'. Because we forego the derivation of Eq. (5-4), some commentary on its terms is needed. An inclination or obliquity factor, which depends upon the direction of s with respect to the wavefront, has been rounded to unity with little inaccuracy for most practical applications. The factor i/λ, resulting from a derivation operation in the development of Fresnel theory, has been retained in traditional form even though our application of the formula will make no use of the $\pi/2$ radian phase advance indicated by i. The total electromagnetic disturbance at P' due to all secondary wavelets originating on a wavefront is found by integrating over that portion of a wavefront S passing through the aperture:

$$\hat{U}(P') = (i/\lambda) \iint_S [\hat{U}(Q)/s] \exp(-iks)\, d\sigma. \tag{5-5}$$

Substituting from Eq. (5-3), we obtain

$$\hat{U}(P') = \frac{iU_0 \exp i\omega t}{\lambda rs} \iint_S \exp[-ik(r+s)]\, d\sigma. \tag{5-6}$$

Moving $1/rs$ out from under the integral sign is an acceptable approximation because even though $k(r+s)$ is large enough that the exponential is a rapidly varying quantity, r and s are relatively constant as the element of area moves over the surface. This expression, Eq. (5-6), is either the Fresnel–Kirchhoff diffraction formula or the Rayleigh–Sommerfeld diffraction formula depending upon the obliquity factor that is ordinarily included. Since we have made this factor unity, we are not distinguishing here between the two formulas.

Fresnel Diffraction

A formula useful in solving Fresnel diffraction problems can be developed from Eq. (5-6) by setting up a particular coordinate system and making further approximations. Let us assume that r is so large that the spherical wave passing through the opening has essentially a plane wavefront that coincides with the plane of the screen as it passes through Q. The origin of a rectangular coordinate system is at a point near the "center" of the aperture, and the z-axis is perpendicular to the plane of the screen. The coordinates at P are (x,y,z), those at Q are $(x_0,y_0,0)$, and those at P' are (x',y',z'). The element of area at Q is $dx_0\,dy_0$. Then

$$r^2 = (x_0 - x)^2 + (y_0 - y)^2 + z^2. \tag{5-7}$$

Let r_0 be the distance from P to the origin.

$$\begin{aligned} z^2 &= r_0^2 - x^2 - y^2, \\ r &= r_0\left(1 + \frac{x_0^2 + y_0^2 - 2x_0x - 2y_0y}{r_0^2}\right)^{1/2}. \end{aligned} \tag{5-8}$$

If r_0 is very large compared with x_0 and y_0, the parenthesis may be expanded into a series by using the binomial theorem. Let us approximate by dropping all terms in the series containing powers of x_0 and y_0 higher than the second to get

$$r \approx r_0\left[1 + \frac{x_0^2 + y_0^2}{2r_0^2} - \frac{x_0x + y_0y}{r_0^2} - \frac{(x_0x + y_0y)^2}{2r_0^4}\right]. \tag{5-9}$$

Similarly:

$$s \approx s_0\left[1 + \frac{x_0^2 + y_0^2}{2s_0^2} - \frac{x_0x' + y_0y'}{s_0^2} - \frac{(x_0x' + y_0y')^2}{2s_0^4}\right]. \tag{5-10}$$

If $1/r_0s_0$ is substituted for $1/rs$ in the coefficient of the integral, Eq. (5-6) can now be put into the following form:

$$\hat{U}(P') = \frac{iU_0 \exp i\omega t \exp\left[-ik(r_0 + s_0)\right]}{\lambda r_0 s_0}\iint \exp\{-ik[f(x_0,y_0)]\}\, dx_0\, dy_0, \tag{5-11}$$

where

$$f(x_0,y_0) = \left(-\frac{xx_0 + yy_0}{r_0} - \frac{x'x_0 + y'y_0}{s_0}\right) + \left[\frac{x_0^2 + y_0^2}{2r_0} + \frac{x_0^2 + y_0^2}{2s_0} - \frac{(xx_0 + yy_0)^2}{2r_0^3} - \frac{(x'x_0 + y'y_0)^2}{2s_0^3}\right]. \tag{5-12}$$

Phenomena in which this formula is applied are called Fresnel diffraction; however, many common types of diffraction problems can be studied with simpler formulas like those of Fraunhofer diffraction.

Fraunhofer Diffraction

Simplification of Eq. (5-12) can be achieved for a large class of problems where the distances r_0 and s_0 are each so large compared with the maximum values of x_0 and y_0 that the terms in the brackets can be neglected. Furthermore, if the source is very small and is on the z-axis so that $x \approx 0$ and $y \approx 0$, the first term in the parenthesis can also be neglected. In the coefficient of the integral in Eq. (5-11), $\hat{U}_0 = [(U_0/r_0) \exp i(\omega t - k r_0)]$ is the amplitude and phase of the incident wavefronts in the aperture, and $i \exp(-i k s_0) = \exp(-i k \varphi_0)$ is a constant phase term. (Later, when we find the radiant flux density at P' by multiplying the complex amplitude by its complex conjugate, this phase factor is lost.) If $k_x = x'/\lambda z_0$ and $k_y = y'/\lambda z_0$ with $z_0 \approx s_0$, Eq. (5-11) can be written

$$\hat{U}(P') = \hat{C} \iint_S \exp [2\pi i(k_x x_0 + k_y y_0)] \, dx_0 \, dy_0, \tag{5-13}$$

where

$$\hat{C} = (\hat{U}_0/\lambda z_0) \exp(-ik\varphi_0). \tag{5-14}$$

When diffraction phenomena are observed at distances from the aperture so great that this formula can be used, we have what is called Fraunhofer diffraction. By applying imaging techniques discussed in Chapter 6, we can use a lens to form the diffraction pattern near the focal point. That is, the focal distance of the lens becomes an equivalent "great distance" between the aperture and the plane of observation. In fact, the lens itself may constitute an aperture that produces the diffraction pattern as indicated in Chapter 8. Figure 5.3 shows a photograph of a magnificent Fraunhofer diffraction pattern (produced without a lens) and the aperture through which it was formed.

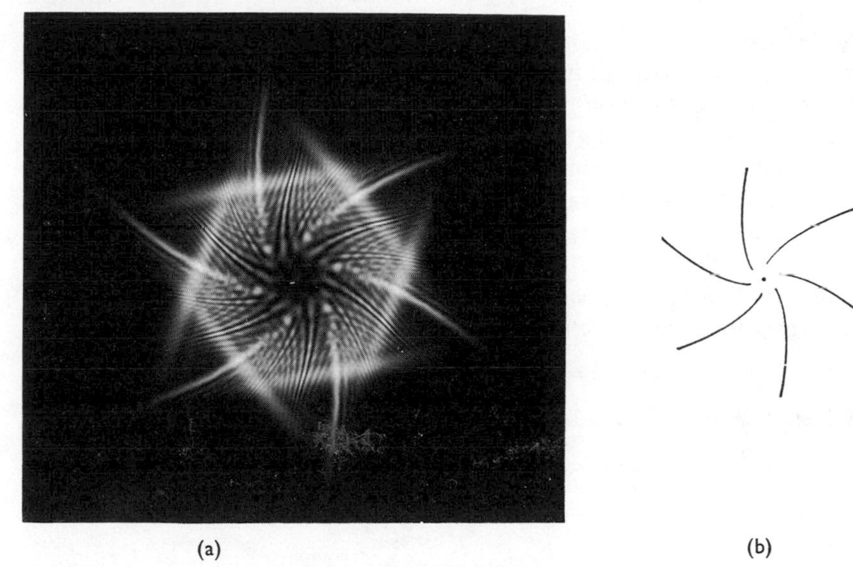

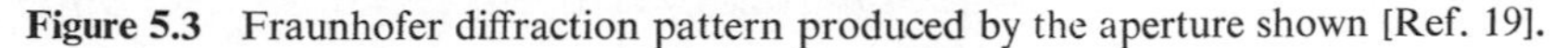

Figure 5.3 Fraunhofer diffraction pattern produced by the aperture shown [Ref. 19].

Rectangular Aperture

Since the integration in Eq. (5-13) is over the wavefront as it passes through the aperture, the size and shape of the aperture determine the region of integration. A factor, called the *aperture function*, can be introduced into the integrand to describe the configuration of the integration region. For the first case, let us consider a rectangular aperture of dimensions $2a$ and $2b$ with the sides parallel to the x_0 and y_0 axes respectively. The related aperture function is defined by the following, called *rectangular functions* in mathematics texts:

$$\begin{aligned} R_a(x_0) &= 1 \qquad \text{when } |x_0| \leq a, \\ R_a(x_0) &= 0 \qquad \text{when } |x_0| > a, \\ R_b(y_0) &= 1 \qquad \text{when } |y_0| \leq b, \\ R_b(y_0) &= 0 \qquad \text{when } |y_0| > b. \end{aligned} \tag{5-15}$$

The aperture function $F_a(x_0,y_0)$ is the product $R_a(x_0)R_b(y_0)$. Integration is accomplished by separating the variables and integrating each from $-\infty$ to $+\infty$:

$$\hat{U}_r(P') = \hat{C}\int_{-\infty}^{+\infty}\int_{-\infty}^{+\infty} [R_a(x_0)\exp(2\pi i k_x x_0)][R_b(y_0)\exp(2\pi i k_y y_0)]\,dx_0\,dy_0. \tag{5-16}$$

As derived, Eq. (5-16) is the diffraction formula for a rectangular aperture; except for the constant $\hat{C}$, it is also the two-dimensional Fourier transform of the rectangular functions. [Actually it is the *inverse* transform according to our definition; see Equations (1-22) and (1-23). However, this distinction is not significant here.] Since the variables are separated, the integration on each—finding the Fourier transform—produces the well-known sinc function, $\text{sinc}\, x \equiv (\sin x)/x$, which is unity where $x = 0$ and passes through zero where $x = q\pi$ $(q = 1,2,3, \ldots)$. The function has maxima, alternating negative and positive, where $x = (2q - 1)\pi/2$, but the maxima become smaller as x increases. The integration of Eq. (5-16) gives

$$\hat{U}_r(P') = 4ab\hat{C}\, \text{sinc}\, 2\pi k_x a\, \text{sinc}\, 2\pi k_y b \tag{5-17}$$

[see Ref. 5, p. 66]. The radiant flux density at points P' is given by

$$\begin{aligned} W_r(P') &= \hat{U}(P') \cdot \hat{U}^*(P') \\ &= W_0\, \text{sinc}^2\, 2\pi k_x a\, \text{sinc}^2\, 2\pi k_y b, \end{aligned} \tag{5-18}$$

where

$$W_0 = (16a^2b^2\, |U_0|^2)/\lambda^2 z_0^{\,2}. \tag{5-19}$$

The total power distributed throughout the diffraction pattern described by Eq. (5-18), that is, the integral of the flux density times an element of area, is finite because even when the limits of integration are extended to infinity,

$$\int_{-\infty}^{+\infty} (\text{sinc}^2\, x)\, dx = \pi. \tag{5-20}$$

The flux density W_0 at the center of the diffraction pattern can be found in terms of the geometry, the wavelength, and the total power Φ_a passing through the aperture. The quantity,

$$W_r(P')\, dx'\, dy' = W_r(x',y')\, dx'\, dy', \tag{5-21}$$

is the differential of the total radiant power in the diffraction pattern. By integrating both variables from $-\infty$ to $+\infty$, the expression for total power in the $x'y'$ plane is obtained, which is also the total power passing through the aperture. Although the approximations restrict x' and y' to small values, the integration to large values can be justified because the integrand, the sinc function squared, will be negligible as x' and y' become large. Thus the power expression, obtained from Eq. (5-18) by transforming variables according to $2\pi ax'/\lambda z_0 = u$ and $2\pi by'/\lambda z_0 = v$, is

$$\begin{aligned} \Phi_a &= W_0(\lambda^2 z_0^{\,2}/4\pi^2 ab) \iint_{-\infty}^{+\infty} (\text{sinc}^2\, u)\, (\text{sinc}^2\, v)\, du\, dv \\ &= W_0 \lambda^2 z_0^{\,2}/4ab. \end{aligned} \tag{5-22}$$

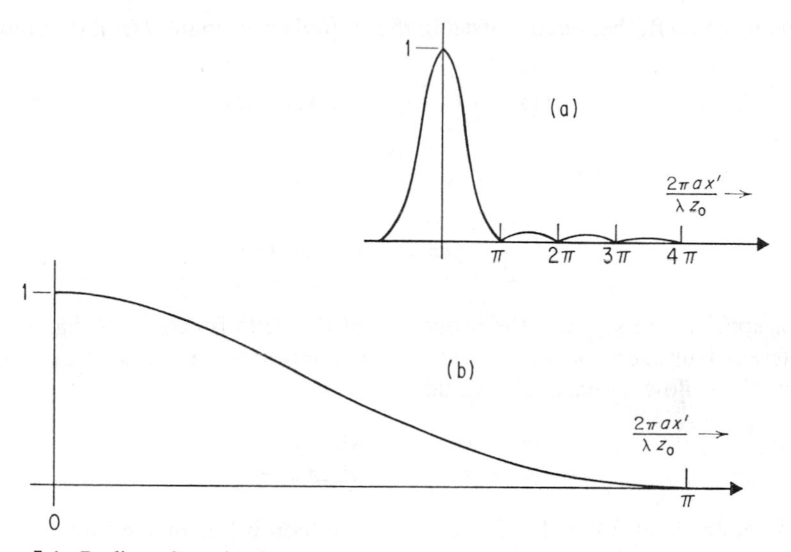

Figure 5.4 Radiant flux density distributed in the diffraction pattern of a slit. [The complete pattern is symmetrical about $(2\pi ax'/\lambda z_0) = 0$.]

Then

$$W_0 = \Phi_a A/\lambda^2 z_0^{\,2}, \tag{5-23}$$

where A is the area of the aperture.

Diffraction by a Slit

An expression for diffraction by a slit can be set up by making $k_\lambda b$ of Eq. (5-18) very large compared with $k_x a$. Then the radiant power distributed in the $x'y'$ plane according to $\text{sinc}^2\, 2\pi k_y b$ is confined to a very small region on each side of the line $y' = 0$. The power distribution in the other direction $(\pm x')$ is given by the following expression for the radiant flux density:

$$W_s(x') = W_0 \,\text{sinc}^2\, (2\pi a x'/\lambda z_0). \tag{5-24}$$

The ratio $W_s(x')/W_0$, the normalized flux density, is plotted in Fig. 5.4 with both condensed and expanded scales for the argument.

The Dirac Delta Function

To treat the diffraction problems of the next three sections, a specialized function or, more correctly, a distribution called the *Dirac delta function* is needed. When the shifted delta function, $\delta(x - x')$, is convolved with some chosen function such as $f(x)$ (which might be, for example, the rectangular function $R_a(x)$), the function itself in the shifted coordinate, $f(x')$, is obtained:

$$f(x') = \int_{-\infty}^{+\infty} \delta(x - x') f(x)\, dx. \tag{5-25}$$

If $f(x)$ is a constant of unit value, this becomes

$$\int_{-\infty}^{+\infty} \delta(x - x')\, dx = 1. \tag{5-26}$$

This special case suggests the properties of the delta function. It has a value different from zero only as the argument approaches zero, so it appears to have the following particular values:

$$\begin{aligned} \delta(x) &= 0 \qquad \text{when } x \neq 0, \\ \delta(x) &= \infty \qquad \text{when } x = 0. \end{aligned} \tag{5-27}$$

A sequence of $2N + 1$ delta functions, a "comb," is defined by

$$\text{III}_0(x_0) = \sum_{q=-N}^{N} \delta(qx_1 - x_0). \tag{5-28}$$

The delta function will have a nonzero value at each position where $x_0 = qx_1$ including $qx_1 = x_0 = 0$. A second comb, which will be useful later, is $\text{III}_e(x_0)$, having $2N$ delta functions, defined by

$$\text{III}_e(x_0) = \sum_{q=1}^{N} \left\{ \delta\left[\left(\frac{2q-1}{2}\right) x_1 - x_0\right] + \delta\left[\left(\frac{2q-1}{2}\right) x_1 + x_0\right]\right\}. \tag{5-29}$$

This becomes a pair of delta functions at $+x_1/2$ and $-x_1/2$ when $N = 1$.

$$\text{III}_2(x_0) = \delta[(x_1/2) - x_0] + \delta[(x_1/2) + x_0]. \tag{5-30}$$

The Fourier transform of the delta function is unity, but the transform of the shifted delta function $\delta(x_1 - x_0)$ is

$$\int_{-\infty}^{+\infty} \delta(x_1 - x_0) \exp(2\pi i k_x x_0)\, dx_0 = \exp(-2\pi i k_x x_1). \tag{5-31}$$

The following relationship, in which the double-ended arrow signifies the Fourier transform in one direction and the inverse transform in the other, shows the transform of the comb $\text{III}_e(x_0)$:

$$\text{III}_e(x_0) \leftrightarrow \sum_{q=1}^{N} \{\exp[-\pi i(2q-1)k_x x_1] + \exp[\pi i(2q-1)k_x x_1]\}. \tag{5-32}$$

The expression in the brace is recognized as twice a hyperbolic cosine function:

$$2\left\{\frac{\exp\left[-\pi i(2q-1)k_x x_1\right]+\exp\left[\pi i(2q-1)k_x x_1\right]}{2}\right\}$$

$$= 2\cosh\left[\pi i(2q-1)k_x x_1\right],$$

$$= 2\cos\left[\pi(2q-1)k_x x_1\right]. \qquad (5\text{-}33)$$

Therefore,

$$\mathrm{III}_e(x_0) \leftrightarrow \sum_{q=1}^{N} 2\cos\left[\pi(2q-1)k_x x_1\right]. \qquad (5\text{-}34)$$

The sum of the series is a ratio of sine functions [Ref. 16, p. 78, No. 420] as given on the right of the following relationship:

$$\mathrm{III}_e(x_0) \leftrightarrow (\sin 2N\pi k_x x_1)/(\sin \pi k_x x_1). \qquad (5\text{-}35)$$

Similarly the Fourier transform of comb $\mathrm{III}_0(x_0)$ can be shown to be [Ref. 5, p. 101]:

$$\mathrm{III}_0(x_0) \leftrightarrow [\sin(2N+1)\pi k_x x_1]/(\sin \pi k_x x_1). \qquad (5\text{-}36)$$

Young's Experiment

Young's experiment, which demonstrates the formation of interference fringes by passing light from a small source through two pinholes, is classical. It was discussed briefly in Chapter 1 (see Fig. 1.4), although it was not identified by name. It will be treated here first by applying our diffraction formulas. In a later section, we return to this experiment and treat it more conventionally as an interference problem.

The convolution of the delta function $\delta(x_1 - x_0)$ with the rectangular function $R_a(x_0)$ [see Equations (1-35) and (5-25)] is

$$\delta(x_1 - x_0) * R_a(x_0) = \delta(x_0) * R_a(x_1 - x_0) = R_a(x_1). \qquad (5\text{-}37)$$

According to this expression, convolving the delta function at x_1 with the rectangular function positions the rectangular function, placing its center at x_1. Thus, a pair of rectangular functions with centers at $-x_1/2$ and $+x_1/2$ is obtained by convolving the pair of delta functions $\mathrm{III}_2(x_0)$ [see Eq. 5-30)] with the rectangular function. The resulting function represents a pair of slits, which take the place of the pinholes, of width $2a$ (height, say, of $2b$) and with separation between centers of x_1. This is written

$$\mathrm{III}_2(x_0) * R_a(x_0) = G_2(x_0). \tag{5-38}$$

In a one-dimensional diffraction formula corresponding to Eq. (5-16), the aperture function $R_a(x_0)$ is replaced by the function $G_2(x_0)$, so

$$\hat{U}_y(x') = 2b\hat{C}\int_{-\infty}^{+\infty} G_2(x_0)\exp(2\pi i k_x x_0)\,dx_0. \tag{5-39}$$

Since this is the Fourier transform of the convolution, represented by $G_2(x_0)$, of two functions, the transform will turn out to be the product of the transforms of the functions themselves [see Ref. 5, p. 74]. The transform of the rectangular function is given in Eq. (5-17), and the transform of the pair of delta functions is given by Eq. (5-34) with $N = 1$. Thus, the expression for $\hat{U}_y(x')$ is obtained:

$$\hat{U}_y(x') = 4ab\hat{C}\cos\pi k_x x_1 \operatorname{sinc} 2\pi k_x a. \tag{5-40}$$

The flux density is given by

$$W_y(x') = W_{y0}\cos^2\pi k_x x_1 \operatorname{sinc}^2 2\pi k_x a, \tag{5-41}$$

in which W_{y0} combines the various coefficients.

If the distance between slit centers is large compared with the slit width—for example suppose $x_1 = 200a$—the sinc function will be nearly unity for many cycles of the cosine factor; and the flux density is nearly constant, at small values of x_1, modulated by a cosine squared factor that varies periodically between one and zero. This pattern is known as Young's interference fringes. A plot of the normalized flux density $W_y(x')/W_{y0}$ is shown in Fig. 5.5. At the 100th fringe away from $x' = 0$ in the example, the fringes disappear because the sinc function becomes zero. Although the flux density in a bright fringe decreases with increasing x', the visibility as defined by Eq. (1-46) remains essentially unity, which should be expected from the assumed perfectly coherent energy beam.

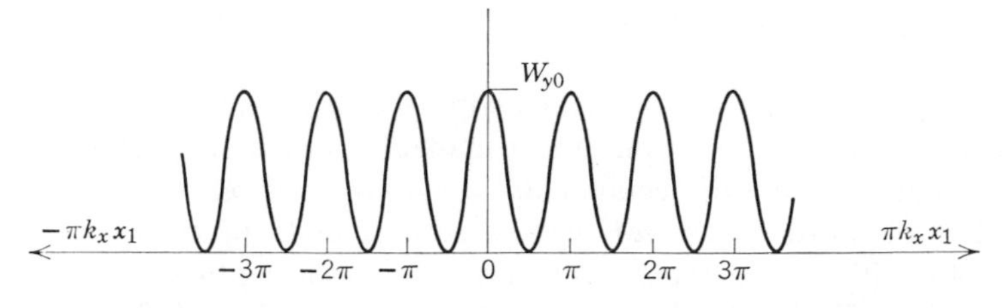

Figure 5.5 Radiant flux density in a Young's interference pattern.

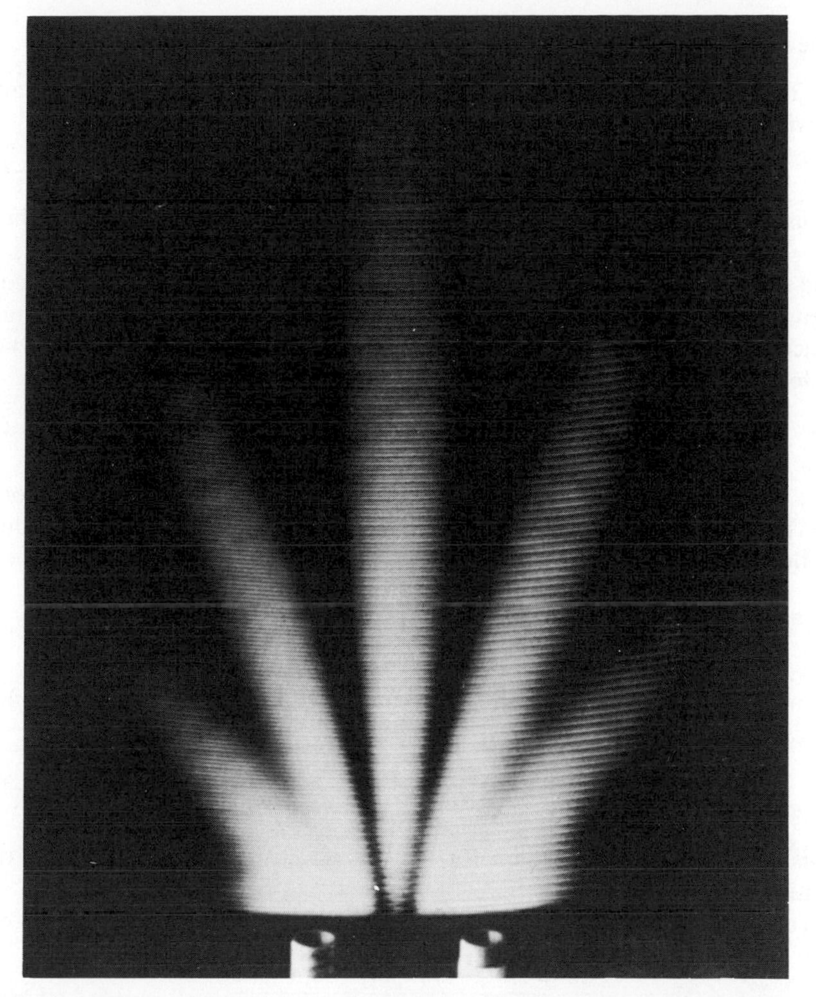

Figure 5.6 Photograph of a Young's experiment—diffraction formed with acoustic waves generated by coherent transducers perpendicular to the figure [Ref. 20]. (This photo also appears in Winston E. Koch, *Lasers and Holography*, Doubleday and Company, Inc., New York, 1968.)

Figure 5.6 is a photograph of a Young's experiment diffraction pattern. The figure is in the plane that is perpendicular to the screen and that includes the pinholes. (The pinholes are actually coherent acoustic transducers.) The photograph shows how the diffraction pattern changes as its distance from the pinholes increases. The regions of bright fringes and regions of dark fringes show clearly.

The Transmission Grating

The operation of Eq. (5-37) can be extended to an array of slits by convolving the comb $\text{III}_0(x_0)$ with $R_a(x_0)$; that is,

$$\text{III}_0(x_0) * R_a(x_0) = G_N(x_0). \tag{5-42}$$

Thus, the center of a rectangular function is positioned at each point where $x_0 = \pm q x_1$ ($q = 0,1,2,3, \ldots ,N$). The separation between slit centers is x_1 and the width of each slit is $2a$. This sequence of slits forms what is called a transmission grating of $2N + 1$ lines. The function $G_N(x_0)$ is an aperture function that can be applied in an equation corresponding to Eq. (5-39) as follows:

$$\hat{U}_N(x') = 2b\hat{C}\int_{-\infty}^{+\infty} G_N(x_0) \exp(2\pi i k_x x_0)\, dx_0. \tag{5-43}$$

Again, this is the Fourier transform of the convolution of two functions. In a manner similar to the evaluation of the integral in Eq. (5-39), the product of the sinc function and, this time, the right side of Eq. (5-36) is obtained:

$$\hat{U}_N(x') = 4ab\hat{C} \operatorname{sinc} 2\pi k_x a \left\{\frac{\sin [(2N + 1)\pi k_x x_1]}{\sin \pi k_x x_1}\right\}. \tag{5-44}$$

By incorporating the $W_s(x')$ already evaluated in Eq. (5-24), the radiant flux density is

$$W_g(x') = W_s(x')\left\{\frac{\sin^2 [(2N + 1)\pi k_x x_1]}{\sin^2 \pi k_x x_1}\right\}. \tag{5-45}$$

This equation is an expression for the diffraction pattern produced by the grating.

To study Eq. (5-45), we first focus our attention on the ratio in the braces:

$$\frac{\sin^2 [(2N + 1)\pi k_x x_1]}{\sin^2 \pi k_x x_1} = \frac{\sin^2 p'\pi}{\sin^2 p\pi}. \tag{5-46}$$

Here p and p' have been defined to group the terms in the arguments of the sine functions. From their definitions we have

$$p' = (2N + 1)p, \tag{5-47}$$

so, when p is zero or an integer, p' is also zero or an integer. Therefore, the numerator in Eq. (5-46) is zero whenever the denominator is zero, but the numerator passes through $2N$ zeros between the common zeros. The ratio is obviously zero for each of the intervening numerator zeros, but the ratio is an indeterminate form at the common zeros. Application of L'Hôpital's rule

[Ref. 17] shows that the ratio approaches $(2N + 1)^2$ when both numerator and denominator approach zero simultaneously. There will be $2N - 1$ ratio maxima between successive denominator zeros; but these maxima turn out to be much smaller than the main peaks that occur when the denominator becomes zero.

The normalized function

$$W_N(x') \equiv \frac{1}{(2N+1)^2}\left\{\frac{\sin^2 [(2N+1)\pi k_x x_1]}{\sin^2 \pi k_x x_1}\right\} \tag{5-48}$$

is a generalized flux density distribution from a transmission grating and is plotted in Fig. 5.7. However, according to Eq. (5-45), to get the actual grating pattern, the variation of $W_s(x')$ must be superposed by combining the relationships expressed in Equations (5-24), (5-45), and (5-48):

$$\begin{aligned} W_g(x')/(2N+1)^2 &= W_s(x')W_N(x') \\ &= W_0 W_N(x') \operatorname{sinc}^2 (2\pi x' a/\lambda z_0). \end{aligned} \tag{5-49}$$

The normalized grating diffraction function, $W_g(x')/[(2N + 1)^2 W_0]$, which is the product of the curves of Figures 5.4b and 5.7, is shown in Fig. 5.8.

The Phase Grating

The characteristics of a phase grating can be investigated by studying an experiment in which a periodic phase variation is produced over a wavefront. The setup is shown schematically in Fig. 5.9. The surface forming the grating is assumed perfectly reflecting and has a sinusoidal profile. The expression,

$$z(x_0) = z_1 \sin 2\pi\sigma x_0, \tag{5-50}$$

describes the surface. In Eq. (5-50), $\sigma = 1/\Lambda$ is the number of crests per unit distance along the surface. Let us assume that the amplitude z_1 is small compared with the distance Λ between crests In fact, z_1 is so small that the maximum slope on the surface $(dz/dx_0)_m = 2\pi z_1 \sigma$ is of the order of milliradians or less. If plane wavefronts are incident on the surface with wavefronts parallel to the mean surface, the grating functions as the aperture, the wavefronts being reflected rather than being transmitted. The surface slopes are so gentle that variations in the direction of rays after reflection are negligible; however, the variations in phase over the wavefront are significant. The reflected wave will have a constant phase term, which is not significant here, and a variable phase term $\varphi(x_0)$ given by

$$\varphi(x_0) = 2\pi[2z_1(x_0)]/\lambda = (4\pi z_1/\lambda) \sin 2\pi\sigma x_0. \tag{5-51}$$

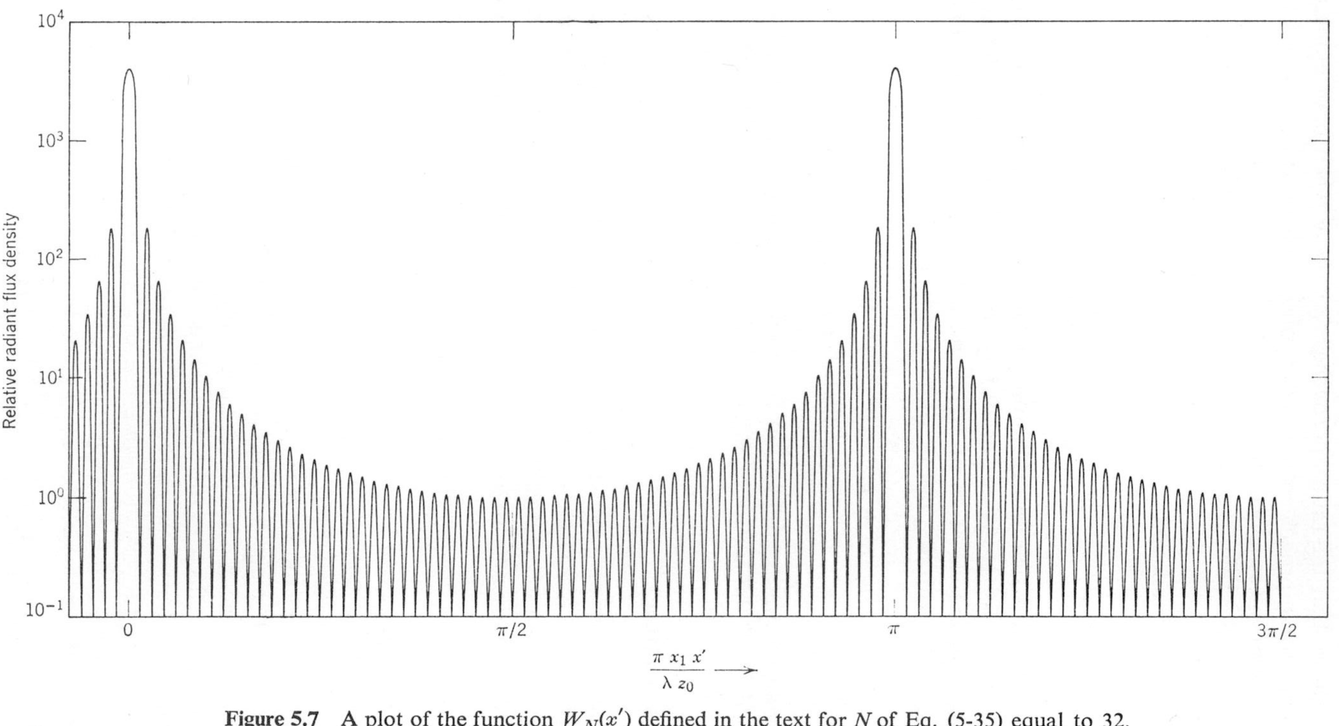

Figure 5.7 A plot of the function $W_N(x')$ defined in the text for N of Eq. (5-35) equal to 32.

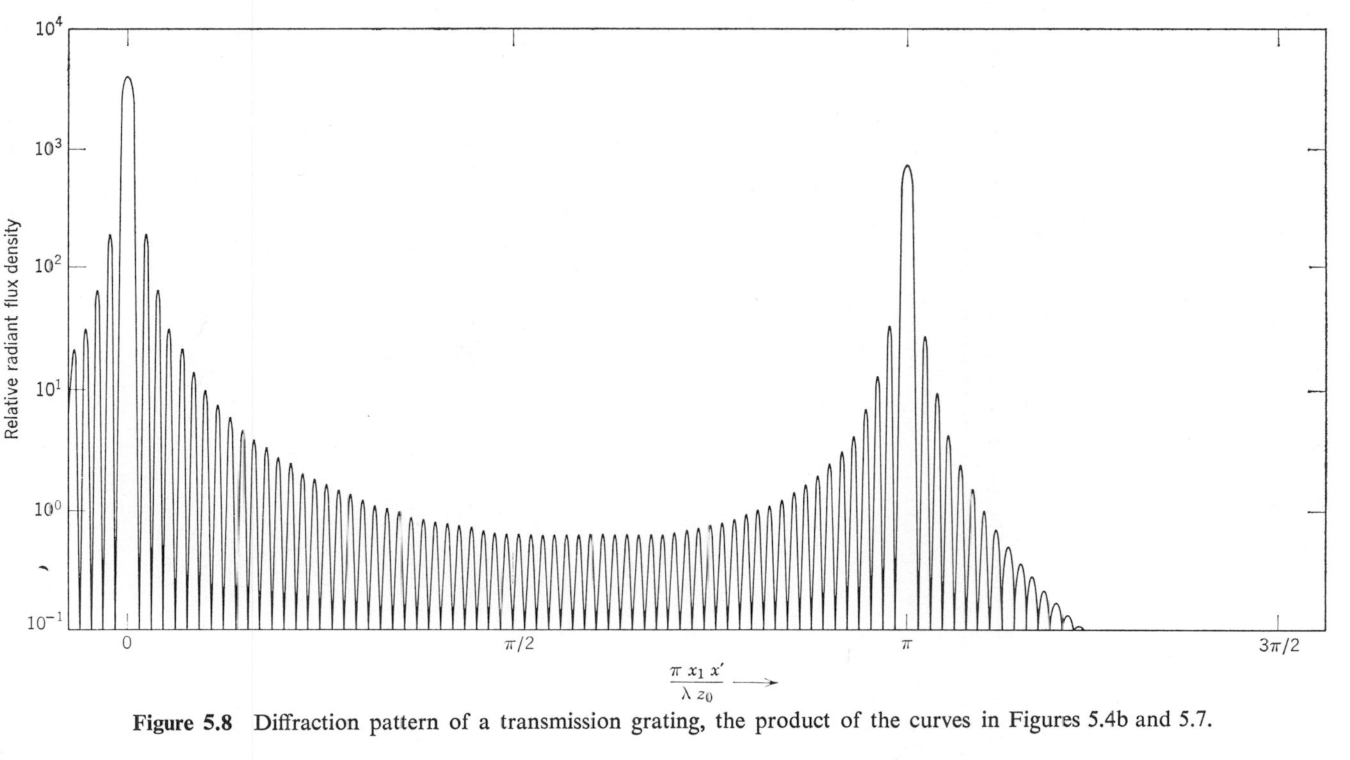

Figure 5.8 Diffraction pattern of a transmission grating, the product of the curves in Figures 5.4b and 5.7.

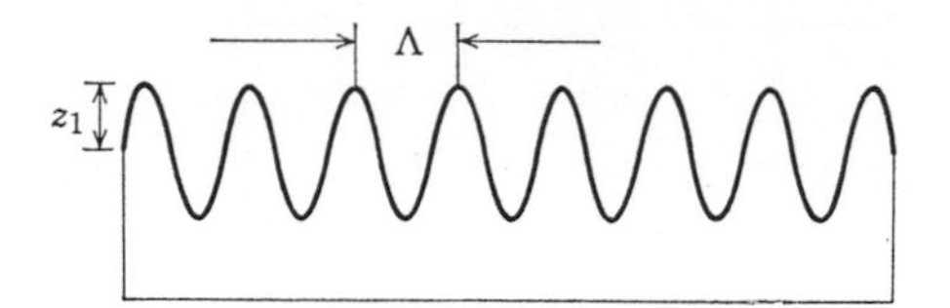

Figure 5.9 Surface profile of a reflecting phase grating. (amplitude z_1 exaggerated)

The variable phase factor in the equation for the wave would, therefore, be

$$\exp\left[-(4\pi i z_1/\lambda)\sin 2\pi\sigma x_0\right]. \tag{5-52}$$

The *slit function* under the assumed conditions is the phase factor for precisely one wave of the profile of the reflected wavefront as follows:

$$\begin{aligned} F_s(x_0) &= \exp\left[-(4\pi i z_1/\lambda)\sin 2\pi\sigma x_0\right] \quad \text{when } (-\Lambda/2) \le x_0 \le (+\Lambda/2), \\ &= 0 \qquad\qquad\qquad\qquad\qquad\qquad\quad \text{when } |x_0| > \Lambda/2. \end{aligned} \tag{5-53}$$

The quantity $(2N + 1)$ is the number of complete surface waves on the grating, so one dimension is $(2N + 1)\Lambda$; the other dimension is $2b$. Only the portion of the wavefront that encounters the grating will be reflected; the rest will pass on by. The value of x_1 in Eq. (5-28) for the comb $\text{III}_0(x_0)$ is Λ, the distance between crests. An aperture function describing the wavefront the moment after it is reflected at the surface is the convolution of the comb with the slit function. Thus

$$\text{III}_0(x_0) * F_s(x_0) = G_p(x_0). \tag{5-54}$$

The diffraction formula is

$$\hat{U}_p(x') = 2b\hat{C}\int_{-\infty}^{+\infty} G_p(x_0)\exp(2\pi i k_x x_0)\,dx_0. \tag{5-55}$$

Once more we have the Fourier transform of the convolution of two functions. To proceed to expressions for the amplitude and flux density distributions, the transform of $F_s(x_0)$ must be found. It is

$$F_s(x_0) \leftrightarrow \int_{-\frac{1}{2}\Lambda}^{\frac{1}{2}\Lambda} \exp\left[-(4\pi i z_1/\lambda)\sin 2\pi\sigma x_0\right]\exp(2\pi i k_x x_0)\,dx_0. \tag{5-56}$$

To facilitate the indicated integration, variables can be changed as follows: $v = 4\pi z_1/\lambda$; $u = 2\pi\sigma x_0$; $2\pi k_x x_0 = (k_x/\sigma)u = nu$ where $n = k_x/\sigma = k_x\Lambda$; $dx_0 = (1/2\pi\sigma)\,du$. Then

$$F_s(x_0) \leftrightarrow \frac{\Lambda}{2\pi}\int_{-\pi}^{+\pi} \exp i[nu - v\sin u]\,du. \tag{5-57}$$

If n is an integer, the integral is a standard form defining a Bessel function of the first kind of order n and argument v [Ref. 18]. That only integral

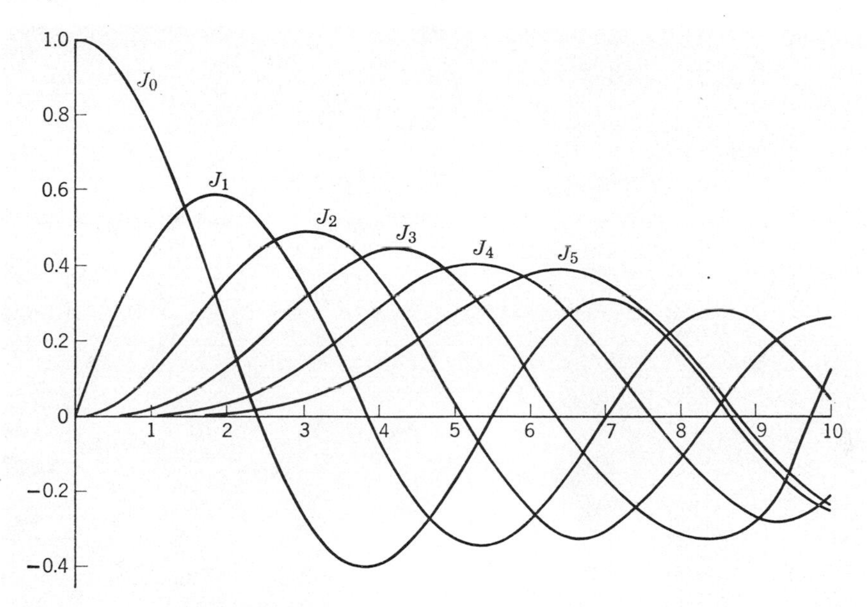

Figure 5.10 The first five orders of the Bessel function.

orders are of interest is shown later. With n an integer:

$$F_s(x_0) \leftrightarrow \Lambda J_n(v). \tag{5-58}$$

Integration of Eq. (5-55) is similar to the integration of Eq. (5-43) with the result,

$$\hat{U}_p(x') = (2\Lambda^2 b\hat{C})J_n(v)\left\{\frac{\sin[(2N+1)\pi k_x\Lambda]}{\sin \pi k_x\Lambda}\right\}. \tag{5-59}$$

The flux density is

$$W_p(x') = W_{po}[J_n(v)]^2\left\{\frac{\sin[(2N+1)\pi k_x\Lambda]}{\sin \pi k_x\Lambda}\right\}^2. \tag{5-60}$$

The quantity in the brace has already been studied. There are main peaks when $\pi k_x\Lambda = q\pi$ ($q = 0,1,2,3,\ldots$). Since the magnitudes of the Bessel functions are always less than one, the orders where the brace has its peak values of $(2N+1)^2$ are the only ones that are significant. The subsidiary peaks may be ignored as in the transmission grating. When $\pi k_x\Lambda = q\pi$, n, which is equal to $k_x\Lambda$, is also equal to q. So, at these peaks the Bessel function is automatically an itegral order. The first few orders of the Bessel function are plotted in Fig. 5.10. The magnitude of a given order function depends upon the argument $v = 4\pi z_1/\lambda$. For example, a maximum proportion of the radiant power appears in the first order (the main peaks where $q = n = 1$) if $4\pi z/\lambda$ is set at 1.841, the value of v for which the first order has

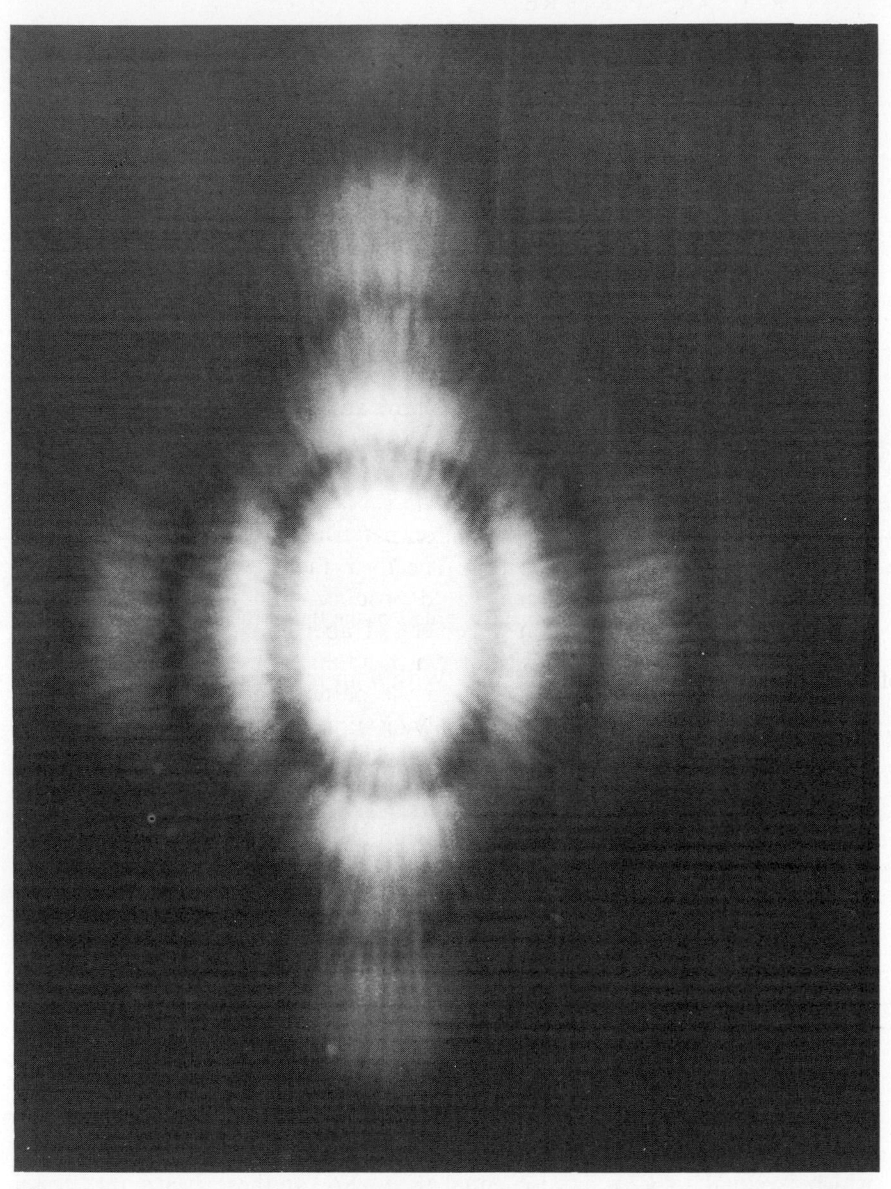

Figure 5.11 A black and white print made from a color transparency showing the diffraction pattern produced by a randomly distributed array of 3000 identical rectangular apertures. (This is a reproduction of a color transparency lent by Dr. R. B. Hoover. A color reproduction appears in Ref. 21.)

its first maximum. That is, the amplitude of the surface profile wave must be $z_1 = 0.146\lambda$. All of the power never appears in any given order, except the zero order where $z_1 = 0$, because when the argument $v = 1.841$, for example, there is also power in the other orders: $[J_0(1.841)]^2 = 0.100$, $[J_1(1.841)]^2 = 0.339$, $[J_2(1.841)]^2 = 0.100$, $[J_3(1.841)]^2 = 0.011$, $[J_4(1.842)]^2 = 0.0006$. The amplitude of the surface wave cannot increase appreciably beyond $z_1 = 0.146\lambda$, for instance, to place a larger proportion in one of the higher orders, because this would be inconsistent with the assumption of small amplitude.

Arrays of Apertures

The comb of delta functions illustrates how an aperture of a given size and shape can be placed at various positions in the aperture plane. With a two-dimensional comb, apertures could be positioned wherever they are desired. The expressions for the diffraction patterns show how a linear array of identical apertures produces a diffraction pattern that can be divided into a form factor and a structure factor. The form factor describes the pattern that one of the apertures alone would produce, and the structure factor describes the effect that the arrangement of apertures has on the pattern. Whenever the apertures are distributed at random over the diffracting plane, the structure factor is not apparent in the pattern formed. However, the flux density in the pattern of N randomly distributed, identical apertures is N times the flux density in the pattern that would be produced by one of the apertures alone. Figure 5.11 is a photograph of the diffraction pattern produced by using a small light source projected through 3000 rectangular apertures randomly distributed. Figure 5.12 is the diffraction pattern produced by a square array of 2916 circular apertures.

The Circular Aperture

If the aperture is circular of radius ρ_m with the z-axis through the center, the diffraction pattern is expected to have circular symmetry, so it is convenient to change Eq. (5-13) by introducing polar coordinates as follows:

$$\begin{aligned} x_0 &= \rho_0 \cos\theta, \qquad & k_x &= (r/\lambda z_0)\cos\varphi, \\ y_0 &= \rho_0 \sin\theta, \qquad & k_y &= (r/\lambda z_0)\sin\varphi. \end{aligned} \tag{5-61}$$

The element of area is $\rho_0\, d\rho_0\, d\theta$. If $2\pi\rho_0 r/\lambda z_0 = u'$, Eq. (5-13) can be put into the following form:

$$\hat{U}_c(P') = \hat{C}\int_{-\pi}^{+\pi}\int_0^{\rho_m} \exp\,[iu'\cos(\varphi - \theta)]\rho_0\, d\rho_0\, d\theta. \tag{5-62}$$

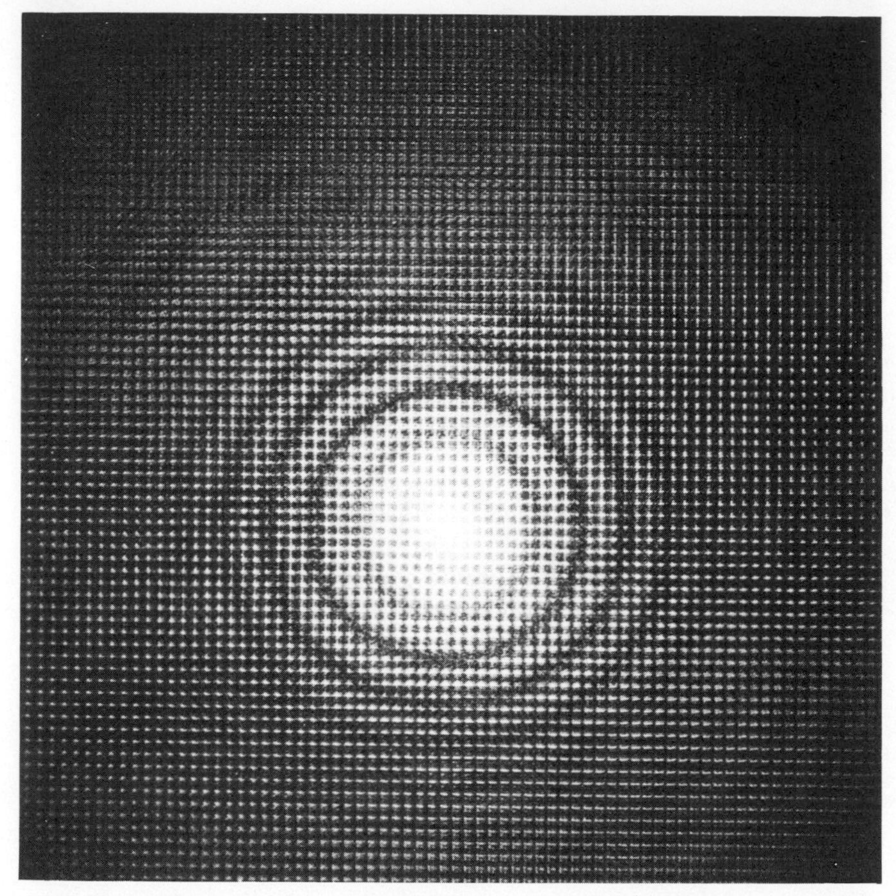

Figure 5.12 Diffraction pattern produced by a square array of circular apertures [Ref. 22].

In preparation for integration on θ, let $\varphi - \theta = (\pi/2) + \alpha$. Then $\cos(\varphi - \theta) = -\sin\alpha$ and $d\theta = -d\alpha$. The integral on α becomes

$$-\int_{-\pi}^{+\pi} \exp[-iu' \sin\alpha]\, d\alpha. \tag{5-63}$$

Since the integration on θ of Eq. (5-62) is over any complete cycle (to go around the whole aperture), the signs of the limits in Eq. (5-63) can be reversed to change sign of the integral. Therefore, if $n = 0$ in Eq. (5-57), the integral is Eq. (5-63), except for notation, and

$$\hat{U}_c(P') = 2\pi\hat{C}\int_0^{\rho_m} J_0(u')\rho_0\, d\rho_0. \tag{5-64}$$

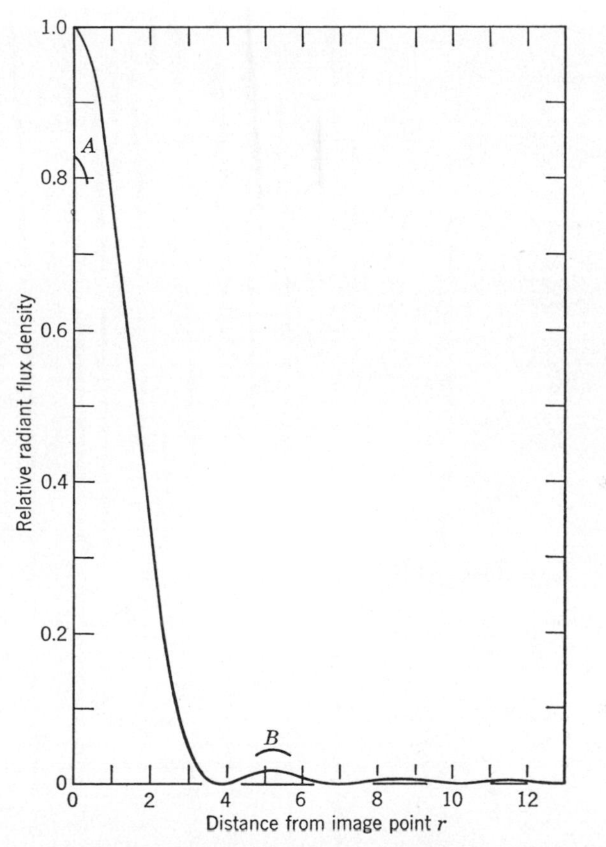

Figure 5.13 Radial distribution of the radiant flux density in a diffraction pattern of a circular aperture. (Arcs at A and B are from Curve 2 of Fig. 8.21.)

Let $u' = \rho_0 u_r$ where $u_r = 2\pi r/\lambda z_0$. Also, by defining a circle function $C_a(\rho_0)$, the aperture function for a circular aperture becomes

$$\begin{aligned} C_a(\rho_0) &= 1 \qquad \text{when } 0 \le \rho_0 \le \rho_m, \\ C_a(\rho_0) &= 0 \qquad \text{when } |\rho_0| > \rho_m. \end{aligned} \tag{5-65}$$

When the integrand of Eq (5-64) is multiplied by C_a, the upper limit of integration can be changed from ρ_m to ∞:

$$\hat{U}_c(P') = 2\pi\hat{C}\int_0^\infty \rho_0 C_a(\rho_0) J_0(\rho_0 u_r)\, d\rho_0. \tag{5-66}$$

This integral is a standard form and is called either a Hankel transform or a Fourier–Bessel transform [see Ref. 5, p. 145]. The following is obtained for $\hat{U}_c(P')$:

$$\hat{U}_c(P') = \pi\rho_m{}^2\hat{C}[2J_1(\rho_m u_r)]/\rho_m u_r. \tag{5-67}$$

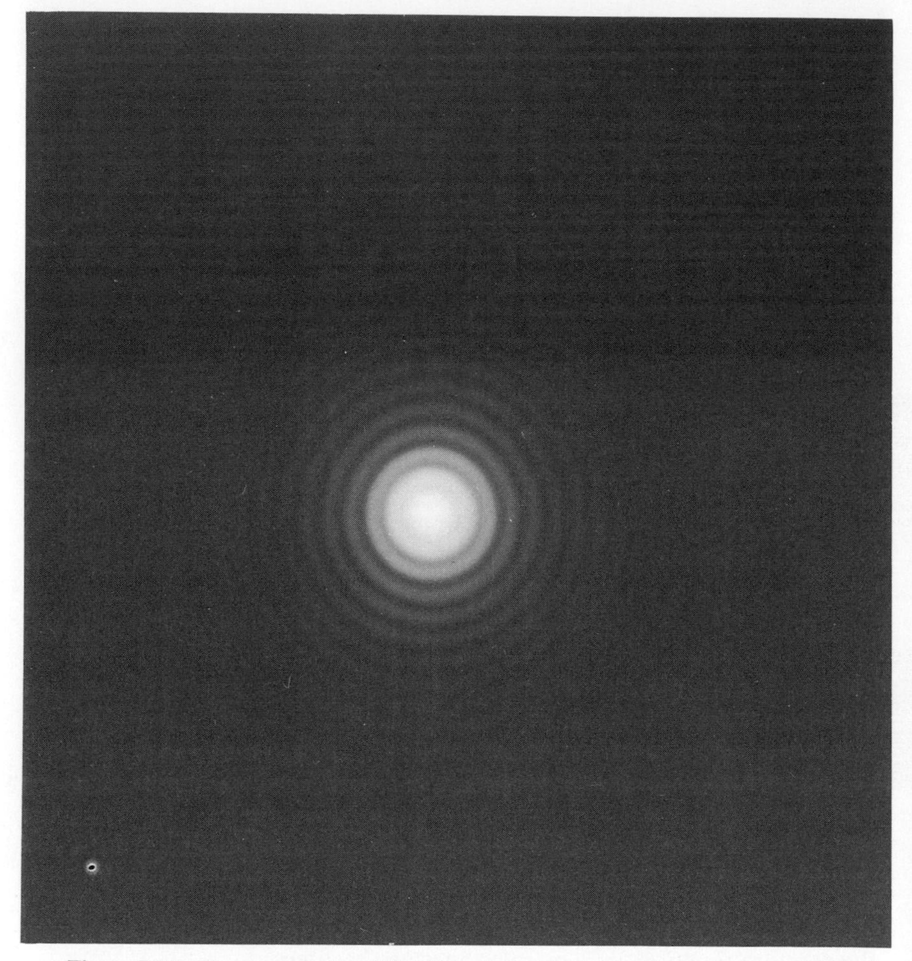

Figure 5.14 Photograph of a diffraction pattern of a circular aperture [Ref. 19].

The radiant flux density obtained from $\hat{U}_c(P')$ is the diffraction pattern for a circular aperture. It is

$$W_c(P') = W_{c0}[2J_1(\rho_m u_r)]^2/(\rho_m u_r)^2, \tag{5-68}$$

where

$$W_{c0} = \pi^2 \rho_m{}^4 \, |U_0|^2/\lambda^2 z_0{}^2. \tag{5-69}$$

The normalized function $W_c(\rho')/W_{c0}$ is plotted in Fig. 5.13. A photograph of a diffraction pattern produced by a circular aperture is shown in Fig. 5.14.

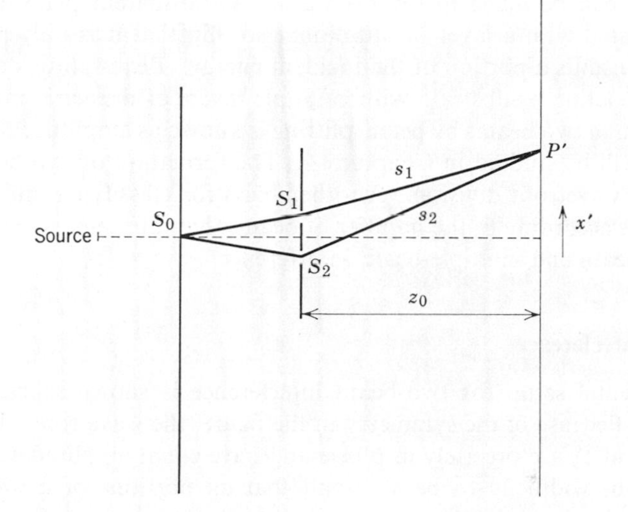

Figure 5.15 Young's experiment: formation of two beams by wavefront division.

Interference Experiments

In conventional interference problems, two or more superposed beams of radiant energy are involved. For instance, in Young's experiment the beam from a source, or from a first slit S_0 as in Fig. 5.15, is divided into two portions by passing radiant energy through two slits like S_1 and S_2 (see also Fig. 1.4). Another way to form two beams from one is illustrated in Fig. 5.16. The beamsplitter, which divides the beam by partial reflection and partial

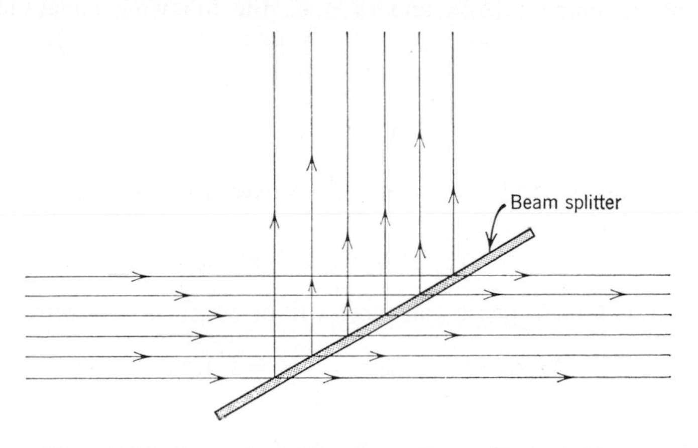

Figure 5.16 Formation of two beams by amplitude division.

transmission, can be made in several ways. A transparent plate, called a substrate, coated with a layer of aluminum so thin that it is only partially reflecting, transmits a portion of the incident energy. Beamsplitters can also be made by coating a substrate with multiple layers of dielectric materials. The formation of two beams by beamsplitting is known as amplitude division; an example will be studied in Chapter 11. The formation of two beams by slits is called wavefront division. Another basis for classifying interference experiments is according to the number of beams that interfere. We consider in turn two-beam and multiple-beam interference.

Two-Beam Interference

The experimental setup for two-beam interference is shown schematically in Fig. 5.15. Because of the symmetry in the figure, the wavefronts that pass through S_1 and S_2 are precisely in phase and have equal amplitudes. Let us assume the slit width $2a$ to be so small that all portions of a wavefront passing through one of the slits and arriving at P' will be essentially in phase. The phase delay in passing from S_1 along s_1 is $\varphi_1 = 2\pi s_1/\lambda$, and the phase delay from S_2 is $\varphi_2 = 2\pi s_2/\lambda$. The distance between S_1 and S_2 is x_1. Then:

$$\begin{aligned} s_1 &= \{z_0^2 + [x' - (x_1/2)]^2\}^{1/2}, \\ s_2 &= \{z_0^2 + [x' + (x_1/2)]^2\}^{1/2}, \end{aligned} \tag{5-70}$$

$$s_2^2 - s_1^2 = (s_2 - s_1)(s_2 + s_1) = 2x_1x',$$

and

$$\Delta s = (s_2 - s_1) = (2x_1x')/(s_2 + s_1). \tag{5-71}$$

Also, by assuming $z_0 \gg x_1$ and $z_0 \gg x'$, the following equations can be written:

$$s_2 + s_1 \approx 2z_0, \tag{5-72}$$

and

$$\Delta s \approx x_1x'/z_0. \tag{5-73}$$

The phase difference between the two waves arriving at P' is, therefore, approximately

$$(\varphi_2 - \varphi_1) = 2\pi x_1x'/\lambda z_0. \tag{5-74}$$

The waves will be in phase and there will be constructive interference when

$$2\pi(x_1x'/\lambda z_0) = 2q\pi \qquad (q = 0,1,2,3,\ldots), \tag{5-75}$$

that is, when

$$x' = q\lambda z_0/x_1. \tag{5-76}$$

The waves will be 180 deg out of phase and there will be destructive interference when

$$2\pi(x_1 x'/\lambda z_0) = (2q - 1)\pi, \tag{5-77}$$

that is, when

$$x' = [(2q - 1)/2]\lambda z_0/x_1. \tag{5-78}$$

Since the distances s_1 and s_2 differ only slightly, the amplitudes of the two waves at P' are essentially equal. Thus, there will be a dark fringe when Eq. (5-78) is satisfied and a bright fringe when Eq. (5-76) is satisfied. This agrees with Eq. (5-41) in which the cosine factor is equal to one when

$$x' = qz_0\lambda/x_1, \tag{5-76}$$

and to zero when

$$x' = [(2q - 1)/2]z_0\lambda/x_1. \tag{5-78}$$

Equation (5-41) gives more information about the pattern of interference because its derivation includes the effects of finite slit width and of diffraction.

Multiple-Beam Interference

To consider an example of multiple-beam interference, we postulate a transparent, nonabsorbing dielectric plate with parallel surfaces P and Q shown in Figure 5.17. The incident ray on P represents monochromatic incident plane wave fronts. Assume both surfaces to have metallic coatings that make them partially reflecting but with a significant portion of each wavefront passing through at each encounter with a surface. If, at each encounter, the fraction of wavefront amplitude transmitted through the surface is τ' and the phase delay caused by the passage through the surface is φ', the complex amplitude transmission coefficient is

$$\hat{\tau} = \tau' \exp(-i\varphi'). \tag{5-79}$$

To get a transmission coefficient that applies to flux densities, the complex coefficient is multiplied by its complex conjugate, which gives the transmittance as defined in Chapter 2:

$$\tau = \hat{\tau} \cdot \hat{\tau}^* = \tau'^2. \tag{5-80}$$

Similarly if $\rho' \exp(i\varphi'')$ is the complex amplitude reflection coefficient, the reflectance is

$$\rho = \hat{\rho} \cdot \hat{\rho}^* = \rho'^2. \tag{5-81}$$

As in Eq. (2-17), assuming the absorptance to be negligible,

$$\rho + \tau = 1. \tag{5-82}$$

To develop a general expression for the infinite series of rays (1,2,3,4, . . .)

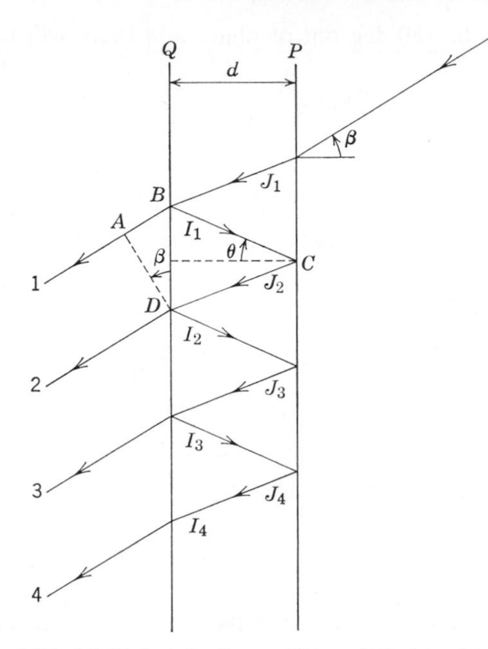

Figure 5.17 Multiple reflections within a dielectric plate.

resulting from the internal reflections in Fig. 5.17, an expression is needed for the delay caused by two passages, two reflections, from B to C to D within the plate, plus the delay from B to A. The total phase delay is

$$\begin{aligned}\delta &= (2\pi/\lambda)[\overline{AB} - n(\overline{BC} + \overline{CD})]\\ &= (2\pi/\lambda)[\overline{AB} - 2n\overline{BC}]. \qquad (5\text{-}83)\end{aligned}$$

In Eq. (5-83) the index of refraction of the plate material is n, and since the wave travels slower in the material, the effective distance (the *optical path*) is $n(\overline{BC} + \overline{CD})$. By Snell's law and from angular relationships in a right triangle,

$$\sin\beta = n\sin\theta. \qquad (5\text{-}84)$$

From Eq. (5-84) and the trigonometric relations in Fig. 5.17:

$$\begin{aligned}n\overline{BC}\cos\theta &= nd,\\ \overline{BD} &= 2\,\overline{BC}\sin\theta,\\ \overline{AB} &= \overline{BD}\sin\beta = n\overline{BD}\sin\theta,\\ &= 2n\overline{BC}\sin^2\theta. \qquad (5\text{-}85)\end{aligned}$$

Then, by applying the relationship of Eq. (5-85) to Eq. (5-83) the equation for the delay is obtained:

$$\begin{aligned}\delta &= 4\pi nd(\sin^2\theta - 1)/\lambda\cos\theta \\ &= (4\pi nd\cos\theta)/\lambda. \qquad (5\text{-}86)\end{aligned}$$

If the incident wave is assumed to have unit amplitude, the amplitudes of several waves between surfaces are as tabulated in Table IX. Complex

TABLE IX AMPLITUDES OF SEVERAL WAVES WITHIN THE PLATE OF FIG. 5.17

Ray	Amplitude
J_1I_1	τ'
I_1J_2	$\tau'\rho'$
J_2I_2	$\tau'\rho'^2$
I_2J_3	$\tau'\rho'^3$
J_3I_3	$\tau'\rho'^4$

amplitudes of the transmitted rays numbered 1, 2, 3, and 4 in the figure are, respectively:

$$\hat{\tau}^2, \quad \hat{\tau}^2\hat{\rho}^2\exp(-i\delta), \quad \hat{\tau}^2\hat{\rho}^4\exp(-2i\delta), \quad \hat{\tau}^2\hat{\rho}^6\exp(-3i\delta). \qquad (5\text{-}87)$$

These rays being parallel, the wavefronts will superpose at infinity; however, the fringes can be localized, by interposing a lens, and can be observed in the focal plane of the lens. The sum of all transmitted waves gives the complex wave amplitude:

$$\hat{U}_m = \hat{\tau}^2 \sum_{q=0} \hat{\rho}^{2q}\exp(-iq\delta). \qquad (5\text{-}88)$$

The infinite series of Eq. (5-88) is a geometric series having a common ratio of $\hat{\rho}^2\exp(-i\delta)$. From the sum formula of such a series,

$$\hat{U}_m = \hat{\tau}^2[1 - \rho^2\exp(-i\delta)]^{-1}. \qquad (5\text{-}89)$$

From this expression, the radiant flux density can be obtained:

$$\begin{aligned}W_m &= \frac{\hat{\tau}^2}{[1 - \hat{\rho}^2\exp(-i\delta)]}\cdot\frac{\hat{\tau}^{*2}}{[1 - \hat{\rho}^{*2}\exp(i\delta)]} \\ &= \frac{\tau^2}{[1 + \rho^2 - 2\rho\cos(\delta + 2\varphi'')]}. \qquad (5\text{-}90)\end{aligned}$$

Let $(\delta + 2\varphi'') = \varphi$. Then from the identity, $\cos\varphi = 1 - 2\sin^2(\varphi/2)$:

$$W_m = \tau^2/[(1 - \rho)^2 + 4\rho\sin^2(\varphi/2)]. \qquad (5\text{-}91)$$

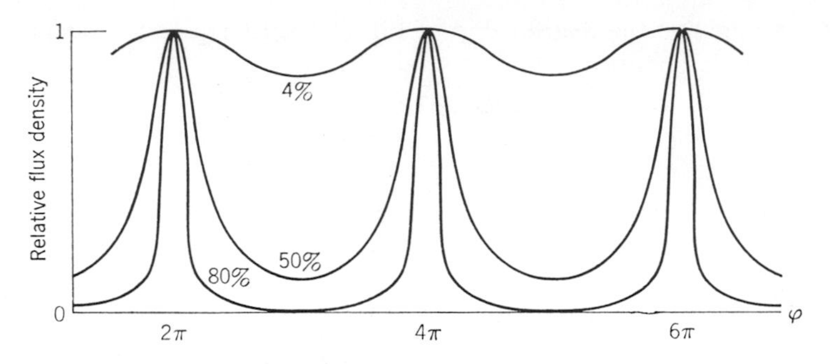

Figure 5.18 Multiple beam interference fringes. (The different curves are for different reflectances of the surfaces in Fig. 5.17.)

From this equation, it is apparent that the maximum flux density will occur when the sine-squared factor is zero, and W_m will have a value

$$W_{m0} = \tau^2/(1 - \rho)^2. \tag{5-92}$$

Let $F = 4\rho/(1 - \rho)^2$. Then

$$W_m = W_{m0}[1 + F\sin^2(\varphi/2)]^{-1}. \tag{5-93}$$

If $(\varphi/2) = (4\pi d\cos\theta/\lambda) + 2\varphi'' = (2q + 1)\pi/2$ (with $q = 0,1,2,3,\ldots$), the sine-squared factor will be unity and W_m will be a minimum:

$$W_{\min} = W_{m0}/(1 + F). \tag{5-94}$$

Since F depends on the reflectance, the visibility of fringes depends upon the reflectance, being higher for higher reflectance. The visibility, from Eq. (1-46), is

$$\mathscr{V} = F/(2 + F). \tag{5-95}$$

Figure 5.18 is a plot of the type of fringes produced in this experiment. The fringes, called Haidinger fringes, are quite sharp. The different curves of the figure show the effect of increasing reflectance.

REFERENCES

1. E. W. Marchand and E. Wolf, "Consistent Formulation of Kirchhoff's Diffraction Theory," *J. Opt. Soc. Am.*, **56**, 1712 (1966).
2. M. Francon, "Interference, Diffraction and Polarization," in *Handbuch der Physik*, Vol. XXIV, p. 268 (in French). Springer-Verlag, Berlin, 1956.
3. M. Born and E. Wolf, *Principles of Optics*, Third Edition. Pergamon Press, London, 1965.
4. E. L. O'Neill, *Introduction to Statistical Optics*, Chapter 5. Addison-Wesley Publishing Company, Inc., Reading, Mass., 1963.

5. A. Papoulis, *Systems and Transforms with Applications in Optics.* McGraw-Hill Book Company, Inc., New York, 1968.
6. J. W. Goodman, *Introduction to Fourier Optics*, Chapter 3. McGraw-Hill Book Company, Inc., San Francisco, 1968.
7. K. Miyamoto and E. Wolf, "Generalization of the Maggi–Rubinowicz Theory of the Boundary Diffraction Wave:" Part I, *J. Opt. Soc. Am.*, **52,** 615 (1962); Part II, *J. Opt. Soc. Am.*, **52,** 626 (1962).
8. E. Marchand and E. Wolf, "Boundary Diffraction Wave in the Domain of the Rayleigh–Kirchhoff Diffraction Theory," *J. Opt. Soc. Am.*, **52,** 761 (1962).
9. E. Wolf and E. W. Marchand, "Comparison of the Kirchhoff and the Rayleigh–Sommerfeld Theories of Diffraction at an Aperture," *J. Opt. Soc. Am.*, **54,** 587 (1964).
10. A. Rubinowicz, "The Miyamoto–Wolf Diffraction Wave," in *Progress in Optics*, Vol. IV. Edited by E. Wolf. North-Holland Publishing Company, Amsterdam, 1965.
11. F. Kottler, "Diffraction at a Black Screen:" Part I. Kirchhoff's Theory, in *Progress in Optics*, Vol. IV. Edited by E. Wolf. North-Holland Publishing Company, Amsterdam, 1965. Part II. Electromagnetic Theory, in *Progress in Optics*, Vol. VI. Edited by E. Wolf. North-Holland Publishing Company, Amsterdam, 1967.
12. J. B. Keller, "Geometrical Theory of Diffraction," *J. Opt. Soc. Am.*, **52,** 116 (1962).
13. J. A. Ratcliffe, "Some Aspects of Diffraction Theory and Their Application to the Ionosphere," in *Reports on Progress in Physics*, Vol. XIX. Edited by A. C. Stickland. The Physical Society, London, 1956.
14. H. Arensault and A. Boivin, "An Axial Form of the Sampling Theorem and its Application to Optical Diffraction," *J. Appl. Phys.*, **38,** 3988 (1967).
15. F. S. Harris, Jr., M. S. Tavenner, and R. L. Mitchell, "Single-Slit Diffraction Patterns: Comparison Experimental and Theoretical Results," *J. Opt. Soc. Am.*, **59,** 293 (1969).
16. L. B. W. Jolley, *Summation of Series.* Dover Publications, Inc., New York, 1961.
17. G. A. Korn and T. M. Korn, *Mathematical Handbook for Scientists and Engineers*, paragraph 4.7-2. McGraw-Hill Book Company, Inc., New York, 1961.
18. G. N. Watson, *A Treatise on the Theory of Bessel Functions*, p. 20. Cambridge University Press, Cambridge, 1958.
19. F. S. Harris Jr., "Light Diffraction Patterns," *Appl. Optics*, **3,** 909 (1964).
20. W. E. Koch, "Acoustics and Optics," *Appl. Optics*, **8,** 1525 (1969).
21. Richard B. Hoover, "Diffraction Plates for Classroom Demonstration," *Am. J. Phys.*, **37,** 871 (1969).
22. J. C. Brown, "Fourier Analysis and Spatial Filtering," *Am. J. Phys.*, **39,** 797 (1971).

6

Ideal Optical Systems

In this chapter we study ideal optical systems and a simple optical element, the lens. Only centered systems are considered, that is, lenses or systems having an axis of symmetry called the *optic axis*. For the most part, we treat only *paraxial rays:* rays forming an angle α with the axis small enough that

$$\sin \alpha \approx \alpha,$$

$$\cos \alpha \approx 1.$$

Usually the rays considered are also *meridional*, that is, rays lying in a plane containing the optic axis. All such planes are called *meridional planes*. Finally, we develop the general formulas for finding the path of any ray between spherical surfaces. These formulas are typical of those programmed for ray tracing by computer. They apply for large as well as small angles and for *skew rays* (rays not in a meridional plane) as well as for meridional rays.

Action of a Lens on a Wavefront

The most common optical element is the lens, illustrated in Fig. 6.1 by a section through the axis. A lens is made of a transparent, optically dense material, usually glass, having an index of refraction greater than one; the two surfaces are conventionally spherical although aspheric surfaces are sometimes used. Aspherics require much hand work in the optics shop and a series of polish-then-test sequences, so, to save cost, they are avoided whenever possible.

If a plane wavefront approaches the lens with rays parallel to the axis and, further, if the lens is "thin" (so that a ray entering the lens on one face at a distance r from the axis emerges at approximately the same distance r on the opposite face), the delays imposed on the various rays of the incident wavefront are proportional to the thickness of the lens at each r. If the maximum

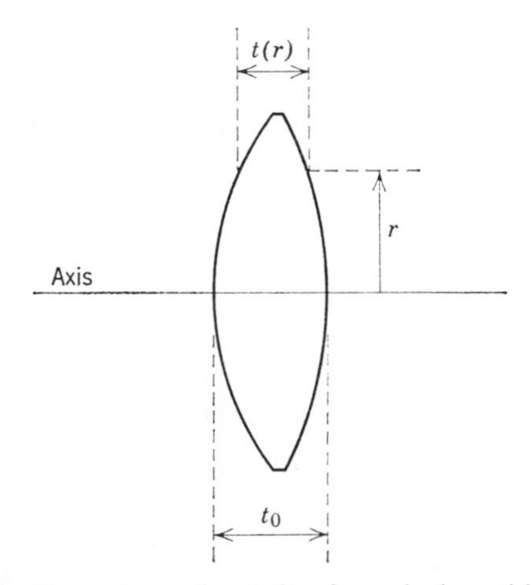

Figure 6.1 Geometry and notation for a single positive lens.

lens thickness is t_0 and the thickness at r is $t(r)$, the phase change resulting from the time taken by a ray or, in general, the wavefront to pass through the lens is

$$\begin{aligned} k\varphi(r) &= knt(r) + k[t_0 - t(r)] \\ &= kt_0 + k(n-1)t(r), \end{aligned} \tag{6-1}$$

where the phase delay caused by the region of free space between the two tangent planes is $k[t_0 - t(r)]$ (see Chapter 4). Since the magnitude of propagation velocity in the material of the lens is lower than that of free space by the factor $1/n$, the free space distance equivalent to the distance in the lens is $nt(r)$ (known as the optical distance), and the phase delay produced by the lens itself is $knt(r)$. The emerging wave, $\hat{U}_2 = \hat{U}_0 \exp ik[\varphi_0 + \varphi(r)]$, at the second tangent plane, is the incident plane wave, $\hat{U}_1 = \hat{U}_0 \exp ik\varphi_0$, at the first tangent plane, multiplied by the phase factor, $\exp ik\varphi(r)$, to account for the delay:

$$\hat{U}_2 = \hat{U}_1 \exp ik\varphi(r). \tag{6-2}$$

The term $k(n-1)t(r)$ of Eq. (6-1) is of particular interest because from it can be deduced the shape of the emerging wavefront.

To find the thickness $t(r)$, let us divide the lens into three parts as shown in Fig. 6.2 so that

$$t(r) = t_1(r) + t_2 + t_3(r). \tag{6-3}$$

From the geometry of circles, t_1 and t_3 are given by

$$t_1(r) = t_{10} - [R_1 - (R_1^2 - r^2)^{1/2}], \tag{6-4}$$

and

$$t_3(r) = t_{30} - [R_2 - (R_2^2 - r^2)^{1/2}]. \tag{6-5}$$

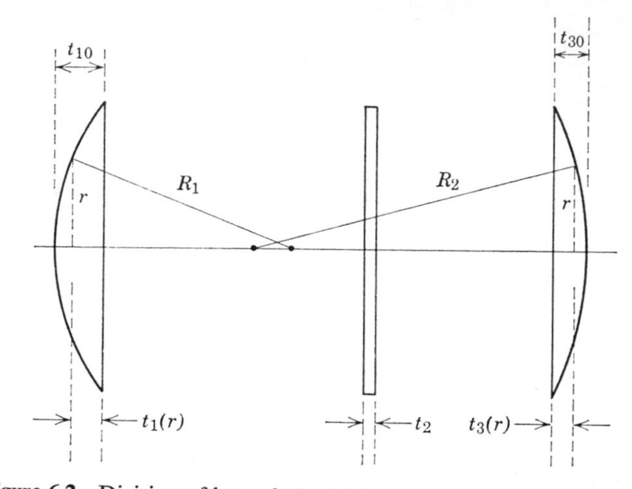

Figure 6.2 Division of lens of Fig. 6.1 into three sections for analysis.

We adopt the convention that a convex surface encountered by the wavefront has, by definition, a positive radius of curvature, and each concave surface a negative radius of curvature; thus, for the lens of Fig. 6.1, the value of R_2 will carry a negative sign in Eq. (6-5). Equations (6-4) and (6-5) can be written in factored form as follows:

$$t_1(r) = t_{10} - R_1\{1 - [1 - (r^2/R_1^2)]^{1/2}\}, \tag{6-4a}$$

$$t_3(r) = t_{30} - R_2\{1 - [1 - (r^2/R_2^2)]^{1/2}\}. \tag{6-5a}$$

Throughout the present discussion, let us assume that $r^2 \ll R_1^2$ and $r^2 \ll R_2^2$; then the expressions in the brackets may be expanded by the binomial theorem and approximated by taking only the first two terms in each series:

$$t_1(r) \approx t_{10} - (r^2/2R_1), \tag{6-4b}$$

$$t_3(r) \approx t_{30} - (r^2/2R_2). \tag{6-5b}$$

When these approximations are substituted in Eq. (6-3), the expression,

$$t(r) = t_0 - \frac{r^2}{2}\left(\frac{1}{R_1} - \frac{1}{R_2}\right), \tag{6-6}$$

results since $t_0 = t_{10} + t_2 + t_{30}$. A quantity $1/f$ is arbitrarily defined in terms of the second expression on the right-hand side of Eq. (6-6) as follows:

$$1/f \equiv (n - 1)[(1/R_1) - (1/R_2)]. \tag{6-7}$$

The symbol f is called the focal length for reasons that become apparent in the discussion following Eq. (6-10). From this definition, the phase of Eq.

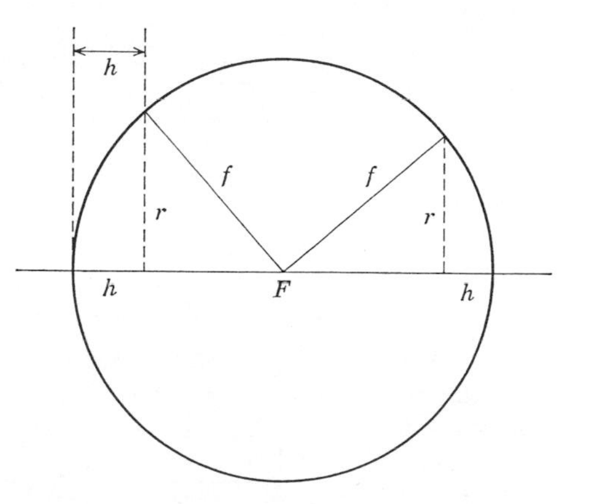

Figure 6.3 Section through a sphere at its center F.

(6-1) or Eq. (6-2) can be written as

$$k\varphi(r) = knt_0 - k(r^2/2f). \tag{6-8}$$

Therefore, the lens introduces a quadratic phase factor of the form $\exp[-ik(r^2/2f)]$ into the expression for the wave. The phase of Eq. (6-8) is made up of a constant phase delay and a variable delay that depends upon the focal length and the radial distance r. By matching the variable phase term $k(r^2/2f)$ with an axial displacement coordinate of a point on a sphere, the sphericity of the emerging wave front can be demonstrated. A great circle section of a sphere is shown in Fig. 6.3 with the coordinate system used to develop the axial displacement coordinate h. Right-triangle relationships give

$$(f - h)^2 = f^2 - r^2,$$

$$h = f\{1 \pm [1 - (r/f)^2]^{1/2}\}. \tag{6-9}$$

If $r^2 \ll f^2$, the same kind of approximation as that following Eq. (6-5a) can be made to get

$$h \approx r^2/2f$$

or

$$h \approx [2f - (r^2/2f)]. \tag{6-10}$$

The two different approximations for h result, of course, from choosing the negative and the positive signs respectively before the brackets in Eq. (6-9). The first approximation applies to the left side of the sphere, the second to the right side. The first approximation is a match with the factor of the variable term in Eq. (6-8), so the emerging wavefront from the lens of Figures 6.1 and 6.2 is a converging spherical one centered on the point F in Fig. 6.3 called the focal point. Also, F is displaced from the origin of h a distance f, which

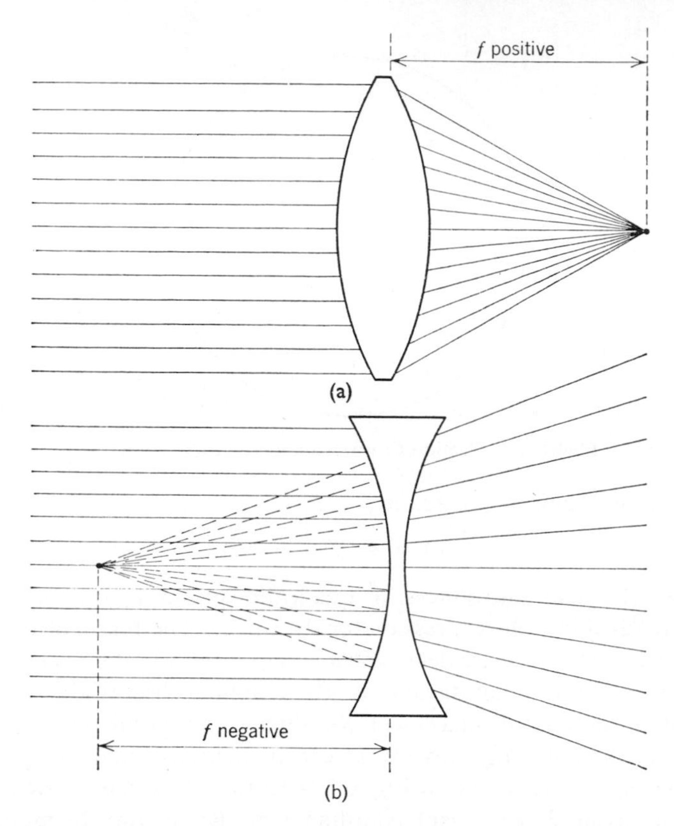

Figure 6.4 Convergence or divergence, respectively, of rays from: (a) a positive lens and (b) a negative lens.

was earlier arbitrarily called the focal length. The origin of h is at the lens.

The optical element discussed above, which produces a converging spherical wavefront from a plane wavefront, is called a positive, or converging, lens. If, instead of the convex–concave (toward the light source) lens, a concave–convex lens (thinnest in the middle) had been used for the illustration, the emerging wavefront would have been an *expanding* spherical one; and the lens would be called negative, or diverging.

A concept of rays goes with the concept of wavefronts, the rays being everywhere perpendicular to the wavefronts. In terms of rays, the positive lens discussed above transforms parallel rays to converging rays that all pass through the focal point, which is a focal length from the lens. The negative lens, on the other hand, converts the bundle of parallel rays to diverging rays that appear to originate from a point on the incident side of the lens. Figure 6.4 shows the ray diagrams for positive and negative lenses.

If a point source of light were placed at the focal point, the positive lens would reverse the process described earlier and produce a parallel bundle of rays. A second lens, identical to the first, could then operate on the bundle of rays to focus them at a second focal point. If a single lens were made "strong" enough, it could conceivably perform the functions of the two lenses and focus the point source at an image point on the other side of the lens. When this capability is extended to off-axis points as well, the lens or lens combination is an ideal imaging system.

An Optical Wave Guide

The explanation of the action of a lens on a wavefront suggests how a series of lenses could simulate a waveguide for a laser beam. The propagation properties of a laser beam can be found by analyzing the laser cavity [Ref. 1]. A laser beam propagates according to one of several mode patterns, the simplest of which produces wavefronts that are spherical with very large radii of curvature; the radiant energy distribution over a wavefront, as a function of r, is Gaussian. The beam spreads, as a beam consisting of spherical wavefronts must, until a thin lens changes a spherical wavefront with a negative radius of curvature to a spherical wavefront having a positive radius of curvature, still with a Gaussian distribution of energy. The beam, after passing through the lens, is a converging beam, but because of the way it propagates, it passes through a region, called the beam *waist*, where its beam width becomes a minimum and its wavefronts become essentially plane. Beyond the waist the beam spreads and the wavefront radius of curvature becomes negative again. A periodic placement of lenses in the beam path limits the maximum beam diameter while it propagates over a great distance.

The same kind of action occurs when a beam reflects from a spherical mirror. Reflection from the mirror is like transmission through a lens and changes a spreading beam to a converging beam. Thus, under proper conditions, a laser beam "bounces back and forth" between two spherical mirrors, which form a resonant cavity.

General Approach to Analyzing Optical Systems—Signs

The remainder of the chapter is given largely to developing a scheme for analyzing an optical system consisting of a number of lenses on a common axis. Passing a ray successively through the several lenses in a system is a repeating computation and invites development of a systematic way to handle the many detailed calculations. From the many approaches to solving this problem, we present a brief form of a matrix method, which is treated in detail elsewhere [Refs. 2–4]. The method was chosen primarily for its conciseness. Also for conciseness, we postulate, rather than derive, a number of the principles commonly applied in optical systems.

Because assignment of algebraic signs is largely arbitrary in setting up optical problems and because no common convention has been adopted regarding what is called positive and negative, explicit rules have to be set up and meticulously followed concerning signs. The rules we use are included in the following definitions and conventions:

(1) Light proceeds from left to right in the optical diagrams unless otherwise stated.
(2) A distance measured in the direction that light is proceeding (usually from left to right) is positive.
(3) A distance is always measured from a refracting surface or from a principal plane (which will be defined later).
(4) The *vertex* of a refracting surface is its intersection point with the axis of symmetry.
(5) A radius of curvature is positive if the direction from the vertex of a surface to the center of curvature is from left to right.
(6) The surfaces are numbered in the order in which light passes through them.
(7) Subscript numbers indicate the surface at which refraction is taking place.
(8) Whenever a distinction must be made, a prime indicates that the quantity applies after the ray has been refracted at a surface.
(9) A reflecting surface requires the use of a negative index of refraction for the medium following the surface, to account for the change in direction or "folding" of the optical system.
(10) *Object space* is that region containing the rays before they enter the lens or optical system.
(11) *Image space* is that region containing the rays after they have passed through the lens or system.

Cardinal Points

Associated with each lens and with each combination of lenses are certain significant points, useful in analyzing optical systems, called the *cardinal points*. We postulate the existence of these points and planes related to them and define them as follows:

(1) Two unit *conjugate planes* (conjugate here meaning that there is an object–image relationship between them), called the first and second *principal planes*, *unit planes*, or *Gauss planes*, are perpendicular to the optic axis; and their points of intersection with the optic axis are called the first and second *principal points* respectively.

(2) The first and second *focal planes* (which, as will develop, can be called infinite conjugates) are perpendicular to the optic axis, and their intersections with the optic axis are called the first and second *focal points*, respectively.
(3) Parallel rays in object space that are incident upon the first principal plane and that pass through the system will reach a common point in the second focal plane; and parallel rays in image space, having passed through the optical system, will have passed through a common point in the first focal plane.
(4) A ray that is incident on the first principal plane at a distance h from the optic axis and that passes through the system will leave the second principal plane at a distance $h' = h$ from the axis.
(5) There are two points on the optic axis called *nodal points* such that a ray passing through the first nodal point will also pass through the second nodal point and its direction in image space will be parallel to its direction in object space.

The principal points, focal points, and nodal points together are called the cardinal points for the system. When the positions of these points are known, the location of the image of an object and its magnification can be determined by a simple ray-tracing procedure. Since a centered system is postulated, only a two-dimensional plot is needed.

If the refractive indices of the media in object space and image space are the same, the nodal points will coincide with the principal points. In the following discussions we shall assume this to be true. In most practical optical systems, the medium is air, which has a refractive index of approximately unity. An important exception is the human eye (see Chapter 14).

Simple Ray Tracing

The principle of simple ray tracing is illustrated in Fig. 6.5a, in which an optical system is shown schematically. The terms H and H' represent the first and second principal planes respectively, and F and F' represent the first and second focal planes. The positions indicated for the cardinal points are not necessarily typical of any particular optical system. Consider any point Q on the object. To say that an image is formed implies that all rays diverging from Q and passing through the system converge to one point somewhere in image space. The image point is designated Q'. We can find Q' graphically by tracing a few rays. Three rays are especially easy to trace. Two are designated in the figure as rays a and b. The third would pass through the first focal point F and would emerge from H' at such a level as to reach Q' by proceeding parallel (for reasons to be developed later) with the optic axis.

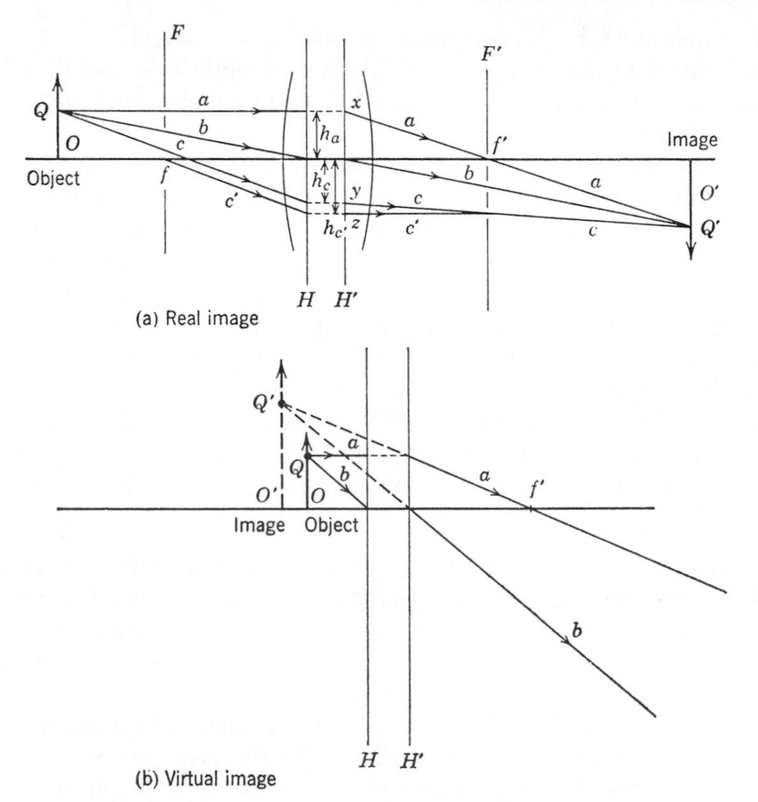

Figure 6.5 Simple ray tracing through a single optical system.

The ray a is parallel to the optic axis. A hypothetical ray a', not from Q, that is coincident with the optic axis will pass through the first principal point and leave the second principal point still coincident with the optic axis. Since ray a' and ray a are parallel in object space, they will, by definition, pass through the second focal point. Ray a is incident on the first principal plane at a distance h_a from the axis and will leave the second principal plane at the same distance, called h_a', at point x. Point x and the second focal point determine ray a in image space. Ray b from Q passes through the first nodal point and leaves the second nodal point. The second nodal point and the requirement to be parallel to the original direction of b determine ray b in image space. Rays a and b will intersect at Q'. A plane through Q' perpendicular to the optic axis is the image plane.

As a check, we can trace a third ray from Q to verify that Q' has been located correctly. The simplest ray would be a ray from Q passing through the first focal point. However, to illustrate some further principles in simple ray tracing, let us arbitrarily choose a ray c that does not pass through f.

We then construct another ray c', not from Q, that does pass through f and is parallel to c. Rays c and c' will pass through the second focal plane at the same point (definition of focal plane). Ray c' intersects the first principal plane at a height $h_{c'}$ and will leave the second principal plane at z, at the same height $h_{c'}$. The point z and the requirement to be parallel to the optic axis determine c' in image space. It will intersect the second focal plane at z'. Ray c will intersect the first principal plane at the height h_c and leave the second principal plane from y, at the same height h_c. The points y and z' determine ray c in image space. Ray c will meet rays a and b at Q'.

Figure 6.5b is a simplified diagram like Fig. 6.5a except that the object has been placed between the focal plane (not shown) and the principal plane H. Under these conditions, the rays from a point on the object diverge in the image space and, at first, appear incapable of forming an image. However, further operation upon these divergent rays by subsequent optical systems (for instance, the eye) interpret the rays as coming from a virtual image located in an imaginary extension of image space as shown.

The reader is advised to perform a few desk-top experiments with a single positive lens of the reading glass variety to verify the relationships between object, lens, and image as indicated in Figures 6.5a, 6.5b, and other similar diagrams with various object-to-lens spacings. The focal length of your lens can be satisfactorily established by noting the spacing between the lens and a piece of paper on which you focus the image of a relatively distant bright object (light bulb, bright window, etc.). Then look through the lens at an object well beyond the focal plane of the lens to get the conditions of Fig. 6.5a. As indicated in the diagram, the image you see is upside down. To establish its position in space, use parallax by wagging your head from side to side (lens stationary) and comparing the movement of the image with the apparent movement of a free finger positioned experimentally between the lens and your eye. Of course, when you have found the place where your finger appears to move exactly with the image as you wag your head, your finger is located at the position of the image. The experiment can be repeated for Fig. 6.5b, the usual mode for a reading glass, where the object is placed between the focal plane and the lens.

Image–Object Relationships

Further definitions are useful in working out image–object relationships:

(1) Object distance s is measured from the first principal plane to the object.
(2) The image distance s' is measured from the second principal plane to the image.
(3) The lateral magnification β is found by dividing the image size by the object size.

Let QO, the distance from Q to the optic axis, represent the object size and $Q'O'$, the distance from Q' to the optic axis, represent the image size. Thus,

$$\beta = Q'O'/QO. \tag{6-11}$$

It turns out that β, by identifying the appropriate similar triangles, is also given by

$$\beta = (s'/s). \tag{6-12}$$

(4) The focal distances f and f' are the distances from the principal points to the first and second focal points, respectively. The classical relationship between s, s', and f', which will be derived later, is

$$-\frac{1}{s} + \frac{1}{s'} = \frac{1}{f'}. \tag{6-13}$$

The Thin Lens

In the thin lens the distance between the principal planes is so small, when compared with other dimensions of the system, that schematically the two planes coincide.

Linear Transformations and Matrices

Our object now is to develop a procedure for finding the principal points, the focal points, and object and image points when positions and radii of curvature of surfaces are known.

An optical system consists of a number of refracting surfaces arranged along an optic axis with varying separations between them. A ray of radiant energy passing through the system will sequentially repeat the process of refraction at a surface and then "translation" to the next surface. The ray is determined in general at any point along its path by the coordinates of the point and the direction cosines at the point. Since only centered systems and meridional rays are considered in particular, the coordinate at a given surface is the distance r from the optic axis and the direction is the angle α that the ray makes with the optic axis. For ray tracing, a method for relating the several r and α at the various stages in the system is needed.

Let us postulate that there is a set of equations which linearly transform r and α at one stage in the system to r' and α' at another stage in the system. The equations can apply for refraction at a surface, for the passage or translation of the ray from one surface to the next, or for passage of the ray through the entire optical system. The transformation has the general form

$$\alpha' = a\alpha + br, \tag{6-14a}$$

$$r' = c\alpha + dr. \tag{6-14b}$$

The transformation can also be written in matrix form, which provides a convenient method for systematizing the computations:

$$\begin{bmatrix} \alpha' \\ r' \end{bmatrix} = \begin{bmatrix} a & b \\ c & d \end{bmatrix} \times \begin{bmatrix} \alpha \\ r \end{bmatrix}.$$

In the work of this chapter, the following standard matrix multiplication rules apply:

$$\begin{bmatrix} a & b \\ c & d \end{bmatrix} \times \begin{bmatrix} w & x \\ y & z \end{bmatrix} = \begin{bmatrix} aw + by & ax + bz \\ cw + dy & cx + dz \end{bmatrix},$$

$$\begin{bmatrix} x' \\ y' \end{bmatrix} = \begin{bmatrix} a & b \\ c & d \end{bmatrix} \times \begin{bmatrix} x \\ y \end{bmatrix} = \begin{bmatrix} ax + by \\ cx + dy \end{bmatrix},$$

so $x' = ax + by$ and $y' = cx + dy$. Ways to determine the matrix elements in terms of refractive indices, radii of curvature, and distances between surfaces will be found.

Refraction and Translation Matrices

We now further postulate that there is a refraction matrix M_{ri} that traces a ray through a given (the ith) surface and takes into account the index of refraction and angle of incidence,

$$M_{ri} = \begin{bmatrix} 1 & -a_i \\ 0 & 1 \end{bmatrix},$$

and a translation matrix M_{ti} that translates a ray to the next surface in the system,

$$M_{ti} = \begin{bmatrix} 1 & 0 \\ \underline{t}_i & 1 \end{bmatrix}.$$

The "power" a_i is given by

$$a_i = (n_i' \cos \alpha_i' - n_i \cos \alpha_i)C_i. \tag{6-15a}$$

Since the angles are assumed small, the small angle approximation applies:

$$a_i = (n_i' - n_i)C_i \tag{6-15b}$$

where n is the refractive index and C is the curvature, equal to the reciprocal of the radius of curvature of the surface. The thickness or distance measured from the general surface i to the surface $i + 1$ is given by $\underline{t}_i$, where the underscore indicates that $\underline{t}_i$ is a "reduced" distance, that is, actual distance divided by the refractive index of the medium along this distance,

$$\underline{t}_i = (t_i/n_i). \tag{6-16}$$

The matrices being used all have the determinant value of unity, so products of any number of them must also have the determinant value of unity. Therefore, a check on the computations is to show that the determinant value of the final matrix is unity.

The power a_i for the surface and the reduced distance $\underline{t}_i$ for translation both depend upon the index of refraction. There will, therefore, be a power and a reduced distance for each color (each wavelength) determined by the refractive index for that color. This important principle must be kept in mind although it is not specifically mentioned in the discussion and problems that follow; monochromatic light is assumed.

The Lens Matrix

The matrix for a lens is found by refracting at the first surface, translating to the second surface, and then refracting at the second surface:

$$M_L = \begin{bmatrix} 1 & -a_2 \\ 0 & 1 \end{bmatrix} \times \begin{bmatrix} 1 & 0 \\ \underline{t}_1 & 1 \end{bmatrix} \times \begin{bmatrix} 1 & -a_1 \\ 0 & 1 \end{bmatrix}$$

$$= \begin{bmatrix} (1 - a_2\underline{t}_1) & [-a_1 - a_2(1 - a_1\underline{t}_1)] \\ \underline{t}_1 & (1 - a_1\underline{t}_1) \end{bmatrix}.$$

The symbols b, d, a, and c are used for elements of the lens or system matrix as follows:

$$M_L = \begin{bmatrix} b & -a \\ d & c \end{bmatrix}.$$

(The choice of symbols and signs in the M_L matrix results from historical usage.)

The matrix for a combination of lenses is found by the product of refraction matrices and translation matrices in the same order as the respective components occur:

$$M_c = M_{r(i+1)}M_{ti} \ldots M_{t2}M_{r2}M_{t1}M_{r1}$$

$$= \begin{bmatrix} b & -a \\ d & c \end{bmatrix}. \qquad (6\text{-}17)$$

The Optical System Matrix

Now that we have a matrix for an optical system, what can we do with it? First, before proceeding, let us make one last postulate: A matrix equation can give a relationship between the angle α' of a ray in image space and the

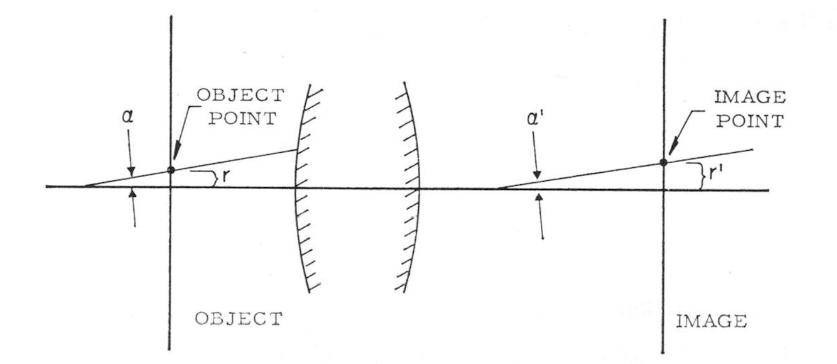

Figure 6.6 Specification of a ray in object and image space.

coordinate r' of the image point on one hand and the corresponding α and r of object space on the other (Fig. 6.6). In the following equation, distances to the object and image planes, $\underline{l}$ measured from the first surface and $\underline{l}'$ measured from the last surface, are used to translate from the object plane to first surface and from last lens surface to image plane respectively:

$$\begin{bmatrix} \underline{\alpha}' \\ r' \end{bmatrix} = \begin{bmatrix} 1 & 0 \\ \underline{l}' & 1 \end{bmatrix} \times \begin{bmatrix} b & -a \\ d & c \end{bmatrix} \times \begin{bmatrix} 1 & 0 \\ -\underline{l} & 1 \end{bmatrix} \times \begin{bmatrix} \underline{\alpha} \\ r \end{bmatrix},$$

$$\begin{bmatrix} \underline{\alpha}' \\ r' \end{bmatrix} = \begin{bmatrix} b + \underline{l}a & -a \\ b\underline{l}' + a\underline{l}\underline{l}' + d - \underline{l}c & c - \underline{l}'a \end{bmatrix} \times \begin{bmatrix} \underline{\alpha} \\ r \end{bmatrix},$$

which is the same as

$$\underline{\alpha}' = (b + \underline{l}a)\underline{\alpha} - ar, \tag{6-18a}$$

$$r' = (b\underline{l}' + a\underline{l}\underline{l}' + d - \underline{l}c)\underline{\alpha} + (c - \underline{l}'a)r. \tag{6-18b}$$

Here the "reduced" angles are $\underline{\alpha}' = n'\alpha'$ and $\underline{\alpha} = n\alpha$.

When the point at r is imaged at r', all rays from the object point focus at the image point regardless of direction α. Therefore the coefficient of $\underline{\alpha}$ in Eq. (6-18b) can be only zero. The lateral magnification is then

$$(r'/r) = c - \underline{l}'a = \beta. \tag{6-19}$$

Also, since the system matrix must have the determinant value of unity,

$$(c - \underline{l}'a) = 1/(b + \underline{l}a) = \beta. \tag{6-20}$$

By inserting expressions in terms of β, as given in Equations (6-19) and (6-20), into the last matrix equation, an equation for the transformation between object and image planes is obtained:

$$\begin{bmatrix} \underline{\alpha}' \\ r' \end{bmatrix} = \begin{bmatrix} 1/\beta & -a \\ 0 & \beta \end{bmatrix} \times \begin{bmatrix} \underline{\alpha} \\ r \end{bmatrix}.$$

For points on axis where $r = 0$,

$$\underline{\alpha}' = \underline{\alpha}/\beta. \tag{6-21}$$

The Lens or System Formulas

The development of the lens formulas in this section assumes the optical system shown in Fig. 6.7. Two general conjugate planes are labeled *object* and *image*, and the principal planes are called H and H'.

If the principal planes are the two conjugate planes, β is unity (see the fourth postulate under "Cardinal Points") and $\underline{l}$ and $\underline{l}'$ are the distances to the principal planes $\underline{l}_H$ and $\underline{l}_H'$ respectively. From Eq. (6-20),

$$c - \underline{l}_H'a = 1; \; \underline{l}_H' = (c-1)/a; \tag{6-22}$$

$$1/(b + \underline{l}_H a) = 1; \; \underline{l}_H = (1-b)/a. \tag{6-23}$$

If now the general system is considered (where s and s' can have values other than zero) and it is noted that $\underline{s}$ and $\underline{s}'$ are measured from principal planes and $\underline{l}$ and $\underline{l}'$ from surfaces, the geometry of Fig. 6.7 gives

$$\underline{l} = \underline{s} + \underline{l}_H, \tag{6-24}$$

$$\underline{l}' = \underline{s}' + \underline{l}_H'. \tag{6-25}$$

By applying the special properties of the principal planes as found in Equations (6-22) and (6-23) to the relationships in Eq. (6-20), one obtains

$$\beta = c - \underline{l}'a = c - \underline{s}'a - \underline{l}_H'a = 1 - \underline{s}'a, \tag{6-26}$$

$$1/\beta = b + \underline{l}a = b + \underline{s}a + \underline{l}_H a = 1 + \underline{s}a. \tag{6-27}$$

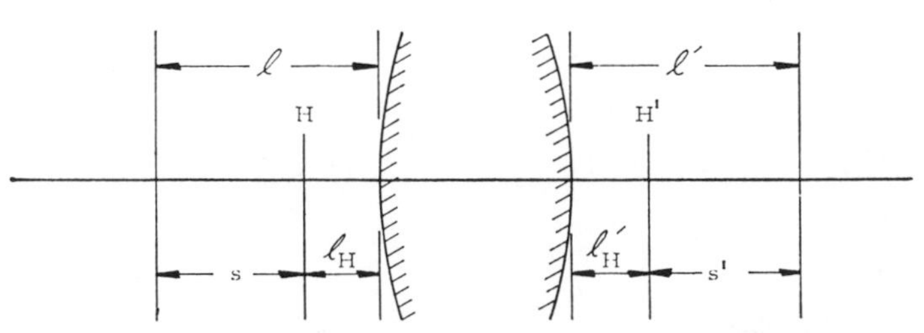

Figure 6.7 Object and image distances for a simple optical system.

The system matrix already found in terms of β becomes, when object and image are measured from the principal planes,

$$\begin{bmatrix} 1 + \underline{s}a & -a \\ 0 & 1 - \underline{s}'a \end{bmatrix}.$$

Because the matrix must have the determinant value of unity:

$$(1 + \underline{s}a)(1 - \underline{s}'a) = 1, \tag{6-28}$$

from which

$$-\frac{1}{\underline{s}} + \frac{1}{\underline{s}'} = a. \tag{6-29}$$

If $\underline{s}$ and $\underline{s}'$ are in turn set equal to infinity, the following expression is obtained to locate the focal planes (since an object at a great distance implies incident parallel rays):

$$a = (1/\underline{f}') = -(1/\underline{f}). \tag{6-30}$$

The classical lens formula follows:

$$-1/\underline{s} + 1/\underline{s}' = 1/\underline{f}'. \tag{6-31}$$

The quantity

$$f' = n\underline{f}',$$

where n is the index of refraction in image space, is commonly known as the focal length of the system.

Example: The Doublet Lens

To illustrate the use of the methods just discussed let us first find the cardinal points for a cemented doublet illustrated in Fig. 6.8. This is a common type of compound lens consisting of a positive biconvex element of crown glass cemented to a negative element of flint glass. This lens will be discussed more in a later chapter. In our example, the second surface of the positive element and the first surface of the negative element have the same radius of curvature so that the two pieces can be cemented together. Thus one surface is common to the two. The dispersions of the two types of glass allow for correction of chromatic aberration, and the curvatures of the three surfaces are adjusted to correct for spherical aberrations. Data on the lens are

$$n_c = 1.511, \qquad r_1 = 4.39 \text{ cm},$$

$$n_f = 1.621, \qquad r_2 = -4.39 \text{ cm},$$

$$t_1 = 0.40 \text{ cm}, \qquad r_3 = -140 \text{ cm}.$$

$$t_2 = 0.25 \text{ cm},$$

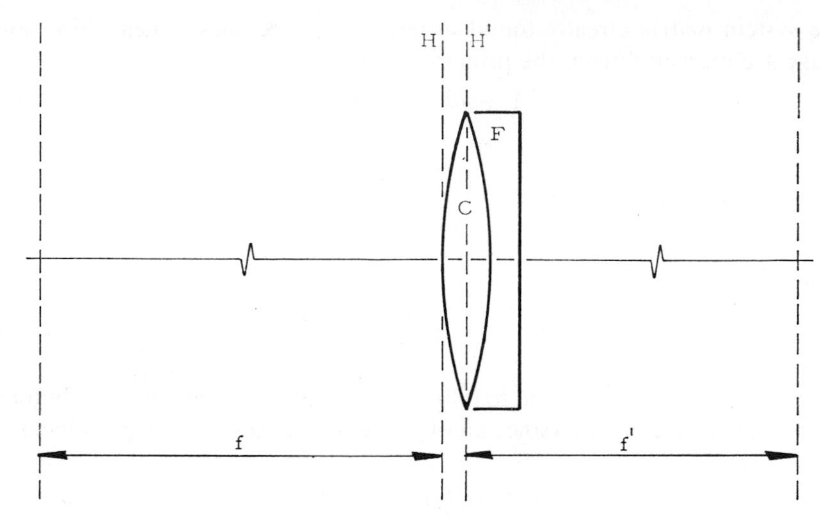

Figure 6.8 A cemented doublet.

The refraction and translation matrices to be multiplied to form the lens matrix are

$$\begin{bmatrix} 1 & -a_3 \\ 0 & 1 \end{bmatrix} \times \begin{bmatrix} 1 & 0 \\ \underline{t}_2 & 1 \end{bmatrix} \times \begin{bmatrix} 1 & -a_2 \\ 0 & 1 \end{bmatrix} \times \begin{bmatrix} 1 & 0 \\ \underline{t}_1 & 0 \end{bmatrix} \times \begin{bmatrix} 1 & -a_1 \\ 0 & 1 \end{bmatrix}.$$

The multiplication gives the following expressions for the elements of the lens matrix (defined in the section on the lens matrix):

$$\begin{aligned} a &= a_1 + a_2 + a_3 - a_1a_2\underline{t}_1 - a_1a_3\underline{t}_2 - a_1a_3\underline{t}_1 \\ &\quad -a_2a_3\underline{t}_2 + a_1a_2a_3\underline{t}_1\underline{t}_2, \\ b &= 1 - a_2\underline{t}_1 - a_3\underline{t}_2 - a_3\underline{t}_1 + a_2a_3\underline{t}_1\underline{t}_2, \\ c &= 1 - a_1\underline{t}_2 - a_2\underline{t}_2 - a_1\underline{t}_1 + a_1a_2\underline{t}_1\underline{t}_2, \\ d &= \underline{t}_2 + \underline{t}_1 - a_2\underline{t}_1\underline{t}_2. \end{aligned}$$

The values for the variables to be substituted into these quantities are found as follows:

$$\begin{aligned} a_1 &= (n_1' - n_1)(1/r_1) \\ &= (1.511 - 1)(1/4.39) = 0.1164, \\ a_2 &= (n_2' - n_2)(1/r_2) \\ &= (1.621 - 1.511)(1/-4.39) = -0.02505, \\ a_3 &= (n_3' - n_3)(1/r_3) \\ &= (1 - 1.621)(1/-140) = 0.004435, \\ \underline{t}_1 &= 0.40/1.511 = 0.2647, \\ \underline{t}_2 &= 0.25/1.621 = 0.1542. \end{aligned}$$

The elements of the lens matrix are then found to be

$$a = 0.964,$$
$$b = 1.005,$$
$$c = 0.955,$$
$$d = 0.420,$$

and the matrix is

$$\begin{bmatrix} 1.005 & -0.0964 \\ 0.420 & 0.955 \end{bmatrix}.$$

The distance $\underline{l}_H$ from the first surface to the first principal plane is

$$\underline{l}_H = (1 - b)/a = -0.005/0.0964 = -0.050.$$

And the distance from the last surface to the second principal plane is

$$\underline{l}_H' = (c - 1)/a = -0.045/0.0964 = -0.467.$$

The focal distances are [see Eq. (6-30)]:

$$f = -(1/a) = -10.38,$$
$$f' = (1/a) = 10.38.$$

Example: The Reflecting Objective

The second example is a reflecting microscope objective shown in Fig. 6.9. The system consists of a large primary spherical reflector and a small aspherical secondary reflector. The primary is concave and the secondary convex.

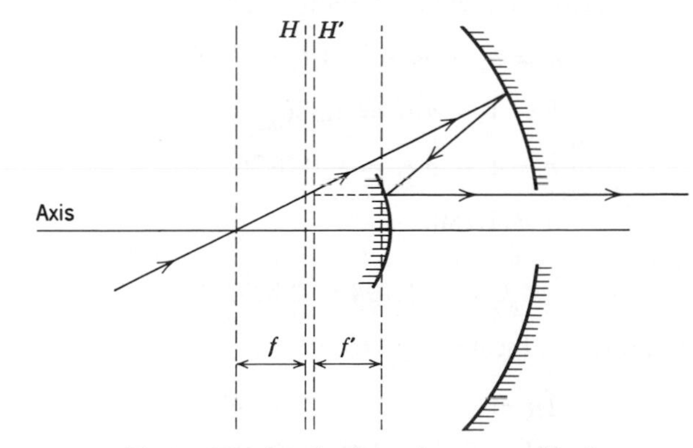

Figure 6.9 A reflecting microscope objective.

Although the secondary is aspherical, an "effective" focal length is used. It is somewhat unfair to use this system as an example of a paraxial system because no paraxial rays pass through. Nevertheless, the problem can be treated as an example in spite of this inconsistency; it will serve as a good example of reflecting optical elements. The data on the system are:

$$\begin{aligned} r_1 &= -2.160 \text{ in.}, \\ r_2 &= +0.6970 \text{ in.}, \\ t_1 &= 1.180 \text{ in.}, \\ n &= 1.000. \end{aligned}$$

At the first surface $n = 1$, $n' = -1$, and

$$\begin{aligned} a_1 &= (n' - n)/r_1 \\ &= (-1 - 1)/-2.16 = 1/1.08 = 0.9281. \end{aligned}$$

The "thickness," t_1, or distance from first surface to second, although from right to left, is positive because that is the direction rays are traveling, the system having been folded. At the second surface,

$$a_2 = (-1 - 1)/0.697 = -2.869.$$

The system matrix is formed by the multiplication of the three matrices,

$$\begin{bmatrix} 1 & -a_2 \\ 0 & 1 \end{bmatrix} \times \begin{bmatrix} 1 & 0 \\ \underline{t}_1 & 1 \end{bmatrix} \times \begin{bmatrix} 1 & -a_1 \\ 0 & 1 \end{bmatrix},$$

to give

$$\begin{bmatrix} b & -a \\ d & c \end{bmatrix} = \begin{bmatrix} 1 - a_2\underline{t}_1 & -a_1 - a_2 + a_1a_2\underline{t}_1 \\ \underline{t}_1 & 1 - a_1\underline{t}_1 \end{bmatrix}.$$

The values of the matrix elements are found by:

$$\begin{aligned} a &= a_1 + a_2 - a_1a_2\underline{t}_1 = 1.194, \\ b &= 1 - a_2\underline{t}_1 = 4.386, \\ c &= 1 - a_1\underline{t}_1 = -0.0926, \\ d &= 1.180. \end{aligned}$$

The focal length is

$$f' = 1/1.194 = 0.8375.$$

The principal planes are located by

$$\begin{aligned} \underline{l}_H &= (1 - b)/a = -2.836 \text{ in.} \\ \underline{l}_H' &= (c - 1)/a = -0.9151 \text{ in.} \end{aligned}$$

Distance to the first principal plane is measured from the first surface of the system, so $\underline{l}_H$, being negative here, is measured from right to left from the primary mirror. Distance to the second principal plane is measured from the last surface in the system. The direction is determined with respect to image space, hence $\underline{l}_H'$ is measured from the secondary mirror from right to left. The resulting locations of the principal planes and focal points are as shown in the figure.

Example: A System of Two Thin Lenses

For this example, two thin lenses with powers a_1 and a_2 separated by a distance t are used. The system matrix is obtained from

$$\begin{bmatrix} 1 & -a_2 \\ 0 & 1 \end{bmatrix} \times \begin{bmatrix} 1 & 0 \\ \underline{t}' & 1 \end{bmatrix} \times \begin{bmatrix} 1 & -a_1 \\ 0 & 1 \end{bmatrix}.$$

Mathematically this is identical with what we had for the reflecting optical system in the second example. The procedure from here on would, therefore, be the same; but we must note that, in contrast with the mirror problem, there is no change in direction of a ray traveling between the elements.

Programming the Matrix Multiplication

A number of computer programs are available for multiplying matrices. Such a program for an accessible computer facilitates the preliminary design of optical systems. Curvatures and thicknesses can be varied to adjust positions of object and image, positions of principal planes, and focal distances until the requirements of the design are approximated. Then, by manually tracing a few paraxial rays on a scale drawing, one can find obvious flaws in the design such as impossible ray heights. All this precedes the major system design by computer to correct for aberrations, which are discussed in Chapter 8.

A program, written in Fortran, that multiplies matrices is shown in Fig. 6.10. Values for the system matrix are printed out at the end together with the distances to the principal planes and the focal length. The example solved in the figure is the doublet lens example given above. (Note that for the refraction matrix, the matrix element a is the negative of the power a_i of the surface.) The value of N, in the unnumbered statement, is the number of surfaces minus one. The program can be used to solve a system of any number of elements. The data cards that follow the program must, of course, be stacked in an order corresponding to the sequence of surfaces and spacings.

```
      DIMENSION A(2,2),B(2,2),C(2,2),D(2,2)
    1 FORMAT(F10.5)
    2 FORMAT(1H1,20X,4E16.6)
    3 FORMAT(21X,3E16.6)
      N=2
    4 M=1
    5 A(1,1)=1.
    6 A(2,1)=0.0
    7 A(2,2)=1.
    8 B(1,1)=1.
    9 B(1,2)=0.0
   10 B(2,2)=1.
   11 READ1,A(1,2)
   12 READ1,B(2,1)
   13 C(1,1)=(A(1,1)*B(1,1))+(A(2,1)*B(1,2))
   14 C(1,2)=(A(1,2)*B(1,1))+(A(2,2)*B(1,2))
   15 C(2,1)=(A(1,1)*B(2,1))+(A(2,1)*B(2,2))
   16 C(2,2)=(A(1,2)*B(2,1))+(A(2,2)*B(2,2))
   17 READ1,A(1,2)
   18 D(1,1)=(C(1,1)*A(1,1))+(C(2,1)*A(1,2))
   19 D(1,2)=(C(1,2)*A(1,1))+(C(2,2)*A(1,2))
   20 D(2,1)=(C(1,1)*A(2,1))+(C(2,1)*A(2,2))
   21 D(2,2)=(C(1,2)*A(2,1))+(C(2,2)*A(2,2))
   22 M=M+1
   23 IF(M-N)24,24,30
   24 READ1,B(2,1)
   25 C(1,1)=(D(1,1)*B(1,1))+(D(2,1)*B(1,2))
   26 C(1,2)=(D(1,2)*B(1,1))+(D(2,2)*B(1,2))
   27 C(2,1)=(D(1,1)*B(2,1))+(D(2,1)*B(2,2))
   28 C(2,2)=(D(1,2)*B(2,1))+(D(2,2)*B(2,2))
   29 GO TO 17
   30 X1=-1./D(1,2)
   31 X2=(D(2,2)-1.)*X1
   32 X3=(1.-D(1,1))*X1
   33 PRINT2,D(1,1),D(1,2),D(2,1),D(2,2)
   34 PRINT3,X2,X1,X3
   35 STOP
      END
0000000000000000000000        END OF RECORD
  -0.1164
   0.2647
   0.02505
   0.1542
  -0.004435
0000000000000000000000


  1.004770E+00    -9.635240E-02    4.199225E-01    9.549837E-01
 -4.672044E-01     1.037857E+01   -4.951074E-02
```

Figure 6.10 A computer program to multiply lens matrices.

Tracing Skew Rays

Thus far in our discussion of optical systems we have traced only paraxial rays, which are a special kind of meridional rays. In this section, the discussion of systems of spherical surfaces is extended to include all possible rays in spherical systems: paraxial rays, meridional rays, and skew rays. The only way to achieve further generality would be to derive formulas that apply for conics and other aspheric surfaces, but these are not treated here.

A number of computer programs have been written, using general formulas, to trace the actual path followed by each of a number of rays through an optical system. The intercept with the image plane is typically found for each ray. Most of the programs allow the computer to adjust the curvatures and separations of surfaces to improve the image.

When the general formulas are reduced to two dimensions, they apply to meridional rays. The paraxial approximation further reduces the formulas to equations that agree with the translation and refraction matrices that are postulated earlier in this chapter.

The problem module upon which most general analysis programs are based consists of two coaxial spherical surfaces and the separation between them. The two surfaces can be any consecutive pair in the sequence of surfaces encountered by a ray. Surface separation is defined as the distance between points where the axis intersects the surfaces. An imaginary plane tangent to the second surface at the axis is used in the development for an intermediate ray-tracing step. The following information is assumed to be known; therefore it would be given as data to the computer:

(1) The coordinates x_j, y_j, z_j where the ray passes through the first surface.
(2) The radius R or the curvature C ($C = 1/R$) of each surface.
(3) The direction of the ray specified by the direction cosines k_j, l_j, m_j, at x_j, y_j, z_j.
(4) The index of refraction n_j in the space following each surface.
(5) The separation t_j.

Ray Path Between Surfaces

To define the ray path between surfaces [Ref. 5], first the intersection of the given ray with the tangent plane is found (Fig. 6.11). The z-axis is the optic axis, and $z = 0$ at the tangent plane. The length of the ray path between the first, the jth, surface and the tangent plane is d. The changes in the coordinates by the passage along this length are

$$\Delta x = x_p - x_j = dk_j, \tag{6-32}$$

$$\Delta y = y_p - y_j = dl_j, \tag{6-33}$$

$$\Delta z = t_j - z_j = dm_j. \tag{6-34}$$

The new coordinates are:

$$z_p = 0, \tag{6-35}$$

$$x_p = x_j + dk_j, \tag{6-36}$$

$$y_p = y_j + dl_j, \tag{6-37}$$

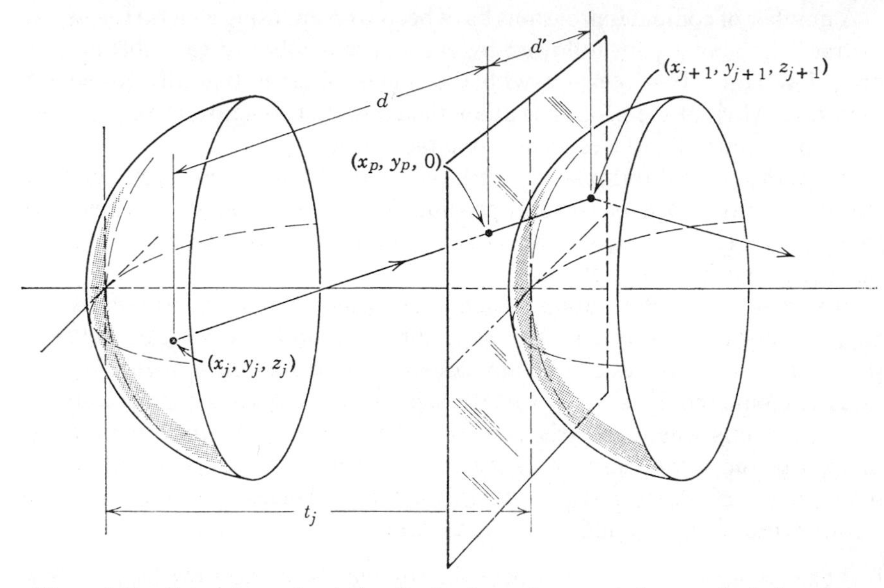

Figure 6.11 Geometry and notation for tracing a skew ray between spherical surfaces.

and the length d is

$$d = (t_j - z_j)(1/m_j). \tag{6-38}$$

The next step is to trace the ray to the second surface. The direction cosines will not change. Let the next increment of the ray path be d'. Then, in a manner similar to the one already used, one can find:

$$x_{j+1} = x_p + d'k_j, \tag{6-39}$$

$$y_{j+1} = y_p + d'l_j, \tag{6-40}$$

$$z_{j+1} = d'm_j. \tag{6-41}$$

To evaluate the ray coordinates at the second surface using these equations, the value of d' is needed. The required evaluating formulas are developed in the following paragraphs.

Figure 6.12 shows a meridional plane containing the ray coordinates at the second surface. The coordinates x_{j+1}, y_{j+1}, z_{j+1} are abbreviated x, y, z. Using the radius of curvature R for the surface, z is expressed in terms of x and y:

$$z = R - [R^2 - (x^2 + y^2)]^{1/2}. \tag{6-42}$$

By transposing and squaring, one obtains

$$\begin{aligned} x^2 + y^2 + z^2 - 2Rz &= 0, \\ C(x^2 + y^2 + z^2) - 2z &= 0. \end{aligned} \tag{6-43}$$

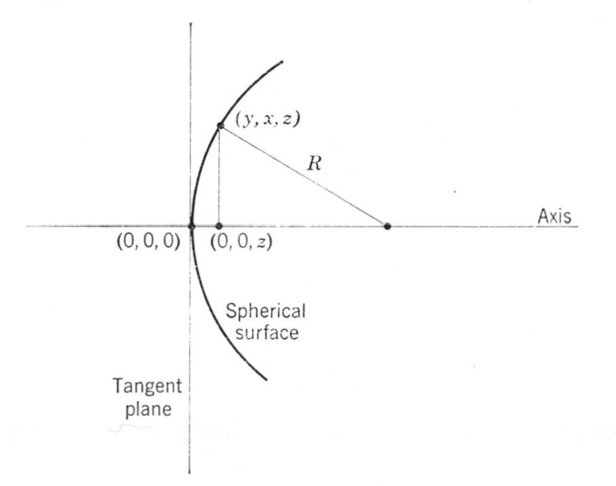

Figure 6.12 Geometry in a meridional plane containing the point of intersection of the ray and the surface.

By substituting for x, y, and z according to Equations (6-39) to (6-41), one obtains

$$d'^2 - (2B/C)\,d' + H = 0 \tag{6-44}$$

in which

$$H = (x_p^2 + y_p^2) \tag{6-45}$$

and

$$B = [m_j - C(x_p k_j + y_p l_j)]. \tag{6-46}$$

By applying the quadratic formula to Eq. (6-44), an equation for evaluating d' is obtained:

$$d' = [B \pm (B^2 - HC^2)^{1/2}]/C. \tag{6-47}$$

According to our sign convention, C and d' must always have the same sign. Also, if the surface were a plane, C would be zero and d' would also have to be zero. Therefore, only the minus sign before the parenthesis in Eq. (6-47) is applicable.

The equations that have been developed thus far are sufficient to find x_{j+1}, y_{j+1}, and z_{j+1}; however, Eq. (6-47) can be put into a more convenient form. In Fig. 6.13 all lines are in the plane of incidence at the $(j + 1)$th surface. The law of cosines gives

$$\begin{aligned} D^2 &= d'^2 + R^2 - 2d'R\cos(\pi - \theta) \\ &= d'^2 + R^2 + 2d'R\cos\theta. \end{aligned} \tag{6-48}$$

The term D^2 can also be written in terms of the coordinates at the ends of the line segment:

$$\begin{aligned} D^2 &= (x_p - 0)^2 + (y_p - 0)^2 + (0 - R)^2 \\ &= x_p^2 + y_p^2 + R^2. \end{aligned} \tag{6-49}$$

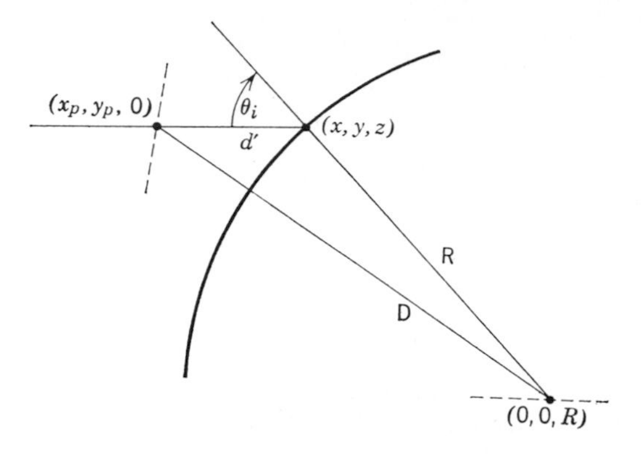

Figure 6.13 Geometry in the plane of incidence at the $(j + 1)$th surface.

By eliminating D between Equations (6-48) and (6-49), an expression for $\cos\theta$ can be found as follows:

$$\cos\theta = (CH - Cd'^2)/2d'. \tag{6-50}$$

If the expression for d' as given in Eq. (6-47) is substituted in Eq. (6-50), the simplified result is

$$\cos\theta = (B^2 - HC^2)^{1/2}, \tag{6-51}$$

which can also be written

$$B - \cos\theta = HC^2/(B + \cos\theta). \tag{6-52}$$

If $\cos\theta$ is substituted in Eq. (6-47) according to Eq. (6-51), and then Eq. (6-52) is used to get an alternate form,

$$\begin{aligned} d' &= (B - \cos\theta)/C \\ &= HC/(B + \cos\theta). \end{aligned} \tag{6-53}$$

Refraction at a Spherical Surface

To derive formulas for refracting a ray at the jth surface, we find it convenient to use vectors [Ref. 5]. Let $\mathbf{n}_j$, $\mathbf{n}_j'$, and $\mathbf{n}$ be unit vectors in the directions of the incident ray, the refracted ray, and the surface normal respectively. The latter is positive in the direction from first medium toward second medium as indicated in Fig. 6.14. To get expressions involving the sine functions needed to apply Snell's law, the following vector products are written:

$$\mathbf{n}_j \times \mathbf{n} = \mathbf{n}_\perp(\sin\theta), \tag{6-54}$$

$$\mathbf{n}_j' \times \mathbf{n} = \mathbf{n}_\perp(\sin\theta'). \tag{6-55}$$

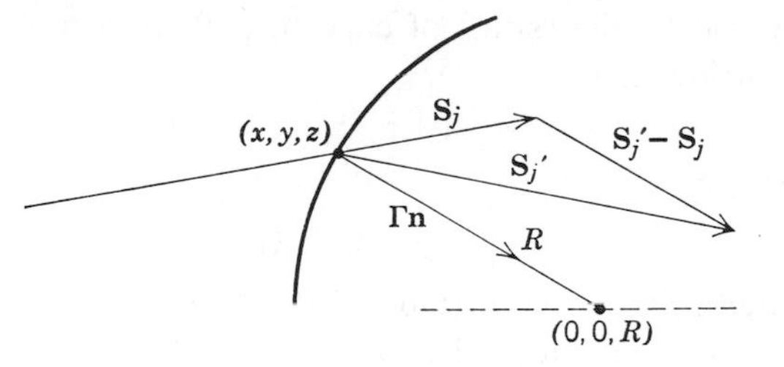

Figure 6.14 Geometry in the plane of incidence showing vector notation for refraction at the surface.

The same unit vector $\mathbf{n}_\perp$ is perpendicular to both $\mathbf{n}_j$ and $\mathbf{n}_j'$ since $\mathbf{n}_j$, $\mathbf{n}_j'$, and $\mathbf{n}$ are coplanar. Snell's law gives the ratio

$$(\sin\theta/\sin\theta') = n_j'/n_j. \tag{6-56}$$

From Eq. (6-54) and (6-55), the vector form of Eq. (6-56) can be written

$$n_j(\mathbf{n}_j \times \mathbf{n}) = n_j'(\mathbf{n}_j' \times \mathbf{n}). \tag{6-57}$$

Let

$$n_j\mathbf{n}_j = \mathbf{S}_j$$

and

$$n_j'\mathbf{n}_j' = \mathbf{S}_j'. \tag{6-58}$$

Then Eq. (6-57) becomes

$$\mathbf{S}_j \times \mathbf{n} = \mathbf{S}_j' \times \mathbf{n} \tag{6-59}$$

or

$$(\mathbf{S}_j' - \mathbf{S}_j) \times \mathbf{n} = 0. \tag{6-60}$$

The only way for this vector product to be zero is for the two vectors $(\mathbf{S}_j' - \mathbf{S}_j)$ and $\mathbf{n}$ to have the same direction or have opposite directions. Therefore, the two can be related by a scalar. This scalar, Γ, called the deviation constant, is defined by the equation

$$(\mathbf{S}_j' - \mathbf{S}_j) = \Gamma\mathbf{n}. \tag{6-61}$$

When the scalar product of each term in Eq. (6-61) and $\mathbf{n}$ is taken, the equation,

$$\Gamma = n_j'\cos\theta' - n_j\cos\theta, \tag{6-62}$$

follows (because of the definitions of $\mathbf{S}_j'$ and $\mathbf{S}_j$). Cos θ' can be eliminated by using Snell's law and a trigonometric identity. The result is

$$\Gamma = -n_j\cos\theta + n_j'\left[\left(\frac{n_j}{n_j'}\cos\theta\right)^2 - \left(\frac{n_j}{n_j'}\right)^2 + 1\right]^{1/2}, \tag{6-63}$$

A vector defined by the radius of curvature R from $(0,0,R)$ to the point (x,y,z) on the surface is

$$\mathbf{R} = x\mathbf{I}_1 + y\mathbf{I}_2 + (R - z)\mathbf{I}_3. \tag{6-64}$$

The unit vector $\mathbf{n}$ is

$$\mathbf{n} = - C[x\mathbf{I}_1 + y\mathbf{I}_2 + (R - z)\mathbf{I}_3]. \tag{6-65}$$

As already stated following Eq. (6-60), the vector $(\mathbf{S}_j' - \mathbf{S}_j)$ is in the same direction as $\mathbf{n}$ and of magnitude Γ, so

$$(\mathbf{S}_j' - \mathbf{S}_j) = -C\Gamma[x\mathbf{I}_1 + y\mathbf{I}_2 + (R - z)\mathbf{I}_3]. \tag{6-66}$$

But

$$\mathbf{S}_j' = n_j'\mathbf{n}_j' = n_j'(k_j'\mathbf{I}_1 + l_j'\mathbf{I}_2 + m_j'\mathbf{I}_3), \tag{6-67}$$

$$\mathbf{S}_j = n_j\mathbf{n}_j = n_j(k_j\mathbf{I}_1 + l_j\mathbf{I}_2 + m_j\mathbf{I}_3); \tag{6-68}$$

therefore

$$(\mathbf{S}_j' - \mathbf{S}_j) = (n_j'k_j' - n_jk_j)\mathbf{I}_1 + (n_j'l_j' - n_jl_j)\mathbf{I}_2 + (n_j'm_j' - n_jm_n)\mathbf{I}_3. \tag{6-69}$$

By equating appropriate coefficients of Equations (6-66) and (6-69) one obtains:

$$(n_j'k_j' - n_jk_j) = -C\Gamma x_j, \tag{6-70}$$

$$(n_j'l_j' - n_jl_j) = -C\Gamma y_j, \tag{6-71}$$

$$(n_j'm_j' - n_jm_j) = -C\Gamma(R - z_0) = -\Gamma(1 - Cz_j). \tag{6-72}$$

Summary of General Formulas

The equations used to translate a ray from one surface to the next are (in some instances transposed from the form used earlier in this chapter):

$$d = (t_j - z_j)(1/m_j), \tag{6-38}$$

$$z_p = 0, \tag{6-35}$$

$$x_p = x_j + dk, \tag{6-36}$$

$$y_p = y_j + dl, \tag{6-37}$$

$$H = (x_p^2 + y_p^2), \tag{6-45}$$

$$B = [m_j - C(x_pk_j + y_pl_j)], \tag{6-46}$$

$$\cos\theta = (B^2 - HC^2)^{1/2}, \tag{6-51}$$

$$d' = HC/(B + \cos\theta), \tag{6-53}$$

$$x_{j+1} = x_p + d'k_j, \tag{6-39}$$

$$y_{j+1} = y_p + d'l_j, \tag{6-40}$$

$$z_{j+1} = d'm_j. \tag{6-41}$$

The equations used to refract a ray at a surface are:

$$\Gamma = -n_j \cos\theta + n_j'\left[\left(\frac{n_j}{n_j'}\cos\theta\right)^2 - \left(\frac{n_j}{n_j'}\right)^2 + 1\right]^{1/2}, \tag{6-63}$$

$$n_j'k_j' = n_jk_n - C\Gamma x_j, \tag{6-70}$$

$$n_j'l_j' = n_jl_j - C\Gamma y_j, \tag{6-71}$$

$$n_j'm_j' = n_jm_j - \Gamma(1 - Cz_j). \tag{6-72}$$

REFERENCES

1. A. G. Fox and T. Li, "Resonant Modes in a Maser Interferometer," *Bell Syst. Tech. J.*, **40,** 453 (1969).
2. W. Brouwer, *Matrix Methods in Optical Instrument Design.* W. A. Benjamin, Inc., New York, 1964.
3. E. L. O'Neill, *Introduction to Statistical Optics*, Chapter 3. Addison-Wesley Publishing Company, Inc., Reading, Mass., 1963.
4. A. Nussbaum, *Geometric Optics: An Introduction.* Addison-Wesley Publishing Company, Reading, Mass., 1968.
5. The development in these two sections follows closely a similar development in: *Optical Design*, MIL-HDBK-141, Defense Supply Agency, Washington, D.C., Oct., 1962.

7

Stops, Pupils, and Image "Brightness"

This chapter is primarily about stops: those elements in an optical system that regulate the amount of light passing through the system and those elements that delineate the portion of the object field actually imaged. Whether an optical system consists of a camera, the eye, or any one of the many possible optical instruments, a common question arises: What proportion of the radiant energy originating at the object can pass through the system and arrive at the image? The camera fan, for instance, discusses this question in his jargon of exposure, shutters, and stops because he is aware that the response of photographic film is related to his shutter and stop settings as well as to the lighting of his subject. A photographic film is an integrating type of detector because it responds to the total accumulated energy, per unit area, received during the time the shutter is open. This time interval is called *shutter speed*. *Exposure* for the film, which is defined in Chapter 13, depends upon the radiant incidance at the film and the shutter speed.

A typical camera, for snapshots, provides for shutter speeds, in discreet steps, from 1/15 to 1/500 sec, a ratio of between 30 and 40 to 1. Unfortunately we do not have space to discuss shutters and their characteristics [Ref. 1].

Besides the shutter, another adjustment that helps control the amount of energy reaching the camera film is the aperture or lens opening. The aperture settings are usually arranged so that changing from one setting to the next changes the flux passing through the lens by a factor of two. "Stopping down" one stop, for example, would be decreasing the flux by a half; three stops would decrease the flux to $\frac{1}{8}$ its original value.

Because our study of stops involves ray tracing, the derived formulas will be based on the laws of geometrical optics; the formulas will, therefore, not apply to coherent beams. Whenever there is an appreciable degree of coherence, interference effects partially determine the distribution of the radiant power, and each instance is a special interference problem. The formulas also do not account for diffraction effects [Ref. 2].

Aperture Stop

As one traces a bundle of rays through an optical system, he discovers that one of the optical components determines the maximum radiant power. This comes about by limiting the maximum ray bundle diameter that can pass through from an element of area on the object to the corresponding element of area on the image. Of course, by deliberate design or by chance, two or more optical components could bound the same bundle of rays, but this is unlikely. The free diameter of the limiting component is called the *aperture stop*. Though called a "stop," it is actually a limiting opening.

The simple lens of Fig. 7.1 forming an image Q' of the object point Q illustrates an aperture stop. The size of the solid angle Ω, with vertex at Q, is determined by the free area of the lens and its distance from Q. Only those rays from Q that are within Ω will pass through the system and contribute to the image at Q'. Hence, the lens constitutes an aperture stop because it limits the size of the bundle of rays. An adjustable iris diaphragm might be placed either before or after the lens to reduce further the size of the bundle; then the diaphragm would be the aperture stop. When the lens, without the diaphragm, constitutes the aperture stop, the "size" of the

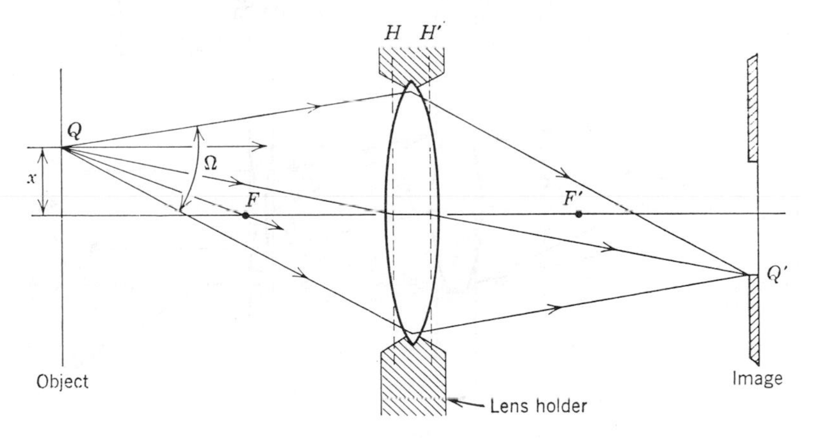

Figure 7.1 Aperture stop and field stop.

solid angle Ω becomes smaller as the distance x of Q from the optic axis becomes greater. This effect will be shown to have an important relationship to the luminance of the image.

Field Stop

A second kind of stop is illustrated in Fig. 7.1. It is the obstruction, with a circular opening centered on the optic axis, located in the image plane. Any object points imaged at distances greater than Q' from the axis will not be "seen." That is, radiant energy from these points will be obstructed and will not pass through the system. The obstruction with its opening constitutes a *field stop* because it limits the field of view that can be "seen."

Determination of Stops

As was demonstrated in Chapter 6, a compound optical system can be reduced to a simple system. Although this can be done by straightforward procedures, the easily obtained results rarely identify which of the several elements of the system is the actual stop. Probably the best way to clarify this problem is to illustrate with a few examples.

In Fig. 7.2, a study of the rays that pass through the first lens reveals the following: All rays from the on-axis point Q_1 and accepted by L_1 pass through L_2. As a matter of fact, L_2 seems to be larger than necessary; thus, it appears that L_1 functions as the aperture stop. However, when off-axis points such as Q_2 are considered, rays accepted by L_1 are not necessarily passed by L_2; for instance, ray d in the figure by-passes L_2 and so is lost

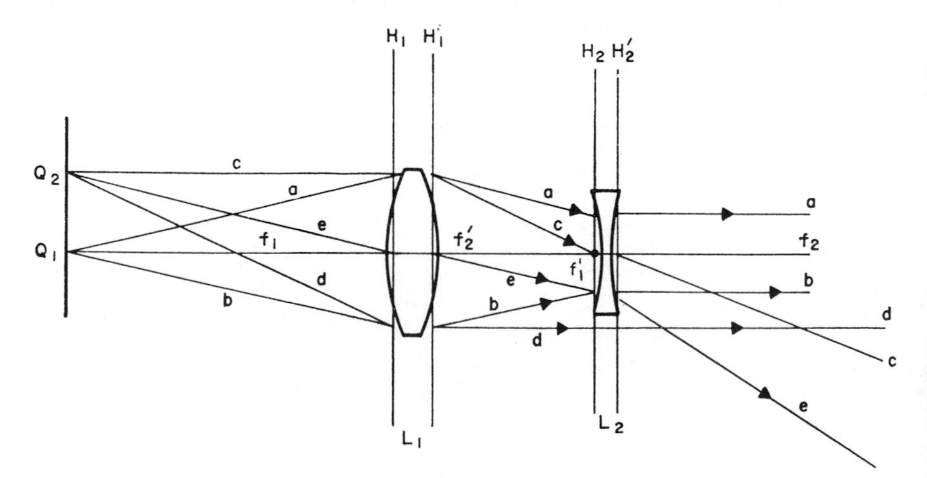

Figure 7.2 Determining stops by ray tracing.

from the beam. Lens L_1 partially limits the "size" of the transmitted beam, and so does L_2. Neither independently can be considered the aperture stop. Also, as Q_2 moves further and further off axis, L_2 will pass a smaller and smaller fraction of the rays accepted by L_1. This effect is called vignetting. Lens L_2 is in effect a field stop.

It is customary to define the limit of the field of view in such a way that the bundle from a point at the edge of the field "transports" half as much radiant power as the on-axis bundle. It is apparent that every element in the system, whether it is one of the lenses or a diaphragm intentionally placed in the system, is likely to be a stop of some kind and must be studied to determine just how it limits the bundle of rays or limits the field. One way to make this study is to trace a number of rays through the entire system as is suggested by Fig. 7.2. Marginal rays for each optical element, one element after another, must be investigated to determine which element actually limits the bundle of rays.

Pupils

With the aid of Fig. 7.3, let us study the effect of the stop S indicated in the figure. This is the aperture stop for the system—at least for the on-axis point of the object. As indicated earlier, this is established by tracing a number of rays through the system to prove that S is the element limiting the bundle of rays. The stop is located between two lenses with the positions of the focal points and principal planes indicated by the several f and H respectively.

The image of S as seen through the first lens is S_0 and is found by ray tracing from S through this lens. Note that the use of "object" and "image"

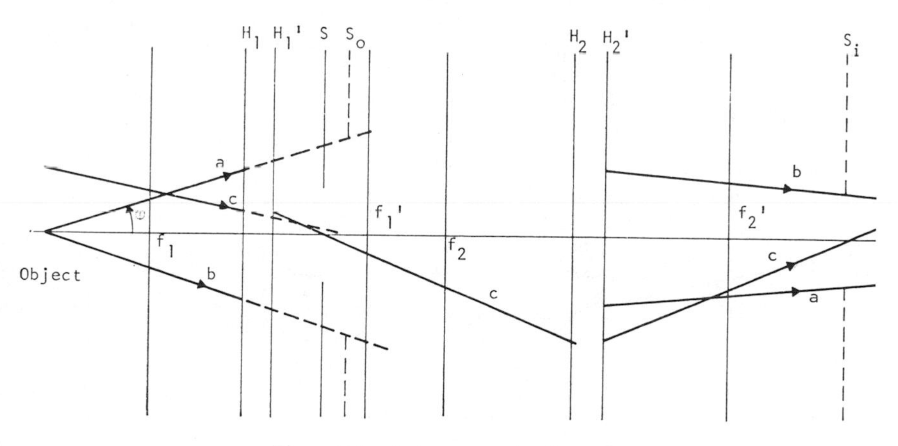

Figure 7.3 Entrance and exit pupils.

is relative to the element being imaged. For instance, for S the object space for the first lens is to the right of the lens, the image space to the left. The image of S will be in object space (considering the whole system); and, in this particular example, it turns out to be a virtual image S_0. (See the discussion of virtual images in connection with Fig. 6.5b.) The next step is to determine the image of S formed by the second lens. This image S_i is real and is in image space. The object image S_0 is called the entrance pupil, and S_i is called the exit pupil for the optical system.

The ray from an off-axis object point that passes through the point of intersection of the entrance pupil and the optical axis is called the *chief*, or *principal ray*, indicated as ray c in the figure. The chief ray also passes through the aperture stop and the exit pupil on-axis.

In a telescope or microscope the entrance pupil usually is the objective itself, but in a camera it is the image of the iris diaphragm in the lens or that part of the lens system, if any, that precedes the iris. In the telescope or microscope the exit pupil is the image of the objective formed by the eyepiece. This image is in the plane of greatest radiant energy density and is thus the place where the entrance pupil of the eye should be located. Telescopes and microscopes should therefore be designed so that the exit pupil is accessible for properly and comfortably positioning the eye and matches the size of the entrance pupil of the eye (see Chapter 14).

Determining Stops by Observing Them as Entrance Pupils

The technique of determining the entrance pupil can be used to determine which elements are functioning as stops. In Fig. 7.4 images of two

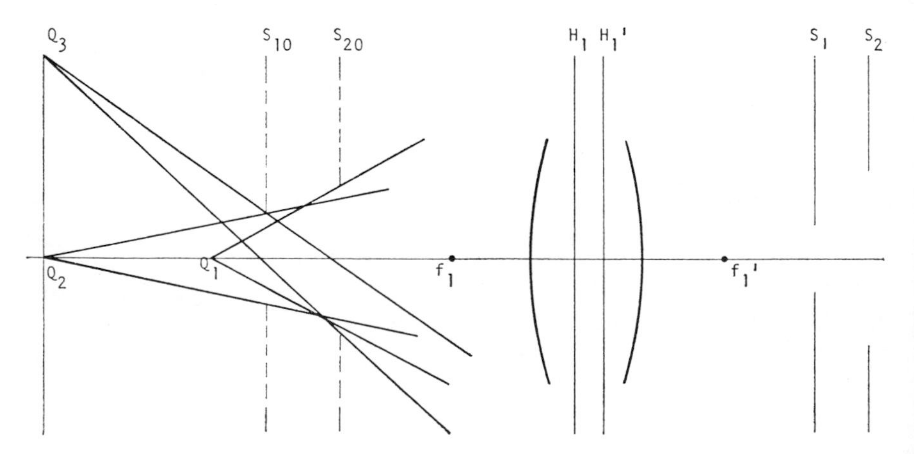

Figure 7.4 Determining stops by imaging to object space.

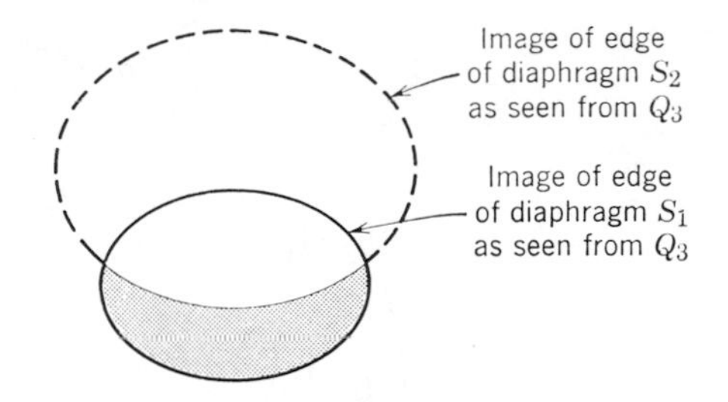

Figure 7.5 Vignetted aperture formed by two potential stops.

diaphragms S_1 and S_2 are shown respectively. The optical system indicated schematically as a simple system includes all optical elements that precede the diaphragms. If S_1 were an optical element—a lens, for instance, instead of a diaphragm—then it would have functioned with the optical system in forming the image of S_2 in object space. If the object is at Q_1, S_{20} is seen to be the entrance pupil and consequently S_2 the aperture stop. The exit pupil would be found by including all elements that follow S_2 in the optical system and finding the image of S_2 in image space. If the object is at Q_2, S_{10} is seen to be the entrance pupil; and S_1 is the aperture stop. For the object point at Q_3, S_{10} is, in effect, the entrance pupil for one side of the beam and S_{20} is the entrance pupil for the other side, so the aperture is the opening through the overlapping portions of S_1 and S_2 [Ref. 3]. (Fig. 7.5.) The combination may also constitute the field stop.

From this discussion, it is seen that the aperture stop of a system may change with a change in object position. The aperture stop of the system is determined by noting which image of the potential stops subtends the smallest angle as seen from the object point.

Angular Field of View

The first step in determining the angular field of view for an optical system is to find the entrance pupil. In Fig. 7.6, S_0 has been found to be the entrance pupil for the system indicated schematically. Then every stop that is to be considered as a potential field stop, such as S_f in the figure, must be imaged through the entire optical system preceding it to object space. The image of S_f is indicated as S_f' in the figure. The image that subtends the smallest angle as seen from the center of the entrance pupil C determines the field stop for the system. The angle, θ in the figure, that this image subtends is the angular field of view.

The field stop usually is an actual diaphragm placed in the system for that purpose. If the edge of the field is to be sharply defined, the stop must be

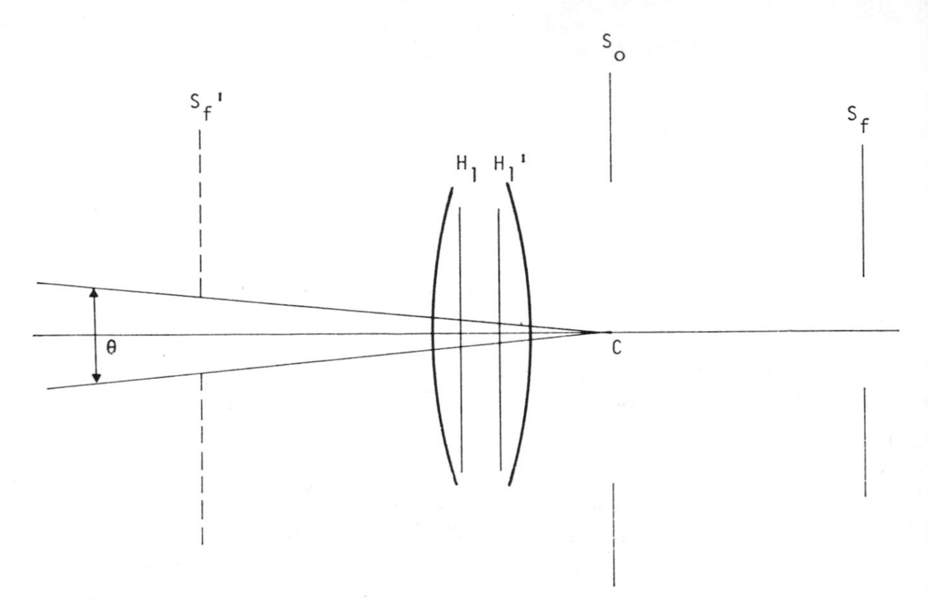

Figure 7.6 Angular field of view—entrance window.

placed in the plane of an image. In a camera, the plate or film holder has an opening just in front of the plate or film to delineate the area that can be exposed. This opening constitutes a simple field stop. A microscope usually has a special field stop placed so that the eye sees the field stop and the object in focus together. To accomplish this, the objective (first lens) forms an image of the object at the field stop; the eyepiece (second lens) in turn forms an image of the first image and the stop. This arrangement is satisfying because the field being viewed has, as a result, a sharp edge.

In Fig. 7.6, the image S_f' of the field stop S_f formed by the lenses in front of it is called the entrance window since this defines the area of the object to be viewed. The image of the field stop formed by the lenses, if any, following the stop is called the exit window. The viewfinder of a camera is designed to show the user the entrance window superposed on the picture that will be taken. "Composing the picture" is arranging the subjects within the entrance window. At the theatre, the screen is the window through which all scenes of the motion picture must be viewed. The screen subtends an equivalent angle θ from each observer in the audience. Whenever the arrangement of projector, viewer, and screen are such that θ differs appreciably from the angular field of view θ that existed at the camera while the picture was being taken, a distorted picture is observed. When one views a reconstruction by a hologram, the hologram plate constitutes the

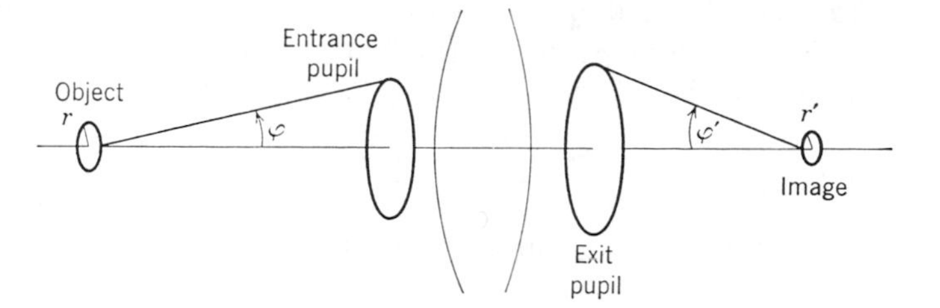

Figure 7.7 Geometry at entrance and exit of an optical system.

entrance window. The restriction of this entrance window to practical sizes of hologram plates is one limitation on holography for displays.

The Sine Condition

In Chapter 6, in Eq. (6-21), it is shown that

$$\underline{\alpha}' = \underline{\alpha}/\beta, \tag{7-1}$$

in which

$$\underline{\alpha}' = n'\alpha' \approx n' \sin \alpha', \tag{7-2}$$

$$\underline{\alpha} = n\alpha \approx n \sin \alpha,$$

and that

$$\beta = r'/r. \tag{7-3}$$

These relationships can be applied to φ, the angle made with the axis by a ray from an on-axis object point to the edge of the entrance pupil. The length r is assumed to be the radius of a small circular element of area of the object centered on the axis. In fact, for later purposes, r is specified so small that the solid angle subtended by the entrance pupil from a point on the object r distance from the axis is essentially the same as the solid angle subtended from an on-axis object point. The symbols φ' and r' are defined in a similar way in image space; however, image instead of object and exit pupil instead of entrance pupil are used in the definitions (Fig. 7.7). From the relationships given in Equations (7-1), (7-2), and (7-3):

$$r'n'\alpha' = rn\alpha \tag{7-4}$$

and, with some approximation,

$$r'n' \sin \alpha' = rn \sin \alpha. \tag{7-5}$$

Equation (7-4) is called *Lagrange's Law* and Eq. (7-5) is known as the *Abbe sine condition*. Because of the way these are derived they seem to be valid only for the paraxial approximation. However, the sine condition is more general than this and applies in optical systems free of certain aberrations for off-axis object points, particularly primary coma. This is discussed again in Chapter 8.

An Optical Invariant

If both sides of Eq. (7-4) are squared and multiplied by π^2, the following result:

$$\begin{aligned} n'^2 \pi r'^2 \pi \alpha'^2 &= n^2 \pi r^2 \pi \alpha^2, \\ n'^2 A' \Omega' &= n^2 A \Omega, \end{aligned} \tag{7-6}$$

where Ω and Ω' are solid angles subtended by the entrance pupil from the object point and by the exit pupil from the corresponding image point respectively. Also, if $n' = n$,

$$A'\Omega' = A\Omega. \tag{7-7}$$

Equation (7-7) is known as *Lagrange's invariant* and indicates that when the paraxial approximation is justified and when the refractive indices in object space and image space are the same, the area–solid-angle product is an invariant in an optical system.

Luminous Flux Transferred From Object to Optical System

From earlier discussions in this chapter, we have established that, for any optical system having a well-defined entrance window and entrance pupil, those light rays that originate on the object within the field of view and that enter the entrance pupil will pass on through the system, out the exit pupil, and finally reach the image. At the outset, the radiating property of the object has to be known, either as a radiance or a luminance in watts or lumens, respectively, per unit solid angle per unit projected area. The following development will use L for luminance, but a parallel development in radiometric, rather than photometric, units would be identical. The object is assumed to be a diffuse radiator.

If an element of area δa_1 is considered at an on-axis point of the object and an element of solid angle with vertex at δa_1 is subtended by an annular ring concentric with the entrance pupil (Fig. 7.8), the indicated solid angle is

$$d\Omega_1 = 2\pi \sin \alpha \, d\alpha. \tag{7-8}$$

Since the projected area of δa_1 in the direction of the annular ring is $\delta \alpha_1 \cos \alpha$, the luminous flux radiated into this solid angle is

$$\begin{aligned} d\Phi &= L_1 \delta a_1 \cos \alpha_1 \, d\Omega_1 \\ &= 2\pi L_1 \delta a_1 \cos \alpha \sin \alpha \, d\alpha. \end{aligned} \tag{7-9}$$

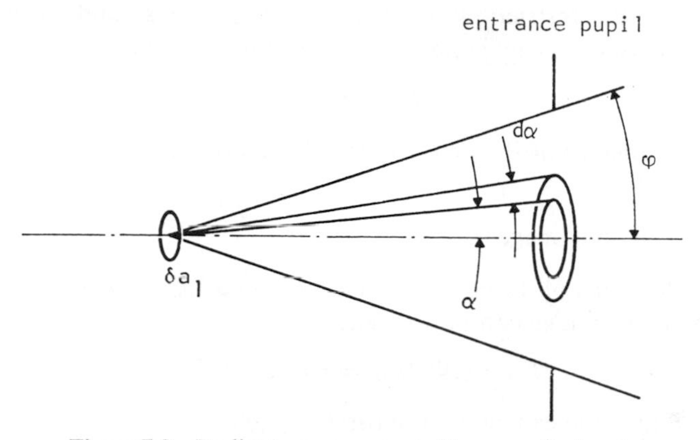

Figure 7.8 Radiant power accepted by an optical system.

The luminous flux radiated into the entrance pupil is found by integrating over the angle φ where φ is the half cone angle subtended by the entrance pupil:

$$\begin{aligned}\Phi &= 2\pi L_1\, \delta a_1 \int_0^{\varphi} \cos\alpha \sin\alpha \, d\alpha \\ &= \pi L_1\, \delta a_1 \sin^2 \varphi. \qquad (7\text{-}10)\end{aligned}$$

Luminance of the Image

If there are no losses in the optical system, the luminous flux Φ given by Eq. (7-10) is also the luminous flux reaching the element of area δa_2 on the image. By a procedure similar to the one followed to get Eq. (7-10), Φ can be found in terms of φ', the half angle subtended by the exit pupil and with vertex at δa_2, as follows:

$$\Phi = \pi L_2 \delta a_2 \sin^2 \varphi'. \qquad (7\text{-}11)$$

Allowances, however, should be made for losses to the beam while passing through the system, the losses consisting of the luminous flux that may be absorbed, scattered out of the beam, or reflected at surfaces. To do this an efficiency factor τ is applied, which we call the transmittance of the optical system:

$$\tau\Phi = \pi L_2 \delta a_2 \sin^2 \varphi'. \qquad (7\text{-}12)$$

By eliminating Φ between Equations (7-10) and (7-12), the following results:

$$\tau(L_1/L_2) = (\delta a_2/\delta a_1)(\sin^2 \varphi'/\sin^2 \varphi). \qquad (7\text{-}13)$$

If the element of area on the object is a circle so that $\delta a_1 = \pi r^2$, Eq. (7-5) can be applied to obtain the following, which is similar to Eq. (2-14):

$$\tau(L_1/L_2) = (n/n')^2. \qquad (7\text{-}14)$$

Finally, within the restrictions used in the derivations, and provided the transmission losses are negligible, ($\tau \approx 1$) and $n = n'$, so

$$L_1 = L_2, \tag{7-15}$$

that is, the luminance of the image is equal the luminance of the object.

Flux Density at the Image

From the definitions of Table V in Chapter 2 and Eq. (7-12), the luminous flux density at the image can be written:

$$F_2 = (\tau\Phi/\delta a_2) = \pi L_2 \sin^2 \varphi'. \tag{7-16}$$

From Eq. (7-13), one can substitute for L_2 to get

$$F_2 = \tau\pi L_1(\delta a_1/\delta a_2) \sin^2 \varphi, \tag{7-17}$$

and from Eq. (7-3),

$$F_2 = (\tau\pi L_1/\beta^2) \sin^2 \varphi. \tag{7-18}$$

Then, from Eq. (7-5), with $\alpha = \varphi$,

$$F_2 = \tau\pi L_1(n'^2/n^2) \sin^2 \varphi'. \tag{7-19}$$

It is apparent from Eq. (7-18), that to have a high flux density at the image one has to start with a source having high luminance as the object. A low magnification and a large entrance window (large $\sin \varphi$) are also needed, but once the system has been determined, these two requirements work against each other. A small magnification requires large s in proportion to s', but making s large moves the object further away from the entrance pupil so that $\sin \varphi$ becomes smaller.

Relative Aperture

The angle φ shown in Figures 7.3, 7.7, and 7.8 has been quite prominent in our expressions for the flux density or luminance of the image. It is often called *angular aperture*. Two other quantities are also used to indicate the relative capacity of an optical system to "gather" light: *numerical aperture* (N.A.) and *f/number*. The first is defined by

$$\text{N.A.} = n \sin \varphi; \tag{7-20}$$

it is discussed further in Chapters 8 and 9 in connection with resolving power. From Equations (7-14) and (2-14), it is apparent that, when there are no losses, the ratio $L_1/n^2 = L_2/n'^2$ is an invariant. Equation (7-18) then can be written

$$F_2 = (\pi L_2/n'^2\beta^2)(n^2 \sin^2 \varphi) \tag{7-21}$$

to show that the square of the numerical aperture is proportional to the flux density in the image.

The f/number is defined as the ratio given by dividing the focal length of the system by the diameter of the entrance pupil:

$$F^* = f/D. \tag{7-22}$$

In telescopes and most cameras, the object is typically a great distance away, in air, and the image is approximately at the focal point. Then the following can be written for Eq. (7-19):

$$F_2 = \tau\pi L_1 (n'/n)^2 (D/2f)^2 \tag{7-23}$$

or, if $n' = n$,

$$F_2 = (\tau\pi L_1/4)(D/f)^2$$
$$= (\tau\pi L_1/4F^{*2}). \tag{7-24}$$

One speaks of a system having a small f/number or a large numerical aperture as being a "fast" system. In common usage, any of the three quantities φ, NA, and F^* might be referred to as relative aperture.

The Cos⁴ Law for Off-Axis Points

To find the flux density at the off-axis image point da_2 in Fig. 7.9, one first writes the expression for the projected area of the aperture stop as seen

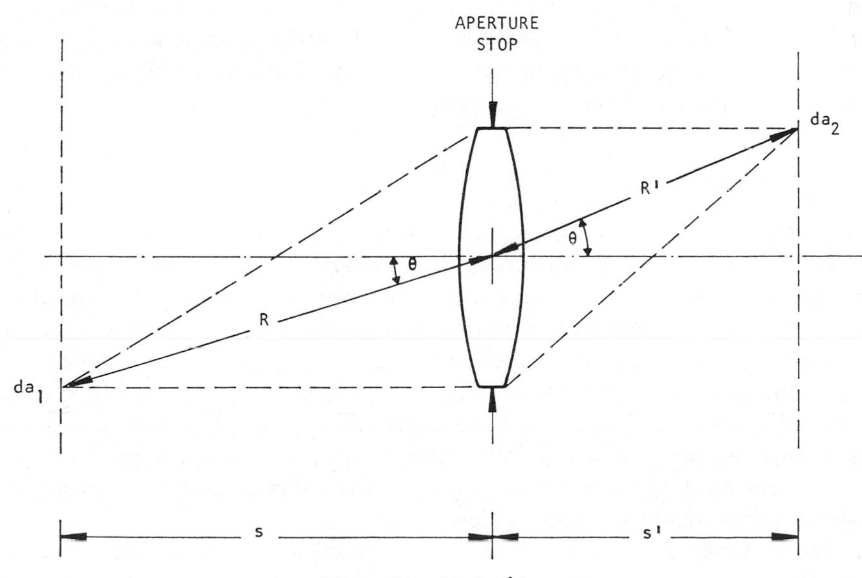

Figure 7.9 Cosine⁴ law.

from da_1, which is

$$A \cos \theta, \tag{7-25}$$

and its distance R is

$$R = s/\cos \theta. \tag{7-26}$$

Then the solid angle subtended by the aperture stop as seen from da_1 is

$$\begin{aligned} \Omega_1 &\approx (A \cos \theta)/R^2 \\ &\approx (A \cos^3 \theta)/s^2. \end{aligned} \tag{7-27}$$

The luminous flux passing through the aperture from da_1 is

$$\begin{aligned} \Phi &= L_1\, da_1 \cos \theta \Omega_1 \\ &= L_1\, da_1 A(\cos^4 \theta)/s^2. \end{aligned}$$

If the transmission losses are neglected, this is also the flux reaching the image area da_2. The flux density is

$$F = \Phi/da_2 = L_1 A(da_1/da_2)(\cos^4 \theta)/s^2.$$

Then,

$$F = (\pi/4)L_1(D/s')^2 \cos^4 \theta \tag{7-28}$$

from Eq. (6-12) and because $A = \pi D^2/4$. If s' is assumed equal to the focal distance,

$$F = (\pi/4)L_1(1/F^*)^2 \cos^4 \theta. \tag{7-29}$$

It should be noticed that this relationship has not included vignetting, the loss of the edge rays of the beam because of the large angle with the axis; thus there may be even more reduction in the illuminance of the off-axis points than the $\cos^4 \theta$ factor indicates.

Radiometers

A particular optical system, called a *radiometer*, is designed to measure radiant energy. If the instrument is calibrated so that the output is in photometric units, the instrument is then called a *photometer*. Typically, these systems have small fields of view and small angular apertures. Often a radiant energy detector is positioned in the image plane centered about the axis; the sensitive area itself functions as the field stop. The optical system in a radiometer requires well-defined stops; this is the reason for the narrow field and aperture. Whenever the paraxial approximation cannot be used, the system must be analyzed to be sure that the sine condition holds for all object points within the field of view.

The accuracy of radiometers, when compared with corresponding instruments in many other fields, is not high. In Chapter 5 it is shown that radiant energy from a "point" source is spread out in a diffraction pattern. Nearly all sources have some coherence; so interference effects are common in radiant energy measurements, that is, the energy is not distributed strictly according to the laws of geometric optics. Lens aberrations, if present, will cause other distortions in the distribution of image energy. To facilitate deriving some of the basic formulas, a diffuse radiator has been assumed, but most sources are actually not truly diffuse radiators. In fact, some projection systems intentionally avoid a diffusely radiating source to reduce the effect of the $\cos^4$ law. The transmittance of the optical system—for example, the lens glass— is a function of wavelength; thus, sources having different spectral characteristics may produce different responses. Standard sources of radiant energy used for calibrating the radiometer are themselves never highly accurate. All these shortcomings of optical systems limit the accuracy of radiometers.

Several writers have discussed radiometers and have given different expressions for the luminance of the image [Refs. 3–7]. These writers also discuss sources of inaccuracies and discuss methods of calibration. Practically, the formulas given are useful primarily during the design of radiometers and for estimating their performance. A radiometer, once designed, depends on a standard source for its calibration.

Baffles for Stray Light

In addition to the stops already discussed in this chapter, at least one more kind is needed: baffles and glare stops to keep stray, extraneous light from reaching the image. Figure 7.10 shows a telescope in which the objective is the aperture stop. Radiant energy from outside the field of view may pass through the objective, then strike the inside wall of the tubing, a lens mount, or a supporting strut; and finally be reflected or scattered into the optical path toward the image. Since the field lens, in addition to erecting the image, forms an image of the objective, a glare stop accurately matching the objective image in size and position can be placed in the system to block

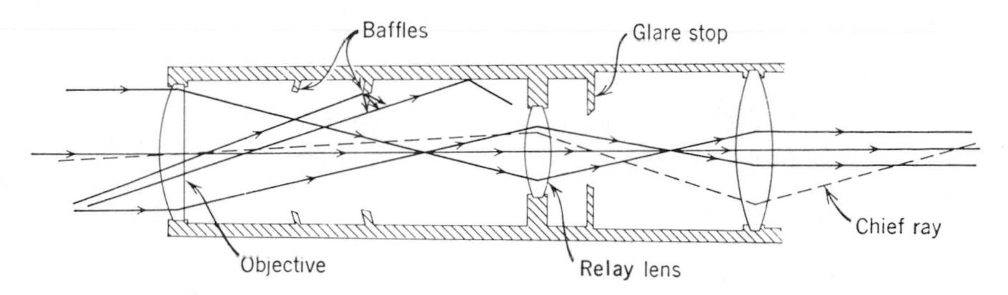

Figure 7.10 Baffles and glare stops to eliminate stray light.

out stray light. Baffles are often used inside the tubing, but outside the path of the normal rays, to serve as light traps. Stops, baffles, lens mounts, and the inside of the tubing should be painted flat black to reduce reflections. If the system is to be used outside the visual range, the "flat black" should be highly absorbent at all the wavelengths used.

REFERENCES

1. A. Schwartz, "Camera Shutters," in *Applied Optics and Optical Engineering*, Vol. IV. Part I. Edited by R. Kingslake. Academic Press, New York, 1969.
2. A. Walther, "Radiometry and Coherence," *J. Opt. Soc. Am.*, **58,** 1256 (1968).
3. W. B. King, "The Approximation of a Vignetted Pupil Shape by an Ellipse," *Appl. Optics*, **7,** 197 (1968).
4. A. Bouwers and A. C. S. Van Heel, "On the Luminosity of Optical Systems," *Physica*, **10,** 714 (1943).
5. V. E. Medvedev and G. G. Paritskaya, "Calculation of Illumination in an Image," *Opt. Spectry.*, **21,** 351 (1966).
6. H. F. Gilmore, "The Determination of Image Irradiance in Optical Systems," *Appl. Optics*, **5,** 1812 (1966).
7. J. W. T. Walsh, *Photometry*, Third Edition. Constable, London, 1958. (Also a reprint by Dover Publications, Inc., 180 Varick Street, New York, N.Y. 10014.)

8

Lens Aberrations

Lens "Imperfections"

The relationships that were either postulated or developed for ideal optical systems in Chapter 6 were based on small angle approximations and on the assumption that the refractive index is constant with variations of wavelength. The small angle approximations, which lead to gross inaccuracies in actual optical systems having finite apertures and finite fields of view, neglect all terms after the first in the series expansions of the sine and cosine:

$$\begin{aligned} \sin \alpha &= \alpha - \alpha^3/3! + \alpha^5/5! - \cdots, \\ \cos \alpha &= 1 - \alpha^2/2! + \alpha^4/4! - \cdots. \end{aligned} \tag{8-1}$$

A typical inaccuracy is illustrated in Fig. 8.1, which shows in cross section a simple positive lens having spherical surfaces. When a few paraxial rays, such as ray *a*, and a few marginal rays, such as ray *b*, are traced exactly through the lens by using the formulas developed at the end of Chapter 6, we find two image points. Paraxial rays image at *c*, and marginal rays at *d*. Rays passing through the lens in zones with radii less than that of the marginal zone will be imaged at axial points between *c* and *d*. (A zone is an annulus of the lens or entrance pupil concentric with the axis.) Figure 8.2 is a plot of focal distance as a function of zone radius for a particular lens.

In Fig. 8.3 a ray of blue light, $\lambda \approx 450$ nm, and a ray of red light, $\lambda \approx 650$ nm, enter at the same point and make the same angle of incidence with the first lens surface; but since the indices of refraction at the two wavelengths are different, the rays will, according to Snell's law, take different directions upon refraction. In the example shown, blue light focuses nearer the lens than red light.

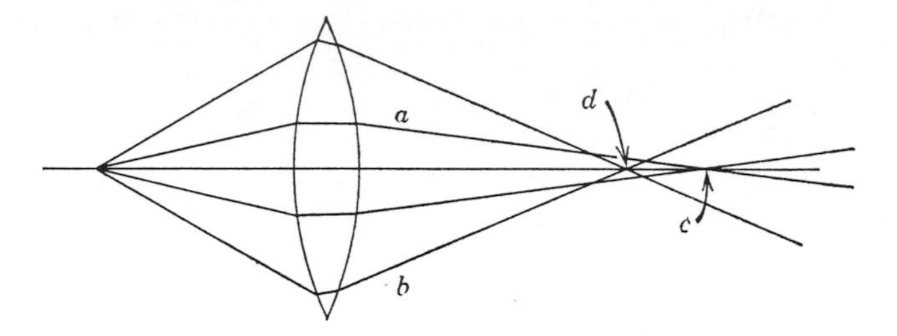

Figure 8.1 Spherical aberration of a simple positive lens.

How shall we define "image point," now that we find that rays originating from an object point do not pass through a common geometrical point in the image? As an additional complication, we show in Chapter 5 that the radiant energy will spread out about an ideal, geometrical image point because of diffraction. The simple definition of image point given in Chapter 6 obviously needs to be liberalized. A more practical question is, "Can anything be done to concentrate all of the energy originating from an object 'point' within a reasonable distance of a defined image point?"

Kinds of Aberrations

This chapter discusses the characteristics of seven conventional classes of image defects, called aberrations. These *aberrations*—which occur when the laws of reflection and refraction are applied to mathematically correct

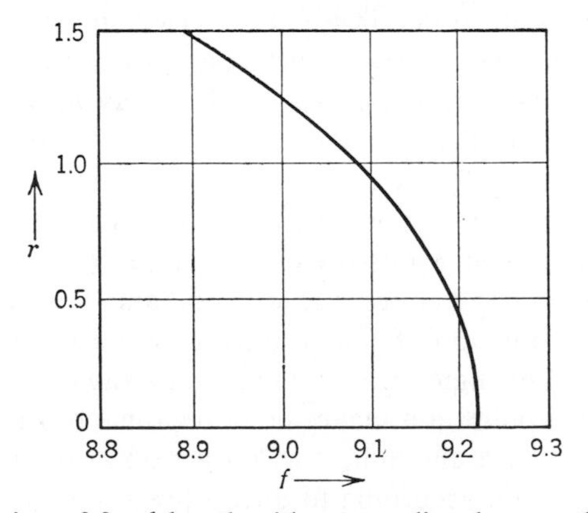

Figure 8.2 Variation of focal length with zone radius due to spherical aberration in a positive lens. (From Jenkins and White, *Fundamentals of Optics*, Second Edition. McGraw-Hill Book Company, Inc., New York, 1950.)

surfaces and which are not a consequence of material inhomogeneity or fabrication errors—are:

(1) Spherical aberration.
(2) Coma.
(3) Astigmatism.
(4) Curvature of field.
(5) Distortion.
(6) Longitudinal chromatic aberration.
(7) Lateral chromatic aberration.

The first five are monochromatic aberrations and were originally defined by L. Seidel in 1856. The importance of this historical classification declines as the system is more highly corrected because higher order aberrations become significant. The higher orders are not as easily visualized as the primaries, and their particular forms differ according to which of the several systems is used for expressing them. Nevertheless, a designer can more easily improve a design if he can identify the predominant kind of aberration present; and the type as well as amount of aberration remaining as a *residual* is still significant in describing the optical quality of most optical instruments.

Because our purpose is to develop an understanding of optical systems and their limitations rather than to train lens designers, we will discuss only the historical classification and chromatic aberration. In the wave theory to be developed, these aberrations will be the fourth-power approximation. In certain systems of small aperture and small field of view only the primaries need be reduced to acceptable amounts, but in most systems higher order aberrations must be taken into account and balanced against the residuals of primary aberrations. The various procedures used by designers to achieve this balance are beyond the scope of this chapter. The effects of diffraction will be discussed later in a section on aberrations and diffraction.

Correcting Optical Systems

The correction of aberrations has always been accomplished during the design of a system by a sequence of trial parameter values changed to improve image quality; this is even the procedure followed by the computer in "automatic" lens design. Although formulas can be derived for the higher order aberrations [Ref. 1], they are too complicated to be of general use; tracing individual rays, by using computer programs to apply the exact formulas, is still the most common way to design optical systems. The following parameters are typically varied during the correction process: curvatures of surfaces, distances between surfaces, indices of refraction, and the dispersions of lens materials. Dispersion is usually expressed as an Abbe number defined for a given material as

$$\nu_d = (n_D - 1)/(n_F - n_C), \tag{8-2}$$

where n_D is the index of refraction at 589.29 nm, the yellow sodium D line; n_F is the index of 486.07 nm, the green hydrogen F line; and n_C is the index at 656.27 nm, the red hydrogen C line. Surfaces can also be aspherized with considerable adjustment latitude by selecting and varying the coefficients of a power series describing the surface. Except for the relatively narrow field systems, one can probably never find enough "degrees of freedom," no matter how many elements are added to the system, to eliminate all measurable higher order aberrations. As suggested earlier, those remaining are called residuals.

Before we can study the behavior of the radiant energy in an "image," we must, rather arbitrarily, define the image plane. This is conventionally done by constructing the image plane perpendicular to the optic axis through the *Gaussian image point*, which is the point image, established by paraxial rays, of an axial object point. The reference image point of an off-axis object point is established on the image plane by tracing the particular ray that passes through the center of the entrance pupil (and, therefore, also through the center of the exit pupil).

Ray Aberrations

In the ray theory of geometrical optics, deviations of rays from the reference image point are the *ray aberrations*. For example, in Fig. 8.1, the distance *cd* is a measure of the spherical aberration. An aberration that is similar in direction and sign to that of a simple positive lens is usually negative and said to be *undercorrected;* so the distance *cd* in Fig. 8.1 is the amount of undercorrected, longitudinal marginal spherical aberration. Figure 8.3 shows another kind of aberration where the distance between the red and blue image points is a measure of the longitudinal chromatic aberration.

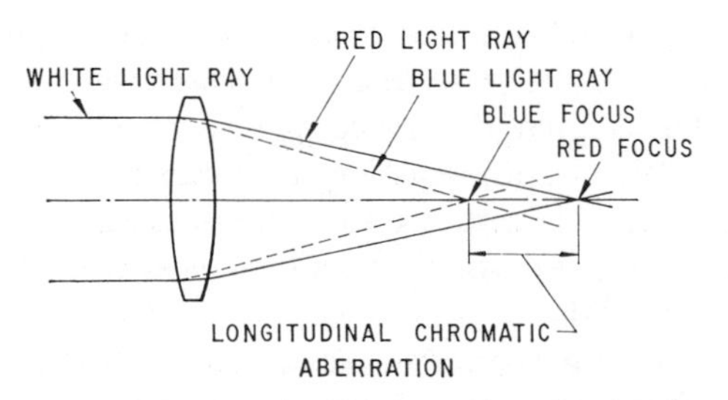

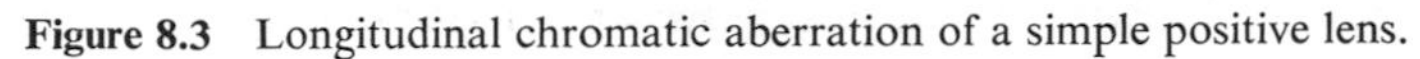

Figure 8.3 Longitudinal chromatic aberration of a simple positive lens.

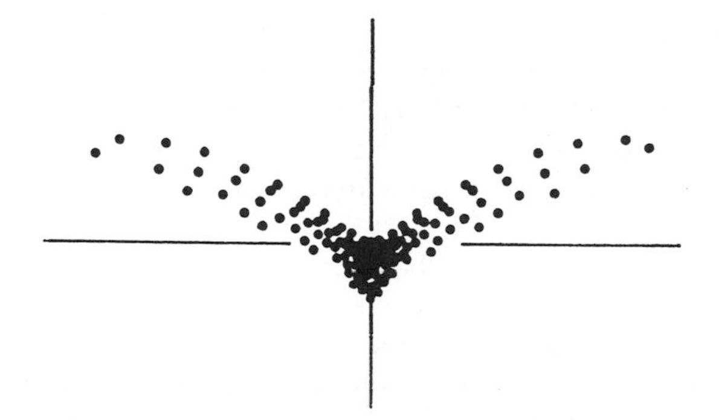

Figure 8.4 Spot diagram of rays incident on the image plane indicating coma and vignetting [Ref. 2].

When a system of rays originating at a single object point is constructed so that the rays are uniformly distributed over the entire entrance pupil, the plot of their consequent intersections with the image plane is called a *spot diagram*. An example of such a diagram is shown in Fig. 8.4 [Ref. 2]. An experienced designer looking at this particular diagram would probably recognize considerable vignetting and coma in the optical system. The size of a spot diagram shows the designer the extent of the energy distribution; the shape is a clue to the type of aberration. When the spot diagram has been reduced to a size comparable to that of the central fringe in a diffraction pattern, ray theory ceases to be as useful as wave theory [Ref. 3].

Wave Aberrations

The discussion of rays and wavefronts in Chapter 6 shows that when all rays of a family converge to a point, the associated wavefront is spherical. From the spreading of rays in Fig. 8.1, for instance, it is evident that the wavefronts emerging from the exit pupil (not shown) cannot be spherical. This simple observation can be extended to a procedure, appropriate for a computer, that will give the shapes of the actual wavefronts. Starting with an on-axis object point, one can find the optical distance D traveled by a ray along the axis from the object to the center of the exit pupil as follows:

$$D = \int n(s)\, ds, \tag{8-3}$$

where $n(s)$ is the index of refraction at the element ds of geometrical path. Then other ray paths can be followed from the same object point and the respective points determined where the optical distances reach the length D. The locus of all such points is the wavefront as it passes through the exit pupil. In Fig. 8.5 a reference sphere is centered on the reference image point P and is of such a radius that it intersects the optic axis where the axis passes through the exit pupil. An "actual" wavefront is also shown coincident with the reference sphere on the axis. Along an actual ray path such as QP' the distance $W = QQ'$ between the actual and reference wavefronts is called *wave aberration* and is positive if Q and P' are on opposite sides of Q'. Positive spherical wave aberration occurs with overcorrected spherical ray aberration. From the example given, it is seen that wave aberration is a measure of the retardation or advancement of the wave with respect to an ideal wave. As the waves arrive in the neighborhood of P, they are out of phase in the image plane by $2\pi W/\lambda$. We show in this and the next chapters that the wave aberration function can describe how the actual flux density distribution in the image plane differs from the distribution of an ideal system. Later the distribution about the reference image point, which is the center of the reference sphere, is discussed. Generally, a judicious choice of reference sphere can be made so that W is everywhere small and can be expressed as a power series of convenient parameters. The ultimate aim is to write the series for the general case by using the minimum number of terms and by using only the required number of parameters. A series that approximates the example described and illustrated in Fig. 8.5 and involves only three parameters is developed in the next section.

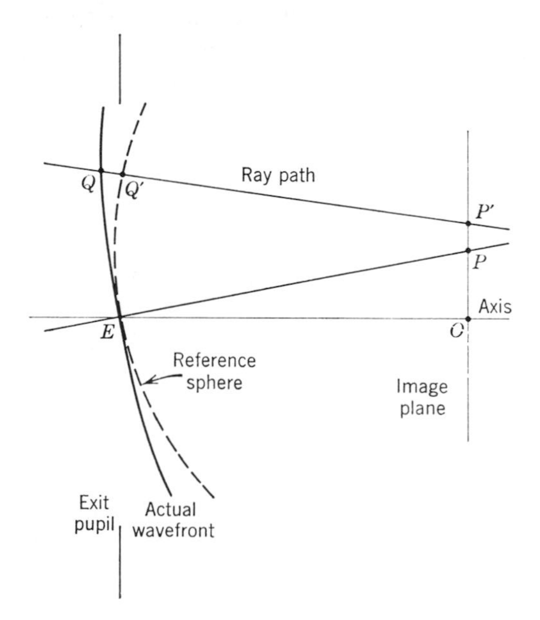

Figure 8.5 Geometry used for defining the wave aberration function.

To facilitate definition of the variables and later discussion of astigmatism, two planes for off-axis points are defined. One is the *tangential plane*, which is the meridional plane containing the object point and the reference image point. The other is the *sagittal plane*, which is perpendicular to the tangential plane and contains the ray from object point to the center of the entrance pupil. Since this ray is refracted at each surface, there is a different sagittal plane for each medium.

The Wave Aberration Function

In the analysis of wavefronts, each from a given object point, the wave aberration varies from ray to ray, which means that W is a function of position in the exit pupil. It is convenient to use the radial distance ρ and an angle φ to specify ray position; φ is measured from the intersection of the tangential plane and the exit pupil plane. Wave aberration is also a function of object position in the field of view; but since there is rotational symmetry, the position can be specified by a single variable in the tangential plane. We use the symbol r for this radial object distance, which can also be used to represent the corresponding distance off-axis of the reference image point since the ratio of the two distances is the magnification β. The parameters ρ and r as used here are normalized distances; the distances from the center to the edge of the exit pupil and from the axis to the edge of the field of view are each taken as unity, so

$$W = W(\rho, r, \varphi). \tag{8-4}$$

(In a more precise analysis, ρ and φ would be polar coordinates in a plane near the exit pupil and symmetrical with the reference sphere; that is, the plane would be perpendicular to both the tangential and sagittal planes.)

When $r = 0$, the aberration function must be symmetrical about the optic axis, and terms containing φ must have zero coefficients. The function must be an even function, $W(\rho) = W(-\rho)$, and so can be represented by terms containing only even powers of ρ. Equation (6-8) and the discussion accompanying it show that terms containing ρ^2 belong to a quadratic phase factor representing a spherical surface. A reference sphere can be chosen such that the coefficient of the term containing ρ^2 only is zero. Since the aberration function is to represent deviations from a reference sphere, only terms containing fourth and higher powers of ρ are, therefore, allowed in terms not containing r and φ. The term containing ($\rho r \cos \varphi$) can be eliminated by a shift of the reference image point in the image plane [Ref. 4, p. 342]. From symmetry, W must be an even function also of φ. Terms that do not contain ρ would represent a constant phase increment over the wavefront in the exit pupil and, therefore, can be omitted in the series.

By suitable choice of the reference sphere, it is found, then, that W may be written in powers of ρ^2, r^2, $\rho r \cos\varphi$ [Ref. 4, p. 345] as follows:

$$W = ({}_0C_{40}\rho^4 + {}_1C_{31}r\rho^3\cos\varphi + {}_2C_{22}r^2\rho^2\cos^2\varphi + {}_2C_{20}r^2\rho^2 + {}_3C_{11}r^3\rho\cos\varphi) + {}_0C_{60}\rho^6 + \cdots. \quad (8\text{-}5)$$

For each term in the parentheses, the sum of powers of ρ and r is four. These represent the primary monochromatic aberrations classified according to Seidel. When these terms are expressed as lateral ray aberrations in the image plane, they are referred to as "third order theory" because they include the second term in the approximation of Equations (8-1) [Ref. 5]. The subscripts of the symbolic C coefficients represent the powers of r, ρ, and $\cos\varphi$, respectively, that appear in the term.

Spherical Aberration

In Eq. (8-5), when $r = 0$ or when the coefficients of terms containing r or $\cos\varphi$ are zero, the expression is purely spherical aberration, and the aberration function is of the form

$$W_s = {}_0C_{40}\rho^4 + {}_0C_{60}\rho^6 + {}_0C_{80}\rho^8 + \cdots. \quad (8\text{-}6)$$

The terms on the right represent primary, secondary, and tertiary spherical aberration, respectively. The three dimensional curve of Fig. 8.6 shows not the actual wavefront shape but the shape of the function W_s for primary spherical aberration. Although an on-axis point has been used to demonstrate spherical aberration, this defect also occurs in general for off-axis as well as on-axis object points.

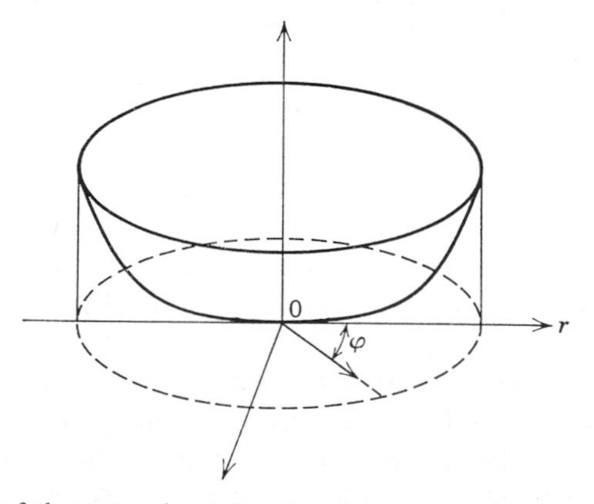

Figure 8.6 A plot of the wave aberration function representing primary spherical wave aberration. (Reproduced from M. Born and E. Wolf, *Principles of Optics*, p. 212. Pergamon Press, Oxford, 1965.)

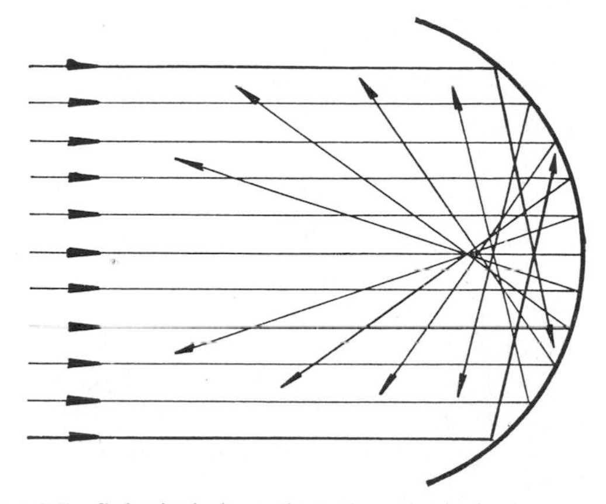

Figure 8.7 Spherical aberration of a spherical mirror.

Figure 8.7 is similar to Fig. 8.1 but shows the spherical aberration of a mirror. It is seen in both figures that when spherical aberration is present, no real focus is possible. In terms of rays, a circular patch of light appears on a screen normal to the axis placed anywhere near the paraxial or marginal focus. There is a position where the size of the patch is a minimum, called the *circle of least confusion;* but we show in the next chapter that this is not the location of best image quality. The light patch may consist of a bright ring with a fainter center or a small bright nucleus with a rather tenuous halo. Then again, it may be a central rather sharp disk and a faint concentric fringe. The image of a distributed object must be formed, of course, by the superposition of the light patches produced individually by the many object points; thus outlines in the image are softened, and details of object structure smaller than the light patches are not visible.

The phrase "superposition of light patches" raises the question of how they will add. With a high degree of coherence among the various light patches, amplitudes of the waves would have to be superposed to form an interference pattern, which could be studied by applying appropriate diffraction formulas. However, here we assume complete incoherence and assume that the density of rays in a small area of the spot diagram is proportional to radiant flux density in the corresponding small area in a light patch. This is true if the spot diagram is appreciably larger than the diffraction pattern central disk (see the later discussion under "Aberrations and Diffraction"); then patches superpose by adding densities.

One way to eliminate spherical aberration is to aspherize one or more of the surfaces. Another way is to balance the undercorrection of a positive lens with the overcorrection of a negative lens. In simple systems, spherical aberration can be adequately minimized by a suitable proportioning of the

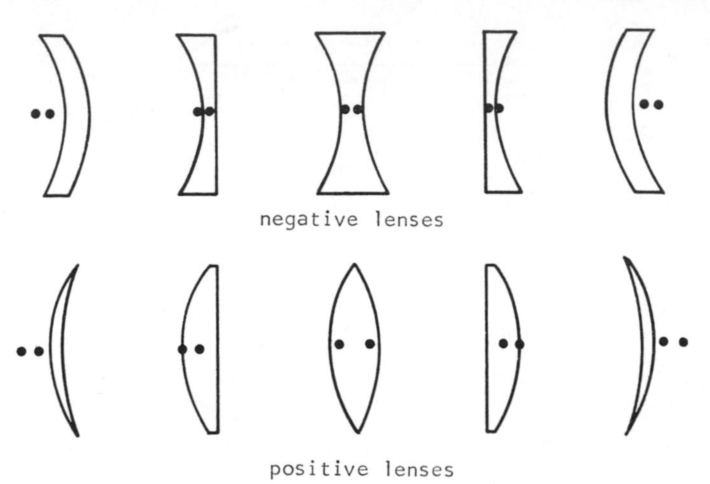

Figure 8.8 Several lens shapes with spherical surfaces showing positions of principal points.

aberration among the several surfaces. It can be minimized in a simple lens, but not completely eliminated, by "bending" the lens, that is by choosing a proper "shape" [Ref. 4, Chapter XVI]. Figure 8.8 shows various shapes of lenses all having spherical surfaces; the plano–convex shape, second from left in the bottom row with the convex side facing the incident rays, is close to the positive meniscus shape that minimizes spherical aberration. A condition often encountered in corrected lenses is shown in Fig. 8.9, which is a plot of focal length against zone radius of a particular lens. This lens has zonal undercorrection and marginal overcorrection; there are instances of the reverse condition, but these are unusual.

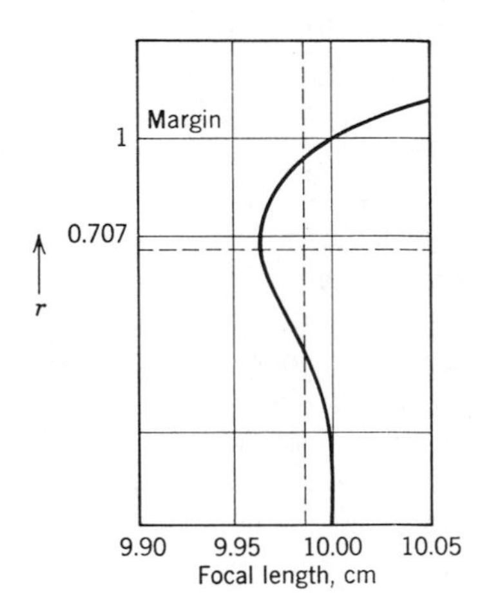

Figure 8.9 Undercorrected zonal and overcorrected marginal spherical aberration for a particular corrected doublet. (From Jenkins and White, *Fundamentals of Optics*, Second Edition. McGraw-Hill Book Company, Inc., New York, 1950.)

As suggested above, the shape and combination of elements for a lens are carefully chosen so that zonal aberration has a reasonable value when compared with the marginal aberration. The choices are made for given angles of incidence and subsequent refraction at the first lens surface. If the object is moved relative to the lens, these angles are changed, and the balance between zonal and marginal aberrations is altered. For this reason, lenses should be used with object distances within the range for which the lens was designed.

Many well corrected lenses have the primary and higher orders of spherical aberration balanced for a specific maximum entrance pupil radius. In cameras and other optical systems, using an aperture larger than this design maximum carries a penalty of decreased fine detail in the image because of high marginal spherical aberration.

Coma

In the general aberration expression of Eq. (8-5),

$$W_c = {}_1C_{31} r \rho^3 \cos \varphi \tag{8-8}$$

represents primary coma. Because of the third power of ρ, it is the first aberration to become noticeable as the object point moves away from the axis. Coma is especially severe for reflecting paraboloids. In a laboratory test of the 200-in. Hale telescope at Mt. Palomar, coma becomes evident only 1 mm off-axis, a field angle of only 13 sec [Ref. 6]. Figure 8.10 shows the distortion of the wavefront in coma. Terms in the power series expression for W containing higher, odd powers of $\cos \varphi$ represent higher order comatic aberration.

In the sagittal plane $\varphi = \pi/2$ or $3\pi/2$, and $W_c = 0$; thus, as far as the coma expression is concerned, the wavefront coincides with the reference sphere. In the tangential plane, $\varphi = 0$ or π, and W_c is proportional to the

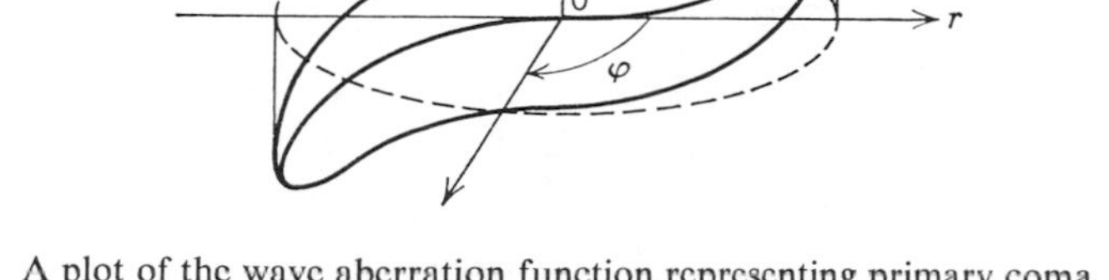

Figure 8.10 A plot of the wave aberration function representing primary coma. (Reproduced from M. Born and E. Wolf, *Principles of Optics*, p. 212. Pergamon Press, Oxford, 1965.)

cube of ρ. If r and ρ are fixed, coma goes through a cycle of values as φ varies from 0 to π and through the same cycle in reverse order when the variation of φ continues from π to 2π. The structure of the comet-shaped image can be visualized by using ray optics. Rays from a point object passing through a particular zone of the lens are imaged in a circle rather than as a point. The center of the circle is further from the axis than the reference image point. These image circles are larger and are centered further from the axis as one moves toward the outer zones of the lens (larger ρ). This is illustrated in Fig. 8.11. Rays passing through a zone shown as a narrow ring in Fig. 8.11a form the circle image shown in Fig. 8.11b (image size is exaggerated) with rays in the tangential section, numbered $+1$ and -1, meeting at 1 on the image circle. The other numbered rays meet in pairs so that the image circle is a double circle. Although $W_c = 0$ in the sagittal section, the wavefront is tilted with respect to the reference sphere so that rays (numbered $+3$ and -3) will not intersect at the reference image point. When the image circles formed by all zones of the lens are superposed, they form the comet-shaped pattern characteristic of coma and illustrated in Fig. 8.11c. The overall image pattern becomes larger as r increases. For crossed lines in the object, with the intersection point off-axis, the precise location of the point of intersection can hardly be determined in the image because the coma pattern is assymmetrical; however, such a point can generally be located in patterns containing spherical aberration.

Coma, as well as spherical aberration, can be reduced by bending the lens; in a simple lens, primary coma can be reduced to zero near the shape that minimizes primary spherical aberration. Both aberrations are corrected for all zones of the aperture when the Abbe sine condition, Eq. (7-5), is fulfilled. For relatively large apertures, higher order coma may become significant although primary coma has been eliminated, but if the sine condition is satisfied for all zones of the wide aperture, all coma must be absent. Thus,

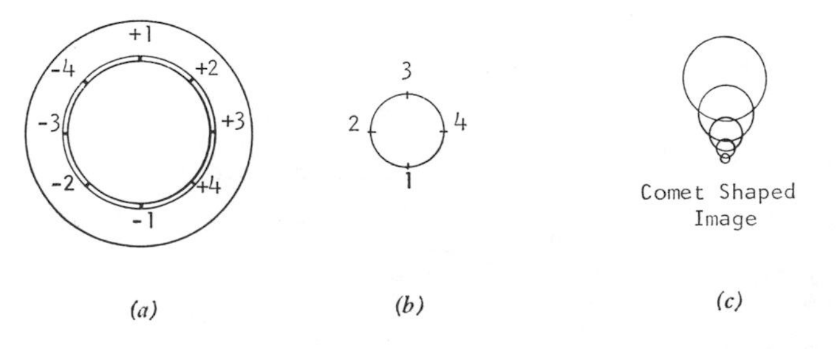

Figure 8.11 Formation of comet-shaped image of an object point. (From Jenkins and White, *Fundamentals of Optics*, Second Edition. McGraw-Hill Book Company, Inc., New York, 1950.)

the satisfaction of the sine condition gives a more highly corrected system than the elimination of primary coma [Ref. 4, p. 352]. Coma is minimized if the optical system is made symmetrical about the aperture stop and if the lateral magnification is unity. In the absence of spherical aberration, coma does not depend on the aperture position; but if spherical aberration is present, a stop position can be found for which coma vanishes, given complete freedom of movement. This procedure applies to the primary aberration only and, of course, is of limited usefulness because the range of practical stop positions is limited.

Astigmatism

In Eq. (8-5), primary astigmatism is represented by the expression,

$$W_a = {}_2C_{22}r^2\rho^2 \cos^2 \varphi; \tag{8-9}$$

higher order astigmatism is represented by terms containing higher, even powers of $\cos \varphi$. Because of the $\cos^2 \varphi$ factor, the wavefront curvature is greater in the sagittal plane where $\cos \varphi = 0$ than in the tangential plane where $\cos \varphi = 1$. Figure 8.12 shows the wavefront distortion. Because of the variation in curvature, a fan of rays in the tangential plane will meet at a point that determines a tangential focus, and a fan of rays in the sagittal plane will meet at another point that determines a sagittal focus. When all rays from an object point are considered, they are found to pass through a line perpendicular to the tangential plane at the tangential focus and through another line perpendicular to the sagittal plane at the sagittal focus. At points between the two foci, the family of rays has an elliptical cross section. The major axis of the ellipse changes from the direction of one focal line to the direction of the other as the ellipse becomes a circle somewhere between the two line images. Fig. 8.13 shows the two focal lines.

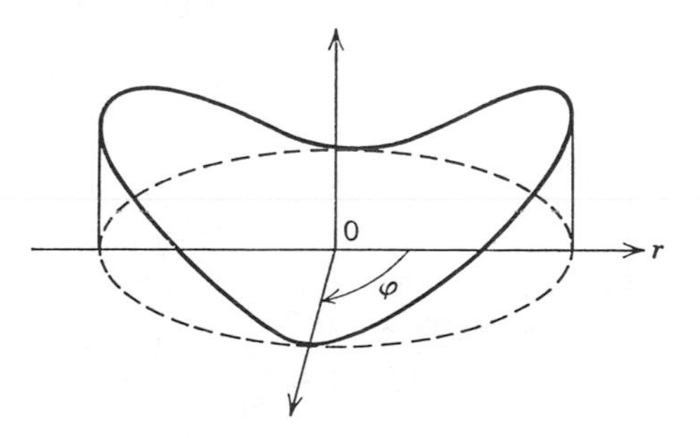

Figure 8.12 A plot of the wave aberration function representing primary astigmatism. (Reproduced from M. Born and E. Wolf, *Principles of Optics*, p. 212. Pergamon Press, Oxford, 1965.)

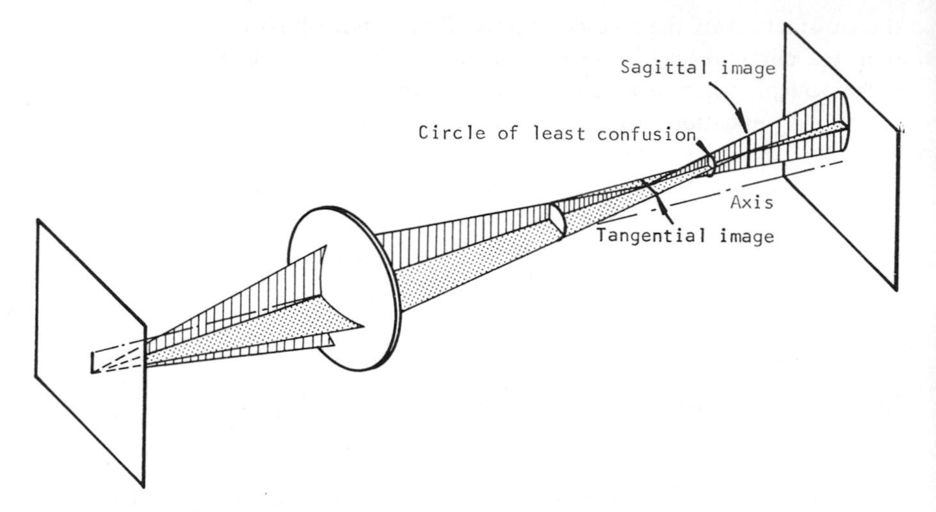

Figure 8.13 Formation of tangential and sagittal images in astigmatism.

Since astigmatism is proportional to r^2, the curvatures of both the tangential section and the sagittal section of the wavefront are greater as the object point moves further off-axis. Thus both foci move toward the lens as the object point moves further off-axis. If the object is a plane perpendicular to the optic axis, a tangential image surface, consisting of the tangential foci of all points on the object surface, and a sagittal image surface are formed. Both surfaces are generally paraboloidal in shape and intersect the optic axis at the paraxial image point. The two surfaces are illustrated in Fig. 8.14. If the two surfaces could be made to coincide, astigmatism would be corrected.

In the presence of astigmatism, circles on the object concentric with the optic axis appear to be sharply imaged in the tangential image surface, and

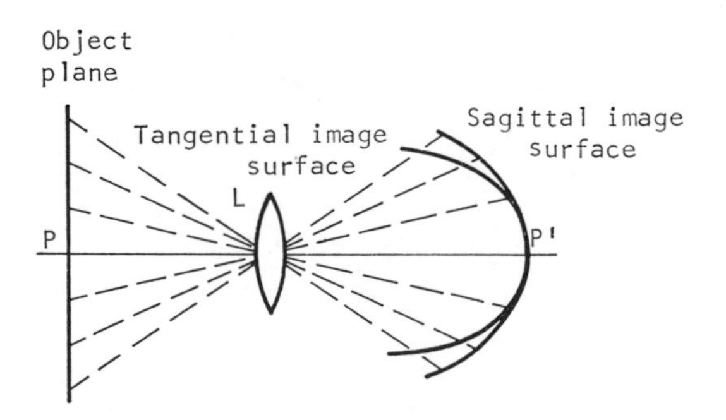

Figure 8.14 Astigmatic images of a plane object.

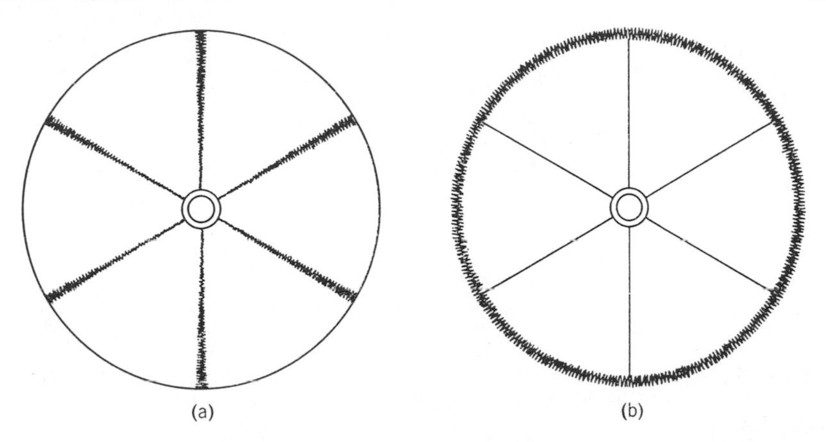

Figure 8.15 Astigmatic images of a spoked wheel: (a) tangential image and (b) sagittal image.

radial lines appear to be sharply imaged in the sagittal image surface. Thus, the image of an object in the general form of a spoked wheel would appear as shown in Fig. 8.15. Astigmatism is a function of both the lens shape and the position of the aperture stop.

Curvature of Field

Every optical system, without specific correction, has curvature of the image "plane." Field curvature, called Petzval curvature, is represented by

$$W_f = {}_2C_{20}r^2\rho^2 \tag{8-10}$$

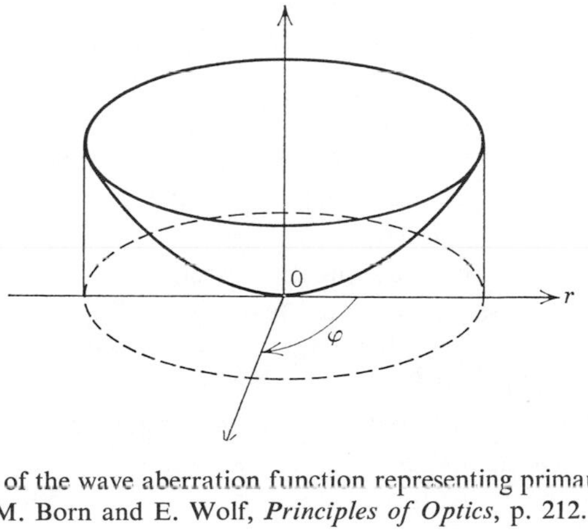

Figure 8.16 A plot of the wave aberration function representing primary field curvature. (Reproduced from M. Born and E. Wolf, *Principles of Optics*, p. 212. Pergamon Press, Oxford, 1965.)

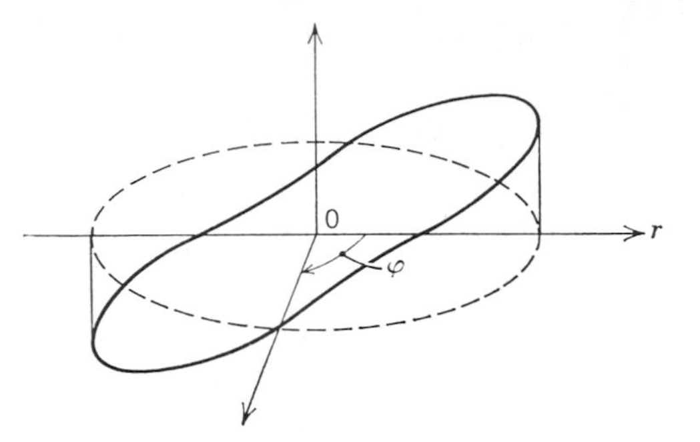

Figure 8.17 A plot of the wave aberration function representing primary distortion. (Reproduced from M. Born and E. Wolf, *Principles of Optics*, p. 212. Pergamon Press, Oxford, 1965.)

in Eq. (8-5) and nearly always accompanies astigmatism. When there is no astigmatism, the sagittal and tangential surfaces coincide and lie on the Petzval surface. The wave distortion is shown in Fig. 8.16. Field curvature is especially objectionable in cameras, enlargers, and projectors because the film plane and the projection screen are typically flat. Correcting for curvature is referred to as "field flattening." Positive lenses introduce inward curvature of the Petzval surface (undercorrection) and negative lenses introduce outward curvature (over correction).

Distortion

If all other aberrations have been eliminated and the term,

$$W_D = {}_3C_{11}r^3\rho \cos \varphi, \tag{8-11}$$

of Eq. (8-5) remains, the image is well defined; but images of points are displaced from the paraxial image by an amount proportional to the cube of the object distance off-axis. The wave distortion is shown in Fig. 8.17. This type of aberration is called distortion; if it is present, the image of any straight line in the object plane that meets the axis will itself be a straight line; but the image of any other straight line will be curved. When the coefficient is positive, the magnification increases in the outer parts of the field to give pincushion distortion illustrated in Fig. 8.18. When the coefficient is negative, it is barrel distortion caused by a decrease in magnification toward the edge of the field.

An aperture stop located between a positive lens and the image increases pincushion distortion, and a stop on the side of the lens remote from the image increases barrel distortion. Distortion is minimized in systems that are symmetrical about an aperture stop.

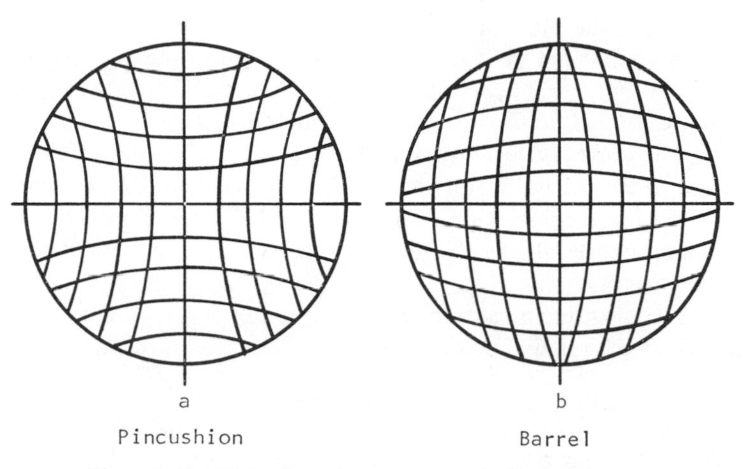

Figure 8.18 Distortion of a square mesh object pattern.

The first three of the aberrations—spherical aberration, coma, and astigmatism—are responsible for a lack of sharpness in the image. The last two—field curvature and distortion—cause geometrical distortion of the image.

Chromatic Aberration

Since the focal length of a lens depends on the refractive index of the lens material and since the index varies with wavelength, the focal length is different for different colors. Consequently, a different image will be found in a different position and of a different size, for each color. The increment of image distance due to refractive index change with wavelength is called longitudinal chromatic aberration, and the increment in image size is called lateral chromatic aberration; these are illustrated as *a* and *b*, respectively, for paraxial optics in Fig. 8.19.

An "achromatic doublet" lens, consisting of two thin lenses (as in Fig. 6.8), can be designed to have the same image position for any two colors.

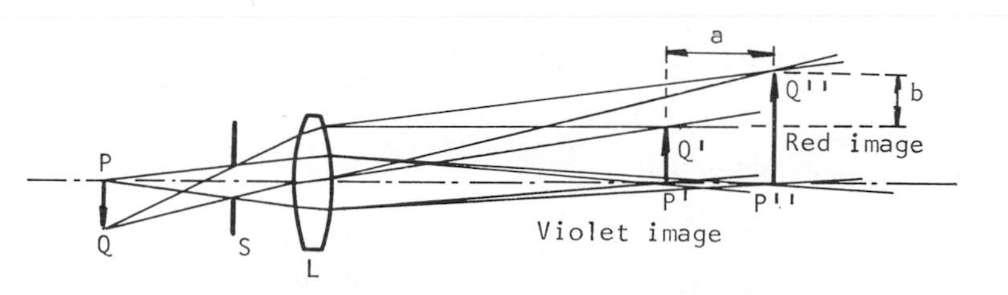

Figure 8.19 Longitudinal and lateral chromatic aberrations.

In such a lens, the materials of the two elements are chosen with different dispersions so that the chromatic aberration of one compensates for that of the other. A lens system corrected in this way is not fully achromatic in the sense that the focal point is the same for all wavelengths. Generally, when the system is corrected by bringing the red and violet images to the same point, the green image is still displaced from the red–violet image. Correcting for a third color, such as green, is called correcting for "secondary color" and requires the use of a lens material having a third dispersion characteristic.

Starting with n_F and n_C as defined following Eq. (8-2), we define a quantity $p(x,y)$, called the relative partial dispersion, for the wavelengths x and y as

$$p(x,y) = (n_x - n_y)/(n_F - n_C). \tag{8-12}$$

The following linear relationship is a good approximation for most glasses, the so-called "normal glasses":

$$p(x,y) \approx a_{xy} + b_{xy}\nu_d, \tag{8-13}$$

where a_{xy} and b_{xy} are constants for the particular glass and wavelengths. The correction for secondary color, that is, for more than two wavelengths, requires at least one glass that does not show normal dispersion.

Unfortunately, correcting longitudinal chromatic aberration by bringing the images to a common point along the axis may also put the principal planes at different positions for different colors. Since magnifications would be correspondingly different, lateral chromatic aberration could still be present.

After the two chromatic aberrations have been corrected for paraxial rays, the corrections must be checked with marginal rays. Each chromatic aberration is made up of a series of terms consisting of a constant (the paraxial term) and terms that are functions of ρ and r. The monochromatic aberrations in general depend on the indices of refraction of the lens materials and, therefore, are functions of wavelength. Thus, when spherical aberration, for instance, has been corrected at one wavelength, it is not necessarily corrected at all wavelengths, and we say that we have chromatic variations of the monochromatic aberrations.

Aberrations and Diffraction

The actual distribution of radiant flux density in the image of a point object formed with some aberration or combination of aberrations present differs, because of diffraction, from the predictions of geometrical optics and spot

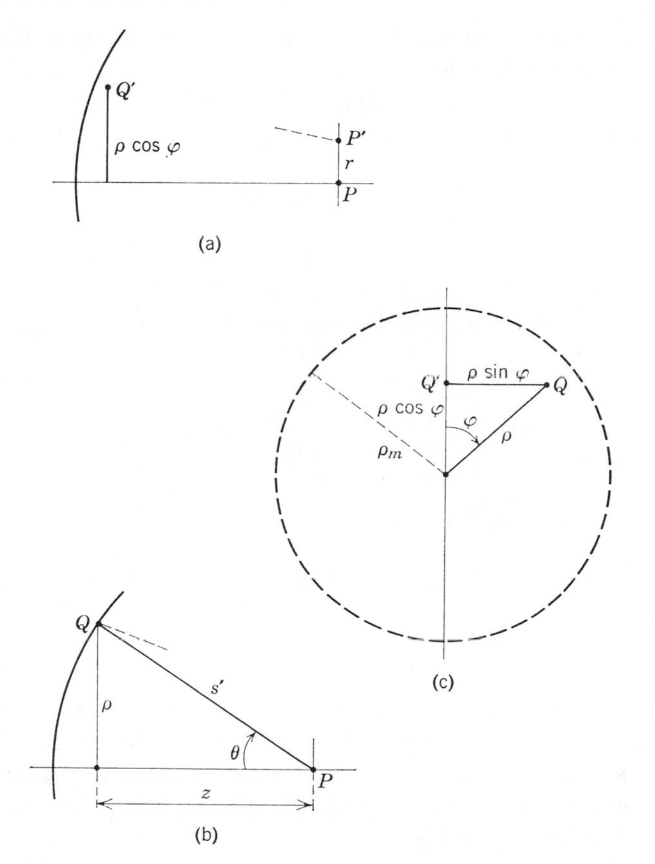

Figure 8.20 Geometry and notation for finding the diffraction formula for a spherical wave.

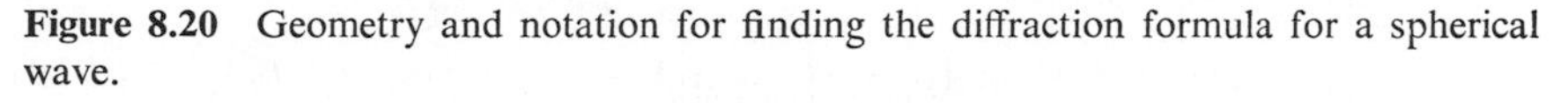

diagrams. In the absence of aberrations, as indicated earlier, a wavefront coincides with the reference sphere in the exit pupil. The diffraction pattern produced is similar to those discussed in Chapter 5 for square and circular apertures where no lenses are involved. The pattern produced by the spherical wavefront emerging from a circular exit pupil of an optical system is derived in the following paragraphs.

The geometry and notation are shown in Fig. 8.20. A meridional plane containing the point of observation P' near the reference image point P is shown in Fig. 8.20a, and another meridional plane containing a point Q on the reference sphere is shown in Fig. 8.20b. A third plane, shown in Fig. 8.20c, is perpendicular to the axis and contains the point Q. The position of the latter plane depends on the choice of Q. A line segment, not shown in the figures, that will be used in our computations is the segment joining Q and P' having a length R'.

An equation, similar to Eq. (5-4), for the disturbance at P' produced by a Huygens wavelet originating from an element of area $d\sigma$ at Q is

$$d[\hat{U}(P')] = (\hat{C}/R') \exp(-ikR')\, d\sigma, \tag{8-14}$$

where $k = 2\pi/\lambda$. Since s' is constant, the phase at P, rather than the phase on the reference sphere, can be taken as zero and $(s' - R')$ used instead of R' in Eq. (8-14). An expression for this phase is as follows:

$$s'^2 = \rho^2 + z^2, \tag{8-15}$$

$$\begin{aligned} R'^2 &= (\rho \cos \varphi - r)^2 + \rho^2 \sin^2 \varphi + z^2 \\ &= \rho^2 - 2r\rho \cos \varphi + r^2 + z^2, \end{aligned} \tag{8-16}$$

$$s'^2 - R'^2 = 2r\rho \cos \varphi - r^2. \tag{8-17}$$

Then, by factoring out $(s' + R')$ and approximating it by $2s'$ on the right-hand side,

$$s' - R' = (r\rho \cos \varphi)/s' - r^2/2s'. \tag{8-18}$$

The second term on the right is a constant phase term; with good approximation, s' can be substituted for R' in the denominator of Eq. (8-14), and these constant factors can be included in the constant $\hat{C}$. The element of area is $d\sigma = \rho\, d\varphi\, d\rho$. Then an equation expressing the total disturbance at P' produced by all wavelets is

$$\hat{U}(P') = 2\pi\hat{C}\int_0^{\rho_m} \rho\, d\rho \left\{\frac{1}{2\pi}\int_{-\pi}^{+\pi} \exp[-ik(\rho r/s') \cos \varphi]\, d\varphi\right\}. \tag{8-19}$$

Except for a change in notation, this is the same as Eq. (5-62), if $u = k\rho_m r/s'$, and Eq. (5-67) follows as before:

$$\hat{U}(P') = \pi\rho_m{}^2\hat{C}[2J_1(u)]/u. \tag{8-20}$$

Also, ρ_m is the radius of the exit pupil. The flux density in the image plane is

$$W_e(p') = W_{e0}[2J_1(u)]^2/u^2. \tag{8-21}$$

where W_{e0} incorporates the appropriate constants and represents the maximum flux density in the image of a point object formed by a diffraction-limited optical system. Because the first zero of the Bessel function, $J_1(u)$, occurs when the argument u is equal to $1.22\pi = 3.832$, the central disk has a radius determined by

$$u_0 = 1.22\pi = 2\pi r_0 \rho_m / \lambda s', \tag{8-22}$$

which, when solved for r_0, is

$$r_0 = 0.61\lambda s'/\rho_m = 0.61\lambda/\sin\theta. \tag{8-23}$$

If the image is being formed in the focal plane so that $s' \approx f$, and if $\sin\theta \approx \rho_m/f$,

$$r_0 = 1.22\lambda F^* \tag{8-24}$$

where F^*, the f/number, is $f/2\rho_m$. About 84% of the radiant power of the pattern is contained in the central disk, Fig. 5.14, and the maximum flux densities in the first three bright rings are, respectively, 0.0175, 0.0042, and 0.0016 times that at the center of the disk [Ref. 7].

As shown above, in the absence of aberrations the flux density is a maximum at the Gaussian image point; with aberrations, this in general no longer holds true. However, if the aberrations are kept small, the maximum flux density will still occur near the Gaussian image point. In terms of the aberration function $W(\rho,r,\cos\varphi)$, Eq. (8-5), whether or not the aberrations are "small" depends both on the maximum of W and the nature of the function (this is discussed further in the next chapter). The popular Rayleigh criterion calls the combined aberrations small if W is nowhere greater than $\lambda/4$. If the aberrations are kept small, aberrations cause a redistribution of the radiant power in the diffraction pattern, but the sizes of the central disk and rings remain essentially unchanged. Figure 8.21 shows graphically how the flux density is redistributed in the image of a point, for example, when the optical system is defocused $\frac{1}{4}$ wavelength at the edge of the pupil. Quantitatively, 17% of the radiant power has been transferred out of the central disk into the rings without significantly changing the size of the disk or the distribution of power in it. If the optical system is a telescope, this defocusing would hardly affect its resolving power. Furthermore, if the object were a double star (two closely spaced point sources) with the eye as a detector, the visual image of the double star would appear almost unaltered by the defocusing, and the relative positions of its components could usually be measured as well as before. However, if the image were explored by some other kind of detector that gave quantitative results, the shift of 17% of power from disk to rings might be a significant change.

If the useful response from a detector is proportional to the ratio of the radiant power in the central disk to the power in the rest of the pattern, the ideal response is 5.2, that is, 84/(100 − 84). In the presence of $\frac{1}{4}$ wavelength of simple wave distortion, the response reduces to $\frac{67}{33}$, or 2. Any obscuration, such as a secondary mirror or a detector array, reduces the signal still further;

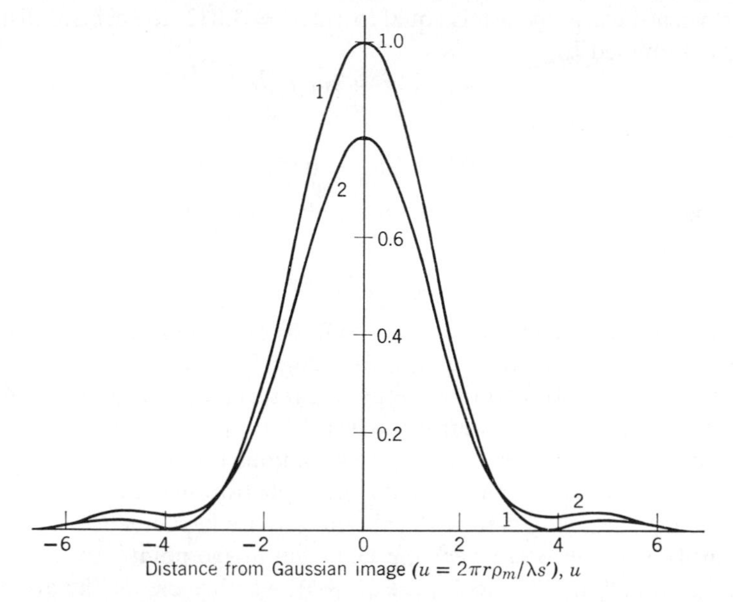

Figure 8.21 The diffraction pattern and the effect of defocusing. (W at the edge of the pupil is $\lambda/4$. Curve 2 is for a defocused image producing a Strehl definition ratio of 0.805.)

moreover, the struts holding the obscuring element alter the shape of the diffraction pattern. For example, in a diffraction-limited system, it turns out that a concentric circular obscuration having a diameter 0.2 that of the aperture reduces the response from 5.2 to 3.9 [Ref. 8]. Four struts, each having a width 0.01 of the aperture diameter, reduce the response further to 2.9. The four struts also change the central disk to a star-like diamond shape.

Another criterion, called the Strehl definition, is based on the ratio of flux density at the center of the diffraction pattern of an aberrated image to that of an aberration-free image. If this ratio is at least 80%, the aberrations are considered small. This criterion turns out to be very close to the Rayleigh criterion for the particular example illustrated in Fig. 8.21.

Large aberrations cause the image from a point object to lose its identity as a diffraction pattern and to become characteristic of the predominant aberration present. A combination of aberrations, misalignment, defocusing, and diffraction can very quickly produce a complicated distribution of radiant power in a pattern much larger than the central disk of a diffraction pattern. Since aberrations must be relatively large before the spot diagram obtained by ray tracing shows significant characteristics, a more sensitive method for discovering defects in well corrected systems is obviously needed. One suitable method, based on the optical transfer function, is discussed in the next chapter.

REFERENCES

1. J. Focke, "Higher Order Aberration Theory," in *Progress in Optics*. Edited by E. Wolf. North-Holland Publishing Company, Amsterdam, 1965.
2. *Optical Design*, MIL-HDBK-141, Oct., 1962. Defense Supply Agency, Washington, D.C.
3. L. C. Martin, *The Theory of the Microscope*, pp. 113–130. American Elsevier Publishing Company, Inc., New York, 1966.
4. R. S. Longhurst, *Geometrical and Physical Optics*. John Wiley and Sons, New York, 1967.
5. W. T. Welford, "Optical Calculations and Optical Instruments, an Introduction," in *Handbuch der Physik*, Vol. 29, p. 15. Springer-Verlag, Berlin, 1967.
6. I. S. Bowen, "Optical Problems at Palomar Observatory," *J. Opt. Soc. Am.*, **42,** 795 (1962).
7. M. Born and E. Wolf, *Principles of Optics*, Third Edition, p. 398. Pergamon Press, Oxford, 1965.
8. E. Everhart and J. W. Kantorski, "Diffraction Patterns Product by Obstructions in Reflecting Telescopes of Modest Size," *Astron. J.*, **64,** 455 (1959).

9

The Evaluation of Optical Systems

An optical system is designed, in general, to make information about the object accessible. Usually this information is expected to appear as an image, but important exceptions immediately come to mind. A hologram, for instance, contains virtually all the information about the object, but it certainly is not an image. However, in our discussion of optical system evaluation, we confine ourselves to systems that do form images, and a judgment of optical quality will be based on how accurately the image represents the object. To enable this judgment, the image must either be readily and repeatedly observed in real time or be accurately stored as on photographic film. Any departures in geometry, radiance, or color (but not scale) from the object are considered image defects.

We would like to define, quantitatively and objectively, a unique generally acceptable measure of image quality. But the purposes of optical systems are so diverse and the consequent image requirements are so different that no single numerical quality index has ever been generally accepted. Instead, each optical system is evaluated in terms of the specific purpose for which its image is formed. Geometrical fidelity, for instance, is essential in a portrait camera. Softness of tone is often desirable, so too much fine detail (too much information!) may be unacceptable. On the other hand, an aerial camera must record small details and small luminance differences; so, according to information theory, maximum information is desired. Geometrical fidelity, though desirable, is not the primary requirement because much information can still be extracted from a distorted photograph by measurement and analysis. When a choice has to be made, the aerial camera recording the most recoverable information is the best. This criterion is carried to the extreme in photographic recording of unusual events, like taking pictures of the moon's surface during a moon walk. Here the most information possible is recorded on the film with secondary consideration given to the resulting analytical difficulties that must be overcome later when the information is recovered.

A Transfer Parameter

The fundamental concept of an image-forming optical system includes a simple correspondence between points (x,y,z) in object space and points (x',y',z') in image space; a linear transformation, accomplished by the optical system, converts from object space to image space. The optical axis is the z-axis; in object space $z = 0$ at the object plane, and in image space $z' = 0$ at the Gaussian image plane. The lateral magnification β determines the x, x' and y, y' relationships:

$$x' = \beta x, \qquad y' = \beta y. \tag{9-1}$$

Also, in the fundamental optical concept we assume that each point in the object plane is weighted by a generalized quantity $Q(x,y)$—representing, for example, radiance or luminance $L(x,y)$ or radiant flux density $W(x,y)$, which in turn may be related to another parameter such as temperature, emissivity, voltage, or current. Each point in image space is then weighted by a quantity $\Gamma(x',y')$, which is not necessarily the same quantity that weights object space but is rather related to the object-space weighting quantity by a transfer parameter $\mathscr{R}$. Thus,

$$\Gamma(x',y') = \mathscr{R}Q(x,y). \tag{9-2}$$

A classical rule, discussed in Chapter 7, for image-forming systems is that the "brightness" of the image equals the "brightness" of the object. Here "brightness" means luminance, which is expressed in lumens per steradian per unit projected area of the source. If Q and Γ represent luminance, the transfer parameter $\mathscr{R}$ is unity according to this rule. To account for losses in the system, in the simple analysis Q and Γ are related by the overall system transmittance τ; thus $\mathscr{R}$ is τ. But, in general, the transfer parameter is more complicated than this.

In practice, the function $\Gamma(x',y')$ of Eq. (9-2) can only approximately represent $Q(x,y)$. In a given system there is a minimum area, $\Delta a' = \Delta x' \, \Delta y'$, from which one can learn significantly about $Q(x,y)$ in the corresponding area $\Delta a = \Delta x \, \Delta y$. Here spatial resolution of the system is involved, and Δa is said to be the resolution limit. Spatial resolution can also be expressed in angular terms, $\Delta x/s$, provided $s \gg \Delta x$, where s is object distance. A corresponding angle occurs in image space and is given by $\Delta x'/s'$, where s' is image distance or, in a system focused at infinity, the focal distance.

Sometimes the image function is made one-dimensional, $\Gamma(x')$. Because this represents a physically realizable function and is, therefore, mathematically well behaved, it has a Fourier transform:

$$\hat{\mathscr{H}}(\kappa') = \int_{-\infty}^{+\infty} \Gamma(x') \exp(-2\pi i \kappa' x')\, dx'. \tag{9-3}$$

By analogy, it is noted that the pair of variables κ' and x' correspond to f and t in the frequency and time domains, respectively. The function $\hat{\mathscr{H}}(\kappa')$ can, therefore, be treated as a complex frequency function, and κ' as a spatial frequency in cycles per unit distance. All information contained in $\Gamma(x')$ is also contained in $\hat{\mathscr{H}}(\kappa')$. Besides this image-space frequency function, there is also an object-space frequency function $\hat{\mathscr{H}}_0(\kappa)$. Since $\Gamma(x')$ only approximately represents $Q(x)$, $\hat{\mathscr{H}}(\kappa')$ can only approximately represent $\hat{\mathscr{H}}_0(\kappa)$. The relative amplitudes of the spatial frequencies in the object function are altered by the intervening optical system to produce the image function. A comparison of the object and image frequency functions, by revealing the kinds and amounts of the alterations, measures the frequency response characteristics of the optical system. A transfer parameter in terms of the frequency response is defined later in this chapter.

Resolution

Ideally, resolution would be defined so that it is generally accepted as the parameter associated with the ability of an optical system to produce images of good quality. There should also be a procedure, implicit in the definition, for measuring an optical system so that the system can be ranked on a scale of values. An approach to this ideal relates resolution, the ability of the optical system to resolve fine detail, with the *point-spread function*.

The spot diagram and the diffraction pattern mentioned in Chapter 8 are examples of point-spread functions, where the radiant energy from a point object is spread out, or distributed, in a small region about the Gaussian image point in the image plane. When two point objects are well separated, there is no apparent overlapping of their spread functions. However, if the two objects are moved toward each other, at some separation the two spread functions obviously overlap. At still closer separations one asks the classical question of resolution: "How near can two point objects be placed before their superposed spread functions cannot be distinguished from the spread function of a single point object?" Before this question can be answered,

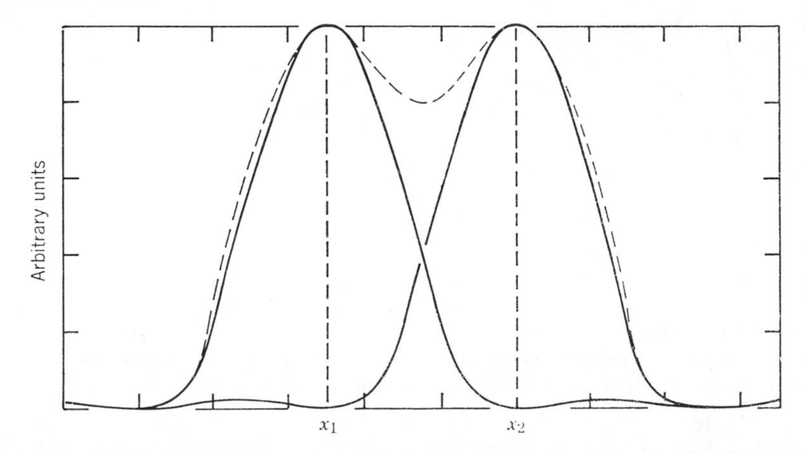

Figure 9.1 Radiant flux density profile (broken line) along a line joining the centers of two overlapping diffraction patterns separated by the Rayleigh criterion distance.

more has to be known about the relative radiant power from the two sources, the nature of the spread function, the degree of coherence between the two sources, and the "noise" characteristics of the detecting system.

In the present discussion we do not undertake the study of noise [Refs. 1–3]. Complete incoherence is assumed, which is justified in considering, for instance, the radiant emissions from different stars. Then the radiant flux densities will add. It is further assumed that the spread function is the familiar diffraction formula, Eq. (8-21), of a circular aperture and that equal radiant power comes from each of two sources. This allows the classical Rayleigh criterion to be stated as follows: When the first minimum (dark ring) of the spread function of one coincides with the peak of the spread function of the other, the two point objects are just resolvable. This is illustrated in Fig. 9.1, which shows the total radiant flux density (broken line in the figure) along a line joining the two Gaussian image points. Thus, according to Eq. (8-23), the theoretical resolution limit is

$$r_0 = 0.61\lambda s'/\rho_m, \tag{9-4}$$

where λ is wavelength, s' is distance from exit pupil to the Gaussian image point, and ρ_m is the radius of the exit pupil. Other criteria have also been used. Conrady [Ref. 4] proposed changing the coefficient of Eq. (9-4) from 0.61 to 0.50 because at the latter separation the eye can still see two regions of maximum flux density. Sparrow [Ref. 5] suggested letting the two sources approach each other until the dip in the density profile disappears, that is, when the second derivative of the profile function becomes zero midway between the two Gaussian image points.

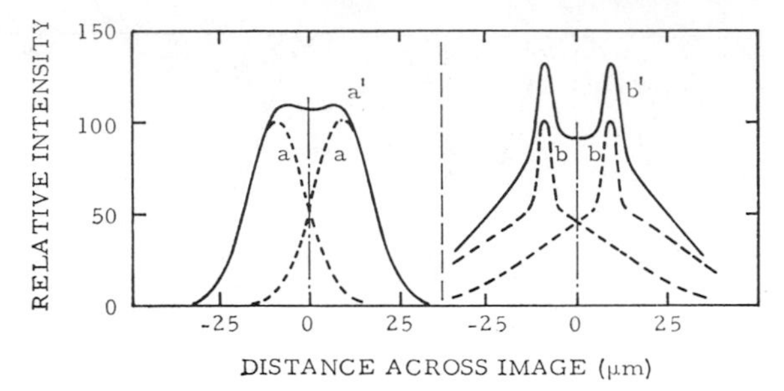

Figure 9.2 Two pairs of identical hypothetical spread functions and the flux density profile formed by the superposition of the two separated functions in each pair. (Separation of centers is near the Sparrow limit for curve a'. Incoherence is assumed) [Ref. 6].

Figure 9.2 shows two hypothetical spread functions which illustrate the dependence of resolution on spread function shape. The flux density profile for a superposed pair of each is also shown. The separation of centers for both examples is chosen so that the density in the minimum of the composite curve b' approximately equals the peak in each individual curve b. The spread functions b are clearly resolved at this separation distance while the spread functions a are not [Ref. 6].

From the foregoing brief discussion, it is evident that the Rayleigh, Conrady, Sparrow, and other similarly defined criteria have serious limitations. Generally, the power from two sources will not be equal; coherence will increase when one considers closer and closer points in a distributed object; noise will be a disturbing factor; and generally insufficient information about the spread function is known. Later it is shown that if the spread function is known, the frequency response function can be found, which tells more than the resolution of the system.

Common methods of testing do not measure resolution in terms of any of the criteria discussed but use resolution test charts, usually called bar charts, or special equipment for measuring the modulation transfer function—both with specified test procedures [Ref. 7]. The test methods are discussed later in this chapter.

Acutance

A mathematically defined function called *acutance* has been found to correlate well with sharpness of detail in an image [Ref. 8]. It can be readily evaluated on a photograph. Again incoherence is assumed. To define acutance, an object field is divided into a bright and a dark region and separated by a straight line. Passing from the dark to the bright region results in a step function in radiant flux density. Under these conditions, the

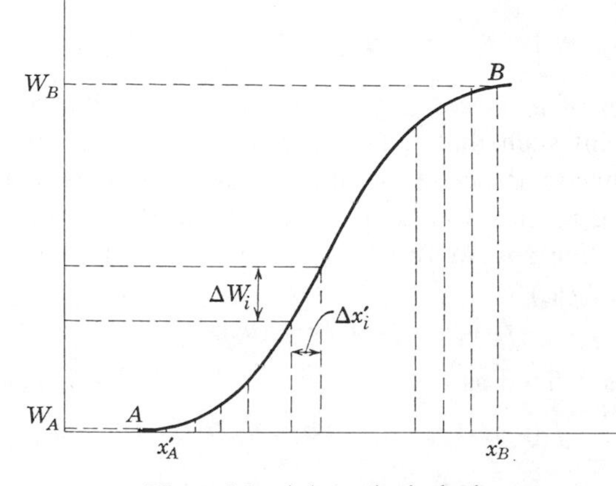

Figure 9.3 A hypothetical edge trace.

image does not have a sharp edge at the boundary; there is actually a gradual change of flux density from the image of the dark region to the image of the bright region. The flux density as a function of displacement along a line normal to the boundary is called an edge trace. One hypothetical edge trace is shown in Fig. 9.3. Acutance can be defined in terms of gradients along this function. But first the edge trace must be developed mathematically.

A *line-spread function*, which is the image of a line object, can be derived from the point-spread function. If the point-spread function is asymmetrical about the Gaussian image point, for example because of coma or astigmatism, then the line-spread function will depend upon the orientation of the point-spread function. However, a symmetrical point-spread function gives a unique line-spread function. For example, the spread function $f_1(u)$ for a diffraction-limited optical system having a circular aperture, is, from Eq. (8-21),

$$f_1(u) = W_{c0}[2J_1(u)]^2/u^2. \tag{9-5}$$

Here W_{c0} is the maximum radiant flux density in the pattern, u is $2\pi r \rho_m/\lambda s'$, and r is radial distance from the Gaussian image point. Let us make the following substitutions in Eq. (9-5): $u = (u_x^2 + u_y^2)^{1/2}$, $u_x = 2\pi\rho_m r_x/\lambda s'$, $u_y = 2\pi\rho_m r_y/\lambda s'$. In these expressions, $r_x = r\cos\psi$ and $r_y = r\sin\psi$ so that r_x and r_y are the x' and y' components respectively of r. Then

$$f_1(u_x,u_y) = W_{c0}\{2J_1[(u_x^2 + u_y^2)^{1/2}]\}^2/(u_x^2 + u_y^2). \tag{9-6}$$

This expression can be integrated with respect to one of the coordinates to obtain the line-spread function. Integration with respect to u_y gives the

Struve function $S_1(2u_x)$ of the first order [Refs. 9–11]:

$$f_2(u_x) = \int_{-a}^{+a} f_1(u_x,u_y)\, du_y = 4W_{c0}S_1(2u_x)/u_x^{\,2}. \tag{9-7}$$

Here a is the value of $u_y = 2\pi\rho_m r_{ya}/\lambda s'$, where r_{ya} is a distance from the Gaussian image point such that $f_1(u_x,u_y)$ is negligible at greater distances. The Struve functions are associated with the Bessel functions, and tabulated Struve functions are usually found in tables of the Bessel functions. The convolution of the line spread function $f_2(u_x)$ with a step function $f_3(u_x)$ gives the edge trace $f_4(u_x)$,

$$f_4(u_x) = f_2(u_x) * f_3(u_x). \tag{9-8}$$

The step function is defined as

$$\begin{aligned} f_3(u_x) &= 1 \qquad \text{when } u_x \geq 0, \\ &= 0 \qquad \text{when } u_x < 0. \end{aligned} \tag{9-9}$$

The Fourier transform of the line spread function gives a one-dimensional frequency function as shown by Eq. (9-3).

Let the edge trace between two points defining the ends of the trace, A and B in Fig. 9.3, be divided into sections by partitioning the distance axis into n equal increments $\Delta x'$, where f_3 is now expressed as a function of x'. The gradient at each $\Delta x_i'$ is then

$$G(x') = \Delta W_i/\Delta x_i', \tag{9-10}$$

and a mean square of the gradients is

$$\overline{G(x')^2} = \sum_{i=1}^{n} (\Delta W_i/\Delta x_i')^2/n. \tag{9-11}$$

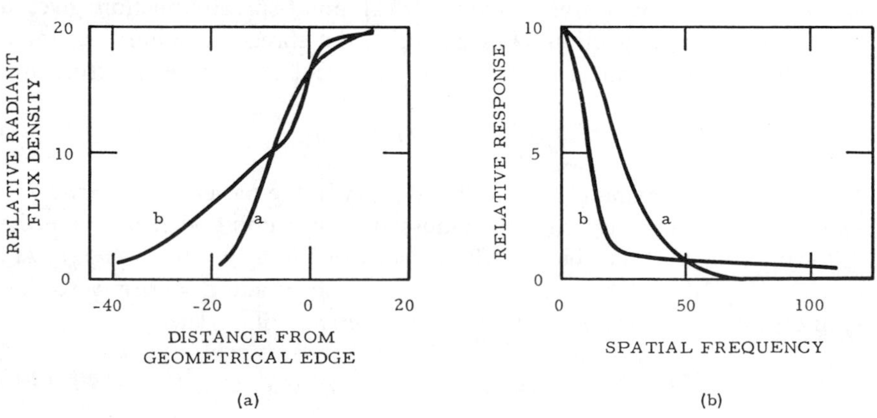

Figure 9.4 Edge trace (a) and spatial frequency function (b) obtained from the spread functions of Fig. 9.2.

The points A and B are chosen by a specified small gradient; for example, points where the gradient reduces to 0.005 watt cm^{-2} cm^{-1} have been found suitable under certain conditions. Acutance, the objective correlate of a subjective sharpness in the image, is defined as

$$\text{Acutance} = \overline{G(x')^2}/(W_B - W_A). \tag{9-12}$$

Acutance is commonly used in evaluating photographs and has been found to be more meaningful than resolution whenever the resolution limit on the photograph is relatively poor when compared with the resolution limit of the eye at a reasonable reading distance. (Acutance does not correlate well with image sharpness whenever the edge trace departs appreciably from the sigmoid shape shown in Fig. 9.3.) Figure 9.4a shows the two edge traces and Fig. 9.4b shows the spatial frequency functions obtained from the respective spread functions of Fig. 9.2.

Resolution Test Charts

Several types of resolution (bar) charts have been designed for testing the ability of an optical system to image the fine detail of an object. One such chart is shown in Fig. 9.5. Usually a chart will have several parallel lines, bars of equal width, in a group with the separation between bars equal to the bar width. The chart contains several groups, all being similar but having different bar widths and spacings. It is desirable that a chart have sharp edges in two directions, requiring bars in two directions, usually mutually perpendicular.

In the image of a bar chart, there will be a smallest group in which the lines are distinctly separate. This group determines the resolution of the

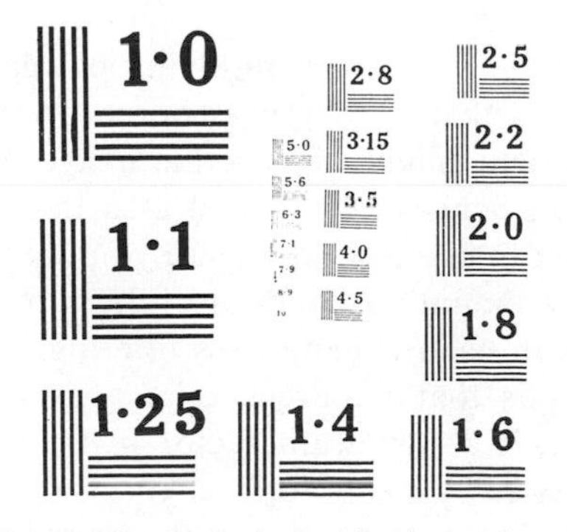

Figure 9.5 A resolution (bar) test chart.

system in lines (line pairs) per millimeter equal to $1/l$, where l is the distance between bar centers in that group, in millimeters. Several tests, with the chart in different positions in the field to test the system for on-axis and off-axis resolution, are usually made with each optical system.

The resolution of an optical system depends somewhat upon how the determinations are made. The image may be recorded, as on photographic film; either the actual image, formed in space (aerial), or the recorded image may be observed with the eye or measured with an instrument such as a microphotometer or a microdensitometer. Methods of measuring resolution, particularly using photographic film, are outlined by Perrin [Ref. 12].

To many users, bar charts are deceptively simple. There is no general agreement among the users nor among the designers of bar charts concerning the most suitable design. The resolving power of an optical system, as determined by using a bar chart, depends on the contrast, which can be defined as

$$C = (L_s - L_b)/(L_s + L_b). \tag{9-13}$$

Here L_s is the luminance of the space between bars and L_b is the luminance of the bars. Thus, charts have been made having low, as well as high, contrast. When the response of an optical system has been determined by using a bar chart, no information is usually obtainable about how the system treats groups having wider bars and spacings. It cannot be taken for granted that all larger groups are imaged perfectly. (This point is discussed further in connection with modulation transfer functions.) However, a bar chart test does provide a single number by which similar lenses can be compared. Also, a lens performing well in a bar chart test, when compared with other lenses, will perform comparatively well in actual use.

Coherent Imaging

An equation like Eq. (9-4), in which resolving power is expressed in terms of wavelength, suggests that a coherent source or the illumination of an object with coherent light might give better image definition than illumination with incoherent light, which has an assortment of wavelengths. To support this idea, we can cite the experience of photographers who usually realize finer detail by mounting band-pass filters on their cameras. But, going to the extreme in narrow band-pass filtering by using coherent light leads to peculiar results that require special analysis. In fact, degree of coherence, in general, is a significant factor in the response of an optical system whether it is tested by resolution test charts or by frequency response.

If an optical system has negligible aberrations and a square aperture, the amplitude of an electromagnetic wave in the image of a point source is

$$\hat{U}_x(P') = \hat{U}_0 \operatorname{sinc} 2\pi k_x a \operatorname{sinc} 2\pi k_y a, \tag{9-14}$$

which is closely related to the similar expression in Equations (5-17) and (8-20)—the latter, however, being for a circular rather than a square aperture. In Eq. (9-14), $k_x = r_x/\lambda s'$ and $k_y = r_y/\lambda s'$.

To study the image of two point sources that are completely coherent with each other, one can superpose their two amplitude diffraction patterns, each of which can be described by an expression like that of Eq. (9-14). If the two Gaussian image points are separated by a distance b, the two spread functions (diffraction patterns) will be centered at $x' = \pm(b/2)$, $y' = 0$; that is, the spread function is shifted by substituting $2\pi a[(b/2) \pm x']/\lambda s'$ for $2\pi k_x a = 2\pi x' a/\lambda s'$ in Eq. (9-14). The amplitude of the combination is

$$\hat{U}_2(P') = \hat{U}_2(x',y') = \hat{U}_0[\operatorname{sinc}(2\pi a y'/\lambda s')][\operatorname{sinc} f_1(x') + \operatorname{sinc} f_2(x')], \tag{9-15}$$

where $f_1(x')$ and $f_2(x')$ represent groups of terms as follows:

$$f_1(x') \equiv \{2\pi a[(b/2) - x']/\lambda s'\}, \tag{9-16}$$

$$f_2(x') \equiv \{2\pi a[(b/2) + x']/\lambda s'\}. \tag{9-17}$$

The radiant flux density is given by

$$W_2(x',y') = \hat{U}_0 \cdot \hat{U}_0^*[\operatorname{sinc}^2(2\pi a y'/\lambda s')][\operatorname{sinc} f_1(x') + \operatorname{sinc} f_2(x')]^2. \tag{9-18}$$

If the two sources were incoherent, the flux density in each image could be found separately and then the densities added to obtain

$$W_2'(x',y') = \hat{U}_0 \cdot \hat{U}_0^*[\operatorname{sinc}^2(2\pi a y'/\lambda s')][\operatorname{sinc}^2 f_1(x') + \operatorname{sinc}^2 f_2(x')], \tag{9-19}$$

a distinctly different result.

In the development above, we chose a square rather than a round aperture for analytical convenience. With the same approach, but with more difficult mathematics, the radiant flux densities in several image configurations have been obtained more generally, including partial coherence [Refs. 13–15]. The results are shown here in several figures.

Figure 9.6 shows that two point sources separated by the Rayleigh criterion distance are resolved with incoherent light but appear as one source with coherent light. In the figure, γ represents degree of coherence. The parameter plotted along the abscissa is the argument, $u = (2\pi a x'/\lambda z)$, of the Bessel function in Equations (8-20) and (8-21).

Figure 9.7 shows two point sources separated far enough to be well resolved when they are either coherent or incoherent, but the two image separations, the distances between image points of maximum flux density, differ by about 15%, the coherent spacing being the greater.

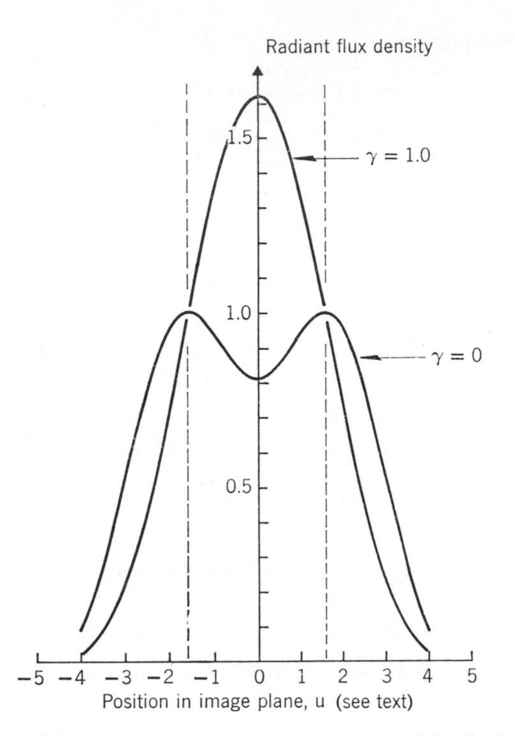

Figure 9.6 Superposed images of two point sources separated by the Rayleigh limit with coherent light, $\gamma = 1.0$, and with incoherent light, $\gamma = 0.0$ [Ref. 13].

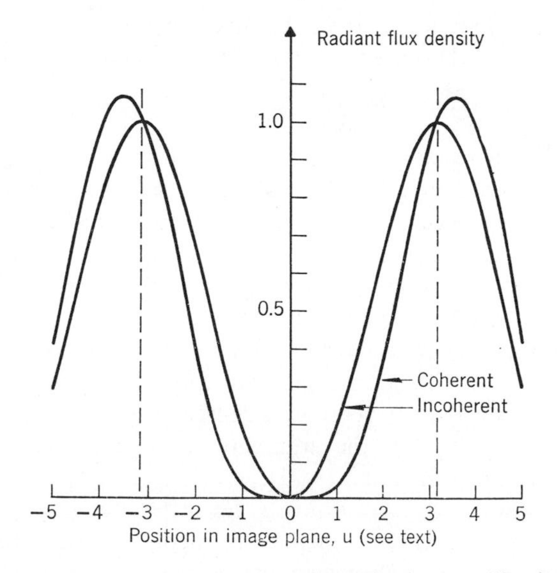

Figure 9.7 Images of two point sources well resolved when illuminated with either coherent or incoherent light. (Broken lines are Gaussian image positions) [Ref. 13].

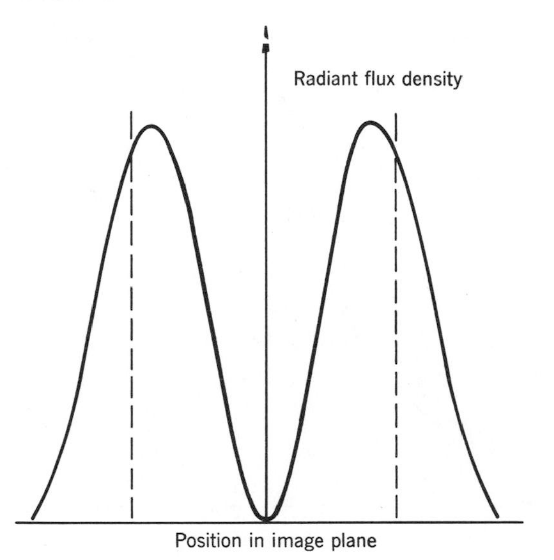

Figure 9.8 Images of the two point sources of Fig. 9.7 illuminated with coherent light but opposite in phase rather than in phase [Ref. 13].

In Figures 9.6 and 9.7, the two coherent point sources were assumed to be in phase with each other. They could in fact have any phase relationship; Fig. 9.8 illustrates what happens when they are opposite in phase. This time the indicated separation is about 15% closer than when the two sources are incoherent.

The image of a coherently illuminated edge is shifted toward the illuminated region, and a "ringing" (closely spaced maxima and minima) appears at the edges of bars and slits. The amount of shift depends on contrast. Figures 9.9 and 9.10 show the flux density distribution near the Gaussian image of a coherently illuminated edge. The numbers on the various curves are values of $n = (\pi \Delta z/\lambda) \sin^2 \varphi'$, which expresses the amount of defocusing [Ref. 14]; Δz is longitudinal displacement of the image plane from the Gaussian image plane, and φ' is the half angle shown in Fig. 7.7. A unit along the abscissa scale is $u = (2\pi/\lambda)x' \sin \varphi'$.

Figure 9.11 is a plot of the ratio R between the indicated and the geometrically correct image separation of two point sources. The various curves correspond to the given degrees of coherence [Ref. 13]. The parameter δ, indicating the actual point separation, is $2\pi ab/\lambda s'$, where $2b$ is separation of the Gaussian image points. Ideally, R should be unity. The curves show that the greater the degree of coherence, the greater the overshoot—reaching as high as 12%. Also, the greater the degree of coherence, the greater the actual separation δ before the ratio approaches unity. The broken lines mark

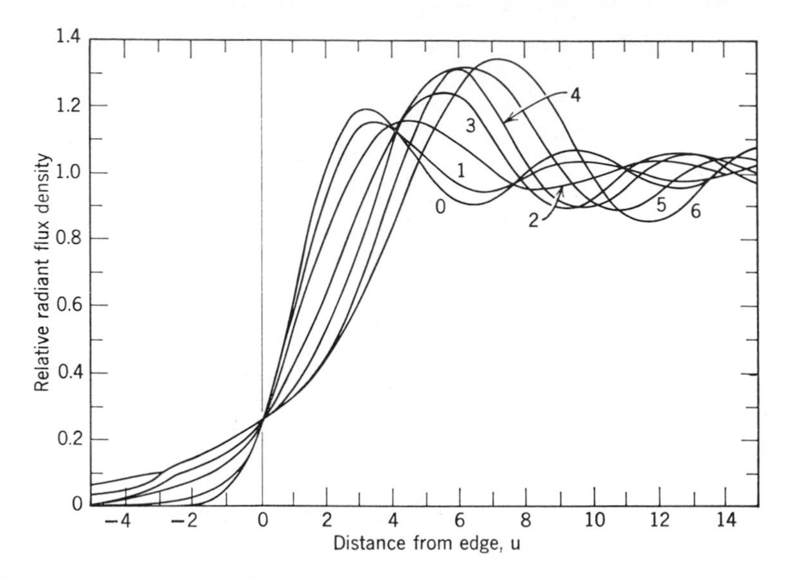

Figure 9.9 Light distribution in the image of a slightly defocused coherently illuminated edge. (The numbers, with increasing value, show the change with increasing defocusing) [Ref. 14].

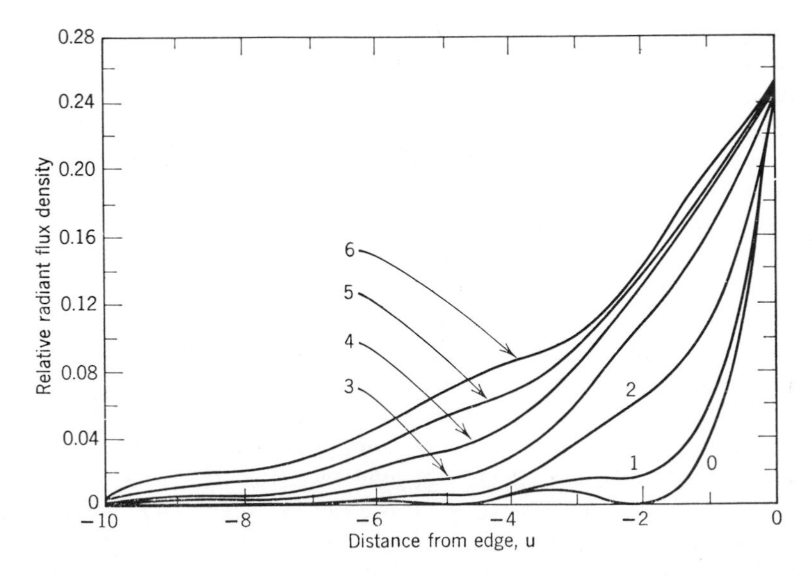

Figure 9.10 Light distribution in the dark side of the image of a coherently illuminated edge [Ref. 14].

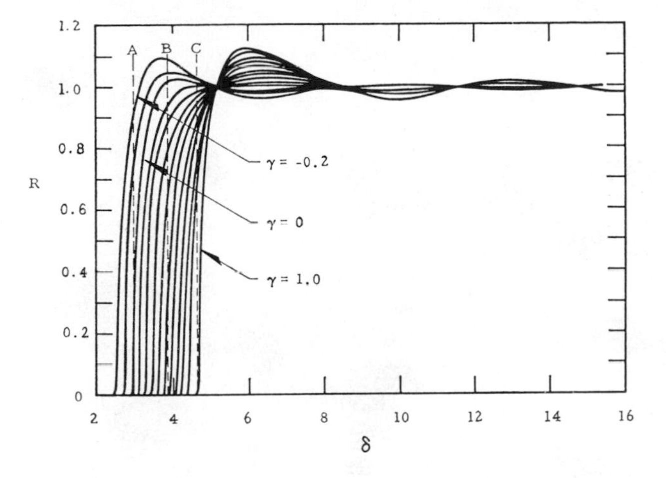

Figure 9.11 Ratio R of the measurable separation between images of two point sources to the geometrical separation δ [Ref. 13].

point separations corresponding to the three criteria: A and C for the Sparrow incoherent and coherent criteria respectively, and B for the Rayleigh criterion.

Figure 9.12 shows photographs of two images of a photographic film which has 16 transparent points casually distributed over its area. The transparent points, when illuminated from behind, act as point sources. Figure 9.12a was made with green light from a laser beam at 514.5 nm; Fig. 9.12b was made with an incandescent source transmitting through a green filter and then through the film. The image formed with laser (coherent) light shows a number of structures (artifacts) formed by interference of the overlapping diffraction patterns; these are not present in the other image. Ernst Abbe predicted the presence of artifacts in microscopy nearly a hundred years ago and had to perform a number of experiments to prove that the artifacts were indeed caused by coherent imaging (coherence occurring spontaneously because objects are very close in microscopy) rather than fine detail in the objects.

An Optical System Always "Sees" Coherent Waves

In our study of a general (incoherent) beam of radiant energy and its representation by the analytic signal $\hat{V}(t)$ in Chapter 1, it was shown that one could think of the real part of $\hat{V}(t)$, written $V^r(t)$, as being synthesized from many component waves [see Equations (1-14)–(1–21)]. It was also

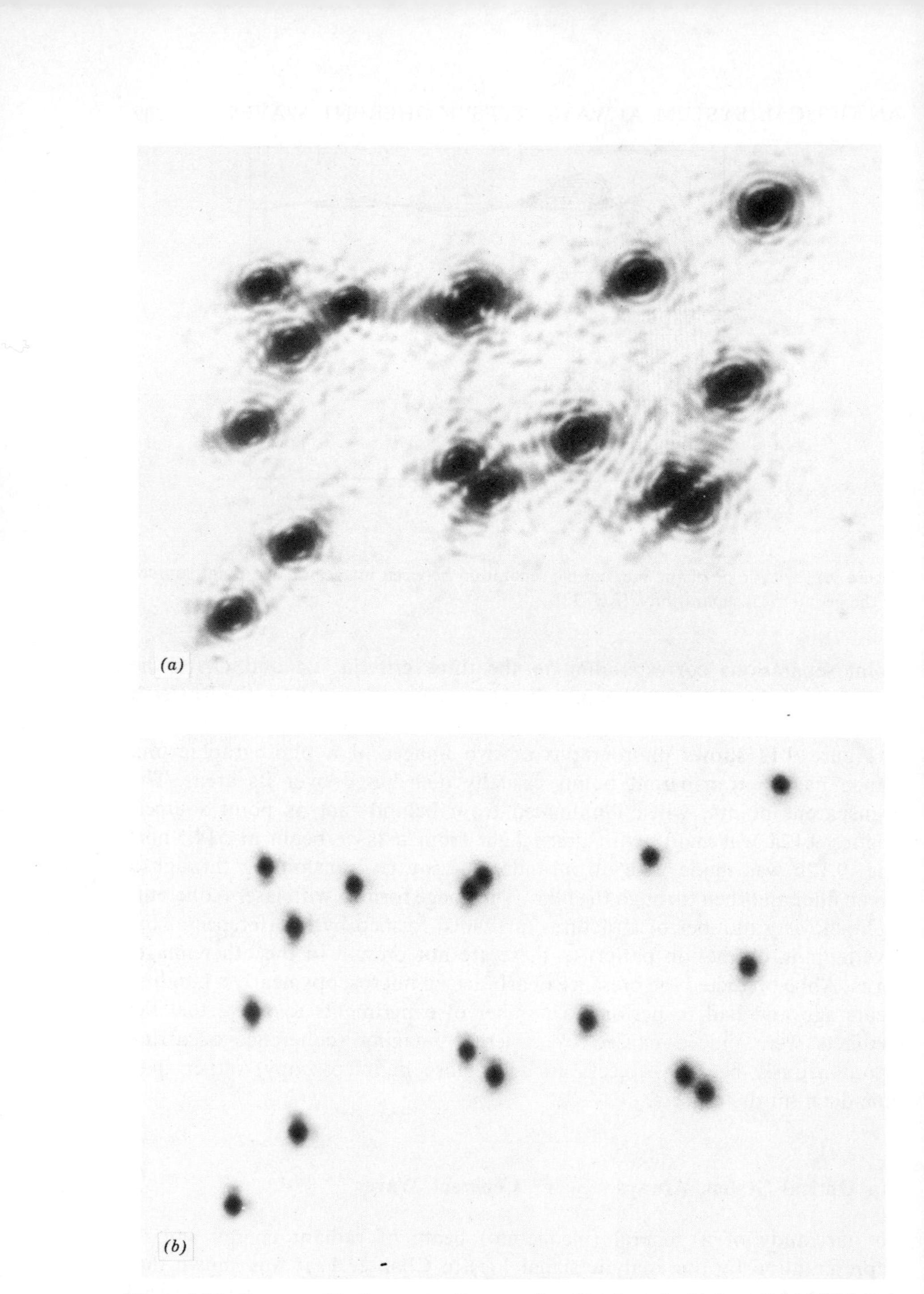

Figure 9.12 Images of "point sources": top, when illuminated with laser light (514.5 nm), and bottom, when illuminated with green filtered white light.

assumed that $V^r(t)$ could be resolved into its frequency components [Eq. (1-23)] by application of the Fourier transform as follows:

$$v(\nu) = \int_{-\infty}^{+\infty} V^r(t) \exp(-2\pi i\nu t)\, dt$$
$$= v^r(\nu) + iv^i(\nu) = |v(\nu)| \exp i\varphi(\nu). \tag{9-20}$$

Here $|v(\nu)|$ and $\varphi(\nu)$ (which are even functions) are expressions for the amplitudes and phases, respectively, of the component waves. The Fourier spectrum of any beam will be composed of monochromatic, pure-frequency components which are assumed to exist mathematically from $t = -\infty$ to $t = +\infty$ but which superpose to give the actual zero beam amplitude outside the time interval of the actual passage of the transient beam through a point in space.

The concept of component waves is useful because the optical system will act on each component wave of a general beam in precisely the same way as it would on a coherent beam of the same frequency. In fact each component is, conceptually, a perfectly coherent beam. We are exercising this concept when we say that the blue light of a white beam is imaged at one point and the red light at another. The independent behavior of the various components is characteristic of "linear" optical materials and is necessary for the application of superposition in wave phenomenon. Where two or more waves cross, they produce a combined effect; but each wave, after it has passed out of the region (for the type of phenomena related to this discussion) is unaffected by its having shared space with the other waves. This is why we can form an image of a wide field of view with waves from diverse parts of the field that must pass simultaneously through a small aperture.

The sometimes expressed idea that there are fundamental differences between incoherent and coherent optical systems is a fiction. Only one additional requirement is imposed on an optical system when it is used with coherent light: Its cosmetic quality must be nearly perfect because every grain of dust and every scratch on the lens surface produce observable diffraction patterns [Ref. 16].

To get some feeling for the transition from coherent to incoherent beams, one can study their relative effects when they are projected through a circular aperture. One specific observation might be the size of the central disk of the resulting diffraction pattern. To make the comparison, an "incoherent" beam can be synthesized by combining three individual, independent beams of slightly different wavelengths, say, 550, 555, and 560 nm, all of equal amplitude. Each beam has a relatively narrow frequency spectrum and a coherence time shorter than the observation time (time constant of the detector or the eye) of the combined beams. Since the optical system acts

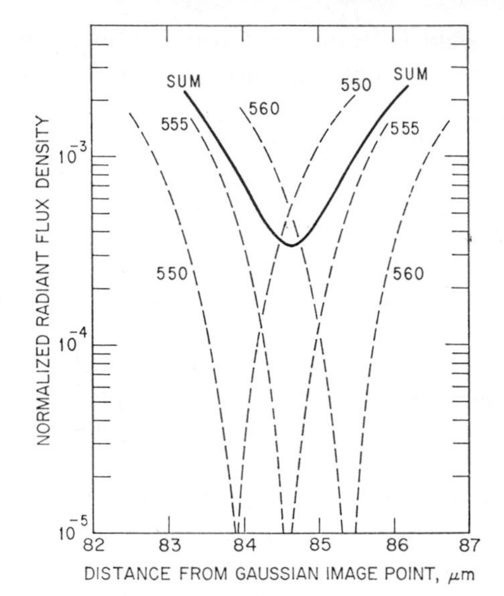

Figure 9.13 Diffraction patterns near the first zero for three incoherent waves at 550, 555, and 560 nm and the sum of their flux densities. (Each curve is normalized to unity at the zero of r.)

on each of these components independently, the diffraction patterns can be found, one by one, and then combined by adding their flux densities. Figure 9.13 shows a magnified portion of the diffraction pattern near the first zero (see Fig. 5.13). The component patterns are indicated by broken lines; the sum represents the whole incoherent beam and is indicated by a solid curve. All curves are normalized to unit flux density at the centers of the patterns. If the detection threshold is set at the minimum of the sum curve, the size of the central disk is hardly different from that of the center wavelength.

Spatial Frequencies

The symbol κ' of Eq. (9-3) has already been referred to as the spatial frequency. Our purpose now is to enlarge this concept. We start with an object that has a luminance variation described by the cosine function,

$$L(x) = a + b \cos(2\pi x/l), \qquad (9\text{-}21)$$

$$0 \leq b \leq a,$$

where a and b are constants in luminance units and l is a constant having the same unit of length as x. Such an object can be realized in a photographic film exposed and developed so that the transmittance varies according to the right-hand member of Eq. (9-21). When illuminated from the back with incoherent light, the film functions as an object with the described luminance characteristic.

Were the film to be illuminated, instead, by coherent light, a complex amplitude transmission factor would apply instead of the transmittance function; and a variable phase, as well as a variable amplitude, might exist over the emerging wavefront.

The object, described by Eq. (9-21), is a striped field, or grating, with bright lines alternating with less bright (or dark) lines, the bright lines gradually fading out to dark lines according to the cosine function. In the following development, we assume complete incoherence. The contrast, as defined in Eq. (9-13), is

$$C = b/a. \tag{9-22}$$

The line spacing is l, which corresponds to the period in time periodic functions; the reciprocal is the frequency κ:

$$\kappa = 1/l. \tag{9-23}$$

The image of this object can be represented by a cosine function [Ref. 17, p. 18] similar to that of the object as follows:

$$L(x') = a' + b' \cos (2\pi\kappa' x' + X). \tag{9-24}$$

The phase X represents a shift of the whole pattern with respect to the optic axis. The units of x' are the units of x multiplied by the magnification β. For example, if β were 2 and κ were 50 waves/mm in object space, then κ' in image space would be 50 waves/2 mm or 25 waves/mm. In the following we will assume unit magnification to avoid the inconvenience of changing frequency. The contrast in image space is b'/a'. The ratio of contrasts is called the frequency response M (a transfer parameter) of the system at the frequency κ.

$$M = (b'/a')/(b/a). \tag{9-25}$$

The frequency response M can have any value between -1 and $+1$; a negative value indicates a reversal of contrast, bright lines of the object appearing as dark lines in the image. A plot of M as a function of spatial frequency is called the modulation transfer function for the optical system.

System Response Functions

To develop the idea of system response function, one can start with an object point source at (x,y) that is represented by the two-dimensional delta function [see Eq. (5-27)],

$$\delta(x,y) = \delta(x)\delta(y). \tag{9-26}$$

The Gaussian image point, of (x,y), is at point (x',y'); so the corresponding image delta function is $\delta(x',y')$.

In the following discussion, it will often be convenient to substitute x'/β and y'/β for the object coordinates x and y, respectively, in expressions for the complex amplitude $\hat{U}_0(x,y)$ or flux density $W_0(x,y)$ to have the object distribution function expressed in image coordinates. The magnification β will usually be assumed unity. This transformation of object to image coordinates gives an image produced by an ideal optical system, which must be modified to include aperture effects and other peculiarities of the actual optical system. This is done by applying the point-spread function introduced earlier. Because this function describes the distribution of intensity in image space resulting from a point source in object space, the whole image of an extended object can be found by superposing the spread functions of all the points in the object. If the distribution of flux density in the object can be described by a function, the convolution of this function with the spread function will give the image distribution function. [See the statement following Eq. (1-36) for the definition of convolution. Also, see Chapter 10 under the *Convolution of Instrument Function and True Spectrum* for an explanation, in one dimension, of a physical meaning of the convolution integral.] If the object is a point source at (x,y), the delta function $\delta(x,y)$ is convolved with the spread function $W_r(x,y)$ to get the image function by translating the object function to image coordinates as follows:

$$W_i(x',y') = \delta(x',y') * W_r(x',y'). \tag{9-27}$$

As in Eq. (5-37), this convolution positions the spread function so that its "center" is at the Gaussian image point because the delta function is at this point. Equation (9-27) holds for all object points where the optical system produces the specific spread function $W_r(x',y')$. Since aberrations actually vary as the source moves over the object field, the spread function is variable. However, the field may generally be divided into finite regions, called isoplanatism patches, so that an appropriate spread function is essentially invariant within each region. An equation like Eq. (9-27) holds, then, within each isoplanatism patch.

The degree of coherence determines the nature of the functions used in the convolution to get the image function. The treatment here is with the two extremes, coherent where both amplitudes and phases are considered and incoherent where only radiant flux densities are used. The general problem, partial coherence, is discussed in Ref. 15.

If negligible aberrations are assumed, the amplitude point-source response of an optical system with a square aperture has already been given in Eq. (9-14). For a circular aperture the point source response, when centered on the optic axis with $r_x = x'$ and $r_y = y'$, is given by Eq. (8-20):

$$\hat{U}_r(x',y') = \hat{U}_r(r) = \pi\rho_m^{\ 2}\hat{C}[2J_1(ru_m)]/(ru_m). \tag{9-28}$$

Here, $u_m = 2\pi\rho_m/\lambda s'$. Let $\hat{U}_0(x,y)$ be the amplitude distribution in the object. Then the amplitude in the image can be found by

$$\hat{U}_i(x',y') = \hat{U}_r(x',y') * \hat{U}_0(x,'y'). \tag{9-29}$$

The flux density point source response for the same system is given by

$$\begin{aligned} W_r(x',y') &= \hat{U}_r(x',y') \cdot \hat{U}_r^*(x',y') \\ &= W_{c0}[2J_1(ru_m)]^2/(ru_m)^2, \end{aligned} \tag{9-30}$$

and the flux density in the image, for a general distribution of flux density $W_0(x,y)$ in the object, can be found by

$$W_i(x',y') = W_r(x',y') * W_0(x',y'). \tag{9-31}$$

Although computations using Eq. (9-31) are fairly complicated, they are not as difficult as those using Eq. (9-29) because W is always positive and independent of phase in the radiant energy wave and $\hat{U}$ generally involves phase of the wave as well as amplitude. However, Eq. (9-31) applies only for complete, or, as an approximation, nearly complete incoherence. Coherence requires Eq. (9-29), and the flux density is found by

$$W_i(x',y') = \hat{U}_i(x',y') \cdot \hat{U}_i^*(x',y'). \tag{9-32}$$

Equation (9-29) accounts for interference effects; Eq. (9-31) does not.

Frequency Response for Diffraction-Limited Systems

As suggested by Eq. (9-3), the image distribution function, either $\hat{U}_i(x',y')$ or $W_i(x',y')$ [instead of $\Gamma(x')$ in Eq. (9-3)], has a Fourier transform which is the two-dimensional frequency function:

$$\hat{\mathscr{H}}_i(\kappa_x',\kappa_y') = \iint_{-\infty}^{+\infty} W_i(x',y') \exp\,[-2\pi i(\kappa_x'x' + \kappa_y'y')]\, dx'\, dy'. \tag{9-33}$$

The expression for $\hat{\mathscr{H}}_i(\kappa_x',\kappa_y')$ can be reached also by first finding the transform individually of the functions on the right of Eq. (9-31) and then forming their product. In this sequence, the operation on $W_r(x',y')$ amounts to taking the transform of the point source response function to obtain the frequency response function of the optical system:

$$\hat{\mathscr{R}}(\kappa_x',\kappa_y') = \iint\limits_{-\infty}^{+\infty} W_r(x',y') \exp\left[-2\pi i(\kappa_x'x' + \kappa_y'y')\right] dx'\, dy'. \quad (9\text{-}34)$$

The frequency response function $\hat{\mathscr{R}}(\kappa_x',\kappa_y')$ is also called the optical transfer (response) function. Its magnitude, $\mathscr{R}(\kappa_x',\kappa_y')$, is known as the modulation transfer function. Figure 9.14 shows $W_r(x',y')$ and $\mathscr{R}(\kappa_x',\kappa_y')$ for an ideal system with circular aperture. Similarly the frequency function, $\hat{\mathscr{H}}_0(\kappa_x,\kappa_y)$, is the Fourier transform of the object distribution $W_0(x,y)$. Then

$$\hat{\mathscr{H}}_i(\kappa_x',\kappa_y') = \hat{\mathscr{H}}_0(\kappa_x',\kappa_y') \cdot \hat{\mathscr{R}}(\kappa_x',\kappa_y'). \quad (9\text{-}35)$$

The amplitude function transforms in similar fashion:

$$\hat{\mathscr{R}}_a(\kappa_x',\kappa_y') = \iint\limits_{-\infty}^{+\infty} \hat{U}_r(x',y') \exp\left[-2\pi i(\kappa_x'x' + \kappa_y'y')\right] dx'\, dy'. \quad (9\text{-}36)$$

The functions $\mathscr{R}_a(\kappa_x',\kappa_y')$ and $U_r(x',y')$ are shown in Fig. 9.15.

According to the preceding development, an optical system is assumed to be a linear, space-invariant system over an isoplanatism patch. The system also functions as a spatial frequency filter system, which alters the amplitude and phase of each frequency according to the frequency response, or optical transfer, function.

If the optical system is centered, including a circular aperture stop, and if it is of such quality as to be diffraction limited, the point source response function and the frequency response function have circular symmetry, and

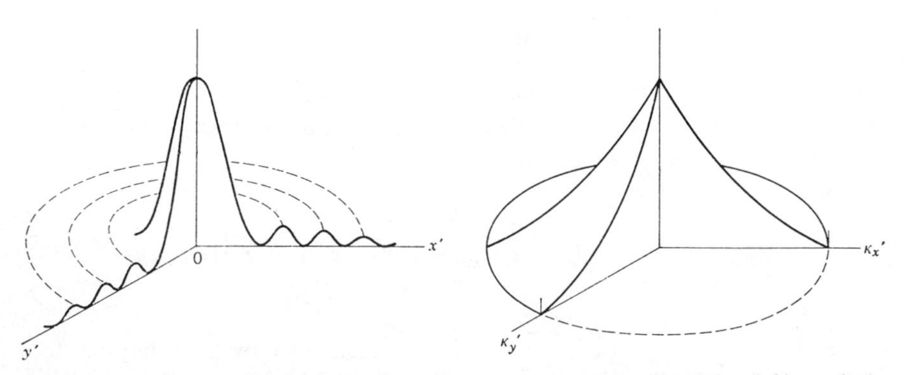

Figure 9.14 The radiant flux density point source response function (left) and the frequency response function (right) for a diffraction limited optical system.

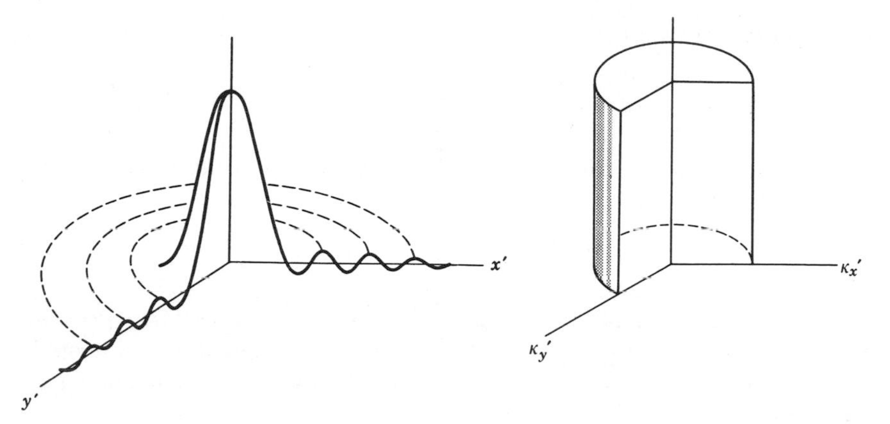

Figure 9.15 The amplitude point source response function (left) and the frequency response function (right) for a diffraction limited optical system.

Hankel transforms are appropriate to find mathematical expressions for these functions. To find $\hat{\mathscr{R}}(\kappa_x',\kappa_y')$, one can begin with $W_r(x',y')$ given by Eq. (9-30). The expression in Eq. (9-34) can be put into a standard Hankel transform by employing the variables ρ, r, s', and φ defined in Fig. 8.20. The coordinates (x_0,y_0) are in the same plane as ρ; ψ is a coordinate angle in the Gaussian image plane and is measured in the same sense as φ. Other variables are defined as follows:

$$u_p = (2\pi\rho/\lambda s') = 2\pi\kappa, \tag{9-37}$$

$$\kappa_x' = (x_0/\lambda s') = (\rho/\lambda s') \cos \varphi, \tag{9-38}$$

$$\kappa_y' = (y_0/\lambda s') = (\rho/\lambda s') \sin \varphi, \tag{9-39}$$

$$(\kappa_x'^2 + \kappa_y'^2) = \kappa^2 = \rho^2/\lambda^2 s'^2, \tag{9-40}$$

$$x' = r \cos \psi, \qquad y' = r \sin \psi. \tag{9-41}$$

Instead of $dx'\, dy'$ for the element of area, $r\, d\psi\, dr$ is used. By accumulating the constants included in W_{c0} from Equations (5-14), (8-19), (9-28), and (9-30), W_r can be written as follows:

$$W_r(r) = Cf_1(r)/2\pi, \tag{9-42}$$

where

$$C = |U_0|^2 \rho_m^{\,2}, \tag{9-43}$$

$$f_1(r) = 2\pi J_1^{\,2}(ru_m)/r^2, \tag{9-44}$$

and

$$u_m = 2\pi\rho_m/\lambda s'. \tag{9-45}$$

Equation (9-34) then becomes

$$\hat{\mathscr{R}}(\kappa) = C\int_0^\infty rf_1(r)\,dr\left\{[1/(2\pi)]\int_{-\pi}^{+\pi}\exp\,[iru_\rho\cos\,(\varphi - \psi)]\,d\psi]\right\}. \quad (9\text{-}46)$$

The integration on ψ can be accomplished by the procedure used on Eq. (5-62) with the result:

$$\hat{\mathscr{R}}(\kappa) = C\int_0^\infty rf_1(r)J_0(ru_\rho)\,dr, \quad (9\text{-}47)$$

Except for the constant, this is the standard Hankel transform [Ref. 18, p. 145]. Then the transform of $f_1(r)$, obtained by using a standard formula, gives the following expression for the modulation transfer function since $u_p = 2\pi\kappa$:

$$\mathscr{R}(\kappa) = C\{2\cos^{-1}(\pi\kappa/u_m) - (2\pi\kappa/u_m)[1 - (\pi\kappa/u_m)^2]^{1/2}\} \text{ when } \pi\kappa \leq u_m, \quad (9\text{-}48)$$

$$= 0 \qquad \text{when } \pi\kappa > u_m.$$

This function, normalized to unit amplitude at $\kappa = 0$, is shown at the right in Fig. 9.14 and also in one dimension as curve (a) in Fig. 9.17. The cut-off spatial frequency is $\kappa_c = 2\rho_m/\lambda s'$. The highest cut-off frequency of an optical system obviously occurs when s' is approximately f, the focal length. Then we can write $\kappa_c = 1/F^*\lambda$ where F^* is the f/number. Although we should generally expect a solution of Eq. (9-34) to produce a complex $\hat{\mathscr{R}}(\kappa_x'\kappa_y')$, it turns out that our solution of Eq. (9-47) [which we obtained from Eq. (9-34)] is real, as shown in Eq. (9-48). Thus, a diffraction-limited optical system produces no phase distortion.

The modulation transfer function of Eq. (9-48) relates expressions of flux density. A corresponding function can be developed to relate amplitudes, as functions of spatial frequency, in the object and image. This is done by beginning with Eq. (9-36) and converting to a standard Hankel transform by the following relationships in addition to those preceding Eq. (9-46):

$$\omega = 2\pi r, \qquad a = \rho_m/\lambda s', \quad (9\text{-}49)$$

$$\hat{C} = 4\pi^2\lambda s'\hat{U}_0, \quad (9\text{-}50)$$

$$f_2(\omega) = a[J_1(a\omega)]/\omega. \quad (9\text{-}51)$$

The complex constant $\hat{U}_0$ includes all constant amplitude and phase factors for the wave at the exit pupil. Applying the variously defined relationships above changes Eq. (9-28) to:

$$\hat{U}_r(r) = \hat{C}f_2(\omega)/2\pi. \quad (9\text{-}52)$$

Then Eq. (9-36) can be written as:

$$\hat{\mathscr{R}}_a(\kappa) = \hat{C}\int_0^{\infty} \omega f_2(\omega)\, d\omega\left\{[1/(2\pi)]\int_{-\pi}^{+\pi} \exp\,[i\kappa\omega\cos(\varphi - \psi)]\, d\psi\right\} \tag{9-53}$$

$$\hat{\mathscr{R}}_a(\kappa) = \hat{C}\int_0^{\infty} \omega f_2(\omega) J_0(\kappa\omega)\, d\omega. \tag{9-54}$$

The indicated transform of $f_2(\omega)$ can be written by applying a standard form (Ref. 18, p. 145), and the amplitude frequency response then becomes

$$\begin{aligned} \hat{\mathscr{R}}_a(\kappa) &= \hat{C} \qquad \text{when} \quad |\kappa| \le a, \\ &= 0 \qquad \text{when} \quad |\kappa| > a. \end{aligned} \tag{9-55}$$

The flux density frequency response $\hat{\mathscr{R}}(\kappa_x{}',\kappa_y{}')$, the Fourier transform of $W_r(x',y')$ in Eq. (9-34), can be found from $\hat{\mathscr{R}}_a(\kappa_x{}',\kappa_y{}')$ by applying the convolution theorem expressed as follows: The Fourier transform of the product of two functions is the convolution of the Fourier transforms of the functions themselves. According to Eq. (9-30), the flux density, as a function of position, is equal to the corresponding position-dependent amplitude function times its conjugate. Equation (9-36) shows how the amplitude function is transformed to $\hat{\mathscr{R}}_a(\kappa_x{}',\kappa_y{}')$. The conjugate is transformed similarly to $\hat{\mathscr{R}}_a{}^*(\kappa_x{}',\kappa_y{}')$. Then, according to the convolution theorem,

$$\hat{\mathscr{R}}(\kappa_x{}',\kappa_y{}') = \hat{\mathscr{R}}_a(\kappa_x{}',\kappa_y{}') * \hat{\mathscr{R}}_a{}^*(\kappa_x{}',\kappa_y{}'). \tag{9-56}$$

Although $\hat{\mathscr{R}}_a(\kappa_x{}',\kappa_y{}')$ of Eq. (9-55) is shown to be complex, the phase factors of $\hat{C}$ are constant, so there is no phase distortion. Also, if Eq. (9-56) is applied for the ideal system, $\hat{\mathscr{R}}(\kappa_x{}',\kappa_y{}')$ will be real.

The functions shown in two-dimensional perspective in Figures 9.14 and 9.15 are point source response functions; the diffraction patterns are on the left, and frequency response functions are on the right. These functions [Equations (9-30) and (9-48) in Fig. 9.14 and Equations (9-28) and (9-55) in Fig. 9.15] are derivable without specifying whether the beam is coherent or incoherent. The functions in Fig. 9.14 are in terms of radiant flux density; those in Fig. 9.15 are in terms of complex wave amplitude. However, the degree of coherence does determine how these equations are applied. In any detection process only radiant flux density expressions have meaning; the complex wave amplitudes are significant only in processes that alter the wave amplitude or phase, such as apodization or spatial filtering. As our derivations indicate, the variation of flux density with frequency is nonlinearly related to the variation of amplitude with frequency.

It is noted that the parameters $\kappa_x' = x_0/\lambda s'$, $\kappa_y' = y_0/\lambda s'$, and $\kappa = \rho/\lambda s'$ serve a dual purpose. They are normalized coordinates on the wavefront in the exit pupil, and they are spatial frequencies in the image plane.

Frequency Response in the Presence of Aberrations

In performing integrations earlier to get diffraction formulas [Chapter 5 and Eq. (8-14)], one tacitly assumed constant amplitude and phase over the wavefront. The region of integration was controlled by an aperture function. By incorporating the wave aberration function developed in Chapter 8 in a similar integration, the frequency response of an optical system with aberrations can now be found.

A phase factor, $\exp(ikW)$, in which W is the wave aberration function, can be introduced into the integrand of the diffraction formula, Eq. (5-13) or Eq. (8-14), to account for an aberration-caused variation of phase over the wavefront.

An amplitude factor $B(x_0,y_0)$ is sometimes also introduced to account for a variable amplitude over the wavefront. The amplitude factor can be manipulated to alter the point-source response, that is, to change, in a predetermined way, the point-spread function. This process is generally called apodization [Ref. 19]. The amplitude factor is also a convenient expression in describing spatial-frequency filtering where it multiplies the amplitude frequency response $\hat{\mathscr{R}}_a(\kappa_x',\kappa_y')$ by an amplitude function to alter the relative amplitudes over the spatial-frequency spectrum [Ref. 20, Chapter 7, and Ref. 21].

To write the diffraction integral containing the aberration phase factor, the dependent variables in the wave aberration function are changed as follows to permit dissymmetry: $x_0 = \rho \cos \varphi$, $y_0 = \rho \sin \varphi$, $x' = r \cos \psi$, and $y' = r \sin \psi$. The general wave aberration function, $W(\rho,r,\cos \varphi)$ of Chapter 8, is actually a function of r; but within an isoplanatism patch, the function is assumed independent of r, or of x' and y', so that $W = W(x_0,y_0)$. Then,

$$\hat{U}_{ra}(x',y') = \hat{U}_{ra}'(k_x,k_y)$$

$$= \hat{C} \iint_S \exp[-ikW(x_0,y_0)] \exp[2\pi i(k_x x_0 + k_y y_0)]\, dx_0\, dy_0. \quad (9\text{-}57)$$

When the variables in Eq. (9-57) are changed as follows

$$k_x x_0 = x' x_0/\lambda s' = x' \kappa_x', \qquad x_0 = \lambda s' \kappa_x', \quad (9\text{-}58)$$

$$k_y y_0 = y' y_0/\lambda s' = y' \kappa_y', \qquad y_0 = \lambda s' \kappa_y', \quad (9\text{-}59)$$

the wave aberration function $W(x_0,y_0)$ becomes $W'(\kappa_x',\kappa_y')$ since $\lambda s'$ is constant in the mathematical processes being considered. An aperture function is defined as

$$\hat{A}_a(\kappa_x',\kappa_y') = \exp\,[-i k W'(\kappa_x',\kappa_y')] \text{ when } (\kappa_x'^2 + \kappa_y'^2) \leq \rho_m^2/\lambda^2 s'^2, \quad (9\text{-}60)$$

$$= 0 \text{ when } (\kappa_x'^2 + \kappa_y'^2) > \rho_m^2/\lambda^2 s'^2.$$

Then the diffraction integral can be written:

$$\hat{U}_{ra}(x',y') = \lambda^2 s'^2 \hat{C} \iint_{-\infty}^{+\infty} \hat{A}_a(\kappa_x',\kappa_y') \exp\,[2\pi i(\kappa_x' x' + \kappa_y' y')]\, d\kappa_x'\, d\kappa_y', \quad (9\text{-}61)$$

and by using the inversion theorem, the following results:

$$\hat{A}_a(\kappa_x',\kappa_y') = (1/\lambda^2 s'^2 \hat{C}) \iint_{-\infty}^{+\infty} \hat{U}_{ra}(x',y') \exp\,[-2\pi i(\kappa_x' x' + \kappa_y' y')]\, dx'\, dy'.$$

$$(9\text{-}62)$$

$\hat{A}_a(\kappa_x',\kappa_y')$ is the amplitude spatial frequency response of an optical system with aberrations. From an equation similar to Eq. (9-56), one can find the flux density frequency response function, which is the optical transfer function:

$$\hat{A}(\kappa_x',\kappa_y') = \hat{A}_a(\kappa_x',\kappa_y') * \hat{A}_a^*(\kappa_x',\kappa_y'). \quad (9\text{-}63)$$

Modulation Transfer Functions for Certain Aberrations

Instruments have been made to measure the optical transfer function of lenses and systems [Refs. 22–26]. A measured modulation transfer function is shown in Fig. 9.16. Since W' of Eq. (9-60) can represent any one of the five monochromatic aberrations, we have a way to calculate the optical transfer function for each kind of aberration [Refs. 27–29]; most of these have been calculated [Refs. 30–37]. Several curves for the modulation transfer function are given in Figures 9.17 and 9.19–9.25. The figures do not show the phase of the optical transfer, which is sometimes significant [Ref. 38] and can generally be found in the references cited. In the figures, except Fig. 9.17, the cut-off spatial frequency is normalized to 2, and the modulation transfer function is normalized to unity at zero spatial frequency.

The curves of Fig. 9.17 are three modulation transfer functions normalized at zero frequency for a lens system [see Ref. 17]. Curve (a) shows the variation for a correctly focused $f/4$ system free of aberrations, transmitting quasi-monochromatic light of a mean wavelength $\lambda = 500$ nm. The effect of defocusing is shown by curve (b), where the modulation transfer function falls off more rapidly with spatial frequency from 0 to 177 lines/mm and becomes negative for frequencies between 117 and 375 lines/mm, and

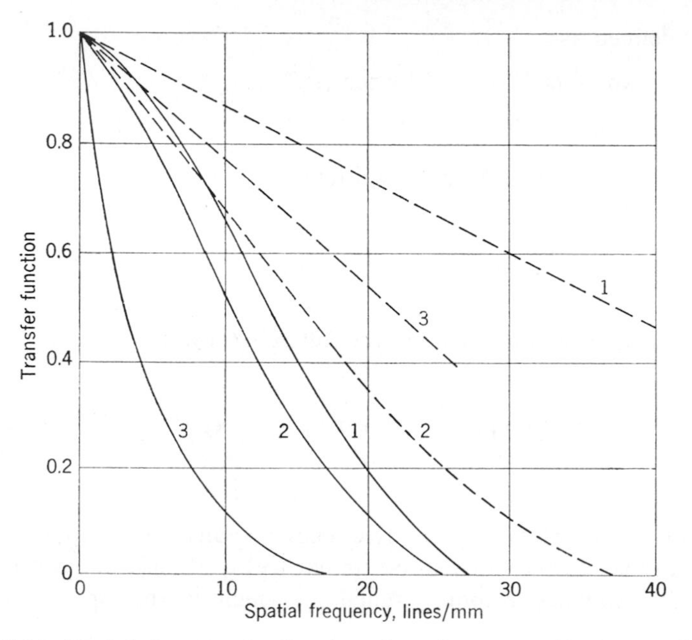

Figure 9.16 Modulation transfer functions for microscope objectives (solid lines are measured values, broken lines are computed values): (1) a 10× objective with $NA = 0.25$, (2) a 25× objective with $NA = 0.50$, and (3) a 45× objective with $NA = 0.65$ [Ref. 26].

positive again for frequencies between 375 and 500 lines/mm, beyond which it is zero. The negative response means that for a grating in object space at a frequency between 117 and 375 lines/mm, points in image space will actually appear dark where geometrically they should appear bright, and bright where they should appear dark. At all frequencies above 117 lines/mm, the resolution of a sine-wave grating of this slightly defocused system is

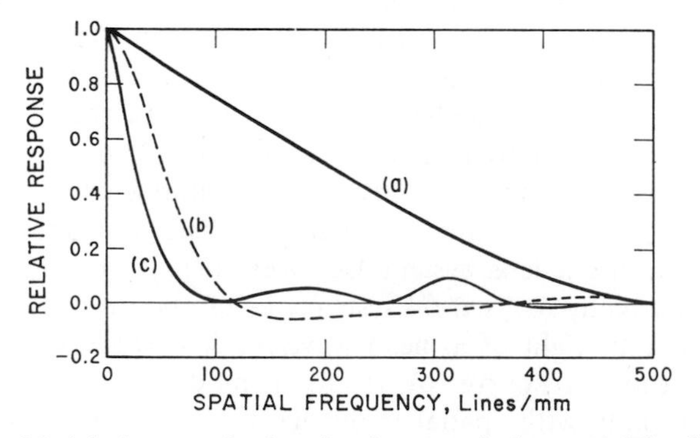

Figure 9.17 Modulation transfer function for an optical system. (The different curves are explained in the text) [Ref. 17].

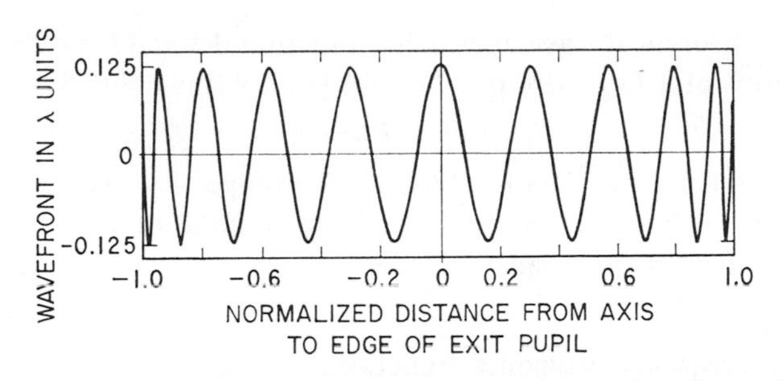

Figure 9.18 An aberration function defined by a Legendre polynomial [Ref. 40].

conventionally described as "spurious." Curve (c) shows the effect of spherical aberration at the focal setting midway between marginal and paraxial focus. The maximum wave distortion, given by the wave aberration function for each of these, is one wavelength.

The shape of the actual wavefront, as well as the magnitude of W', influences the distribution of the radiant power in the image plane. $W_{max} = (\frac{1}{4})\lambda$, referred to as a quarter wavelength aberration, has been considered an acceptable amount of wave aberration. This is another Rayleigh criterion. It is not sufficiently restrictive, however, for certain high quality systems, and it ignores the influence of the wavefront shape.

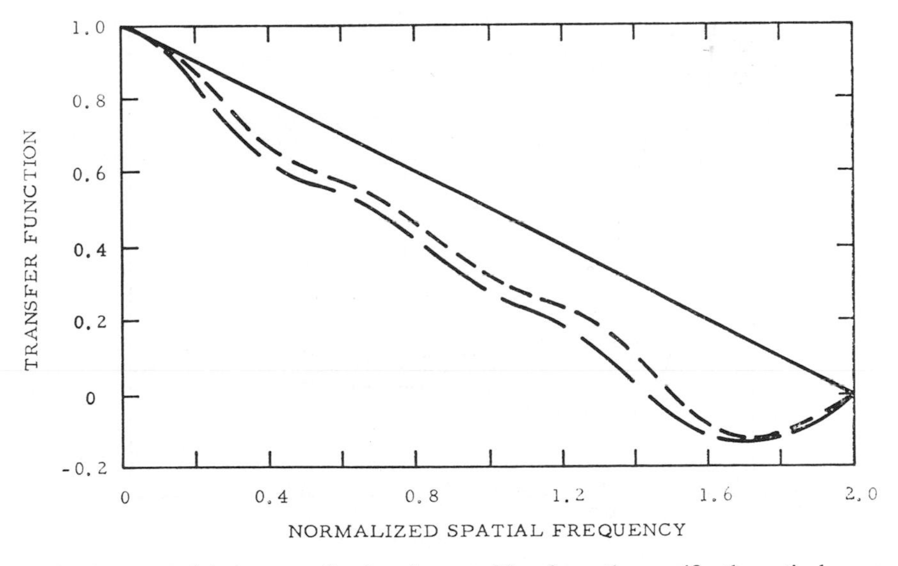

Figure 9.19 Modulation transfer function resulting from the specific theoretical wave aberration function shown in Fig. 9.18 where the amplitude is well within the one-fourth wavelength criterion. (Note negative values just below cut-off) [Ref. 40].

A particular theoretical wave aberration function defined for a rectangular aperture in terms of a Legendre polynomial [Ref. 40] is defined as

$$W(x_0) = A''P_{20}(x_0), \tag{9-64}$$

where $P_{20}(x_0)$ is the polynomial of order 20. The shape of the wave aberration function is shown in Fig. 9.18, and the resulting modulation transfer function for two values of A'' is shown in Fig. 9.19.

Tolerances on Frequency Response Functions

There is no single modulation transfer function (MTF), either determined theoretically during the design and ray trace stages or measured after the system is manufactured, that is sufficient for an optical system evaluation.

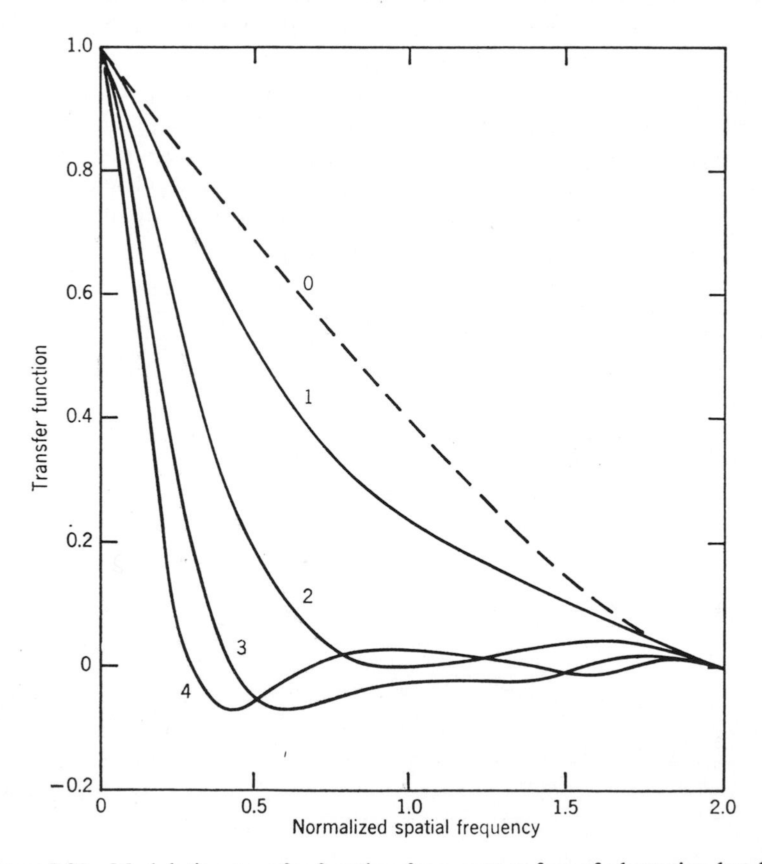

Figure 9.20 Modulation transfer function for a system free of aberration but having a defect of focus. (Maximum wave distortion is n/π wavelengths where n is the number on the curve) [Ref. 30].

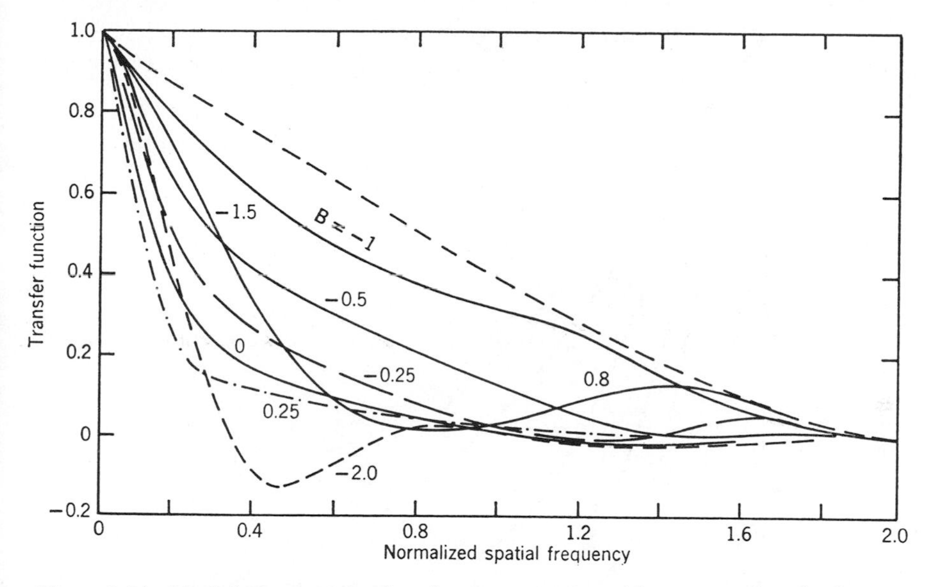

Figure 9.21 Modulation transfer function for a system with one wavelength of wave aberration at the margin of the exit pupil caused by fourth power spherical aberration. (There is, in this case, no correct focusing point. The different curves are for different settings between the marginal focal plane where $B = -2$ and the paraxial focal plane where $B = 0$. The curve labeled $B = -1.0$ is for a setting midway between the two focal planes) [Ref. 33].

There is rather an MTF for each isoplanatism patch: one for an object area near the axis, one for each of several areas at varying distances off-axis, and one each for different directions of the spatial frequency lines when asymmetrical aberrations are present. Within each of these patches, there is an MTF for each radiant energy wavelength region. Therefore, an assortment of MTFs must be in hand when the optical system is appraised. As we have said earlier, the appraisal must consider the intended use. Included here will be the setting of tolerances on the kinds and amounts of degradation in each MTF.

A tolerance on the degradation of an MTF by aberrations is best determined in the systems analysis and design stage during which a maximum detectable spatial frequency κ_m is set. A sufficient signal-to-noise ratio at κ_m for the overall detection process must also be known, which in turn requires that κ_m have a minimum detectable amplitude that will keep the signal sufficiently above all sources of noise. These levels also are determined by the systems analysis. The cut-off frequency κ_c is then set at a high enough frequency by adjusting $F^*\lambda$ (since $\kappa_c = 1/F^*\lambda$) that κ_m is well below κ_c. Then, during design and manufacture, aberrations are kept small enough that the MTF

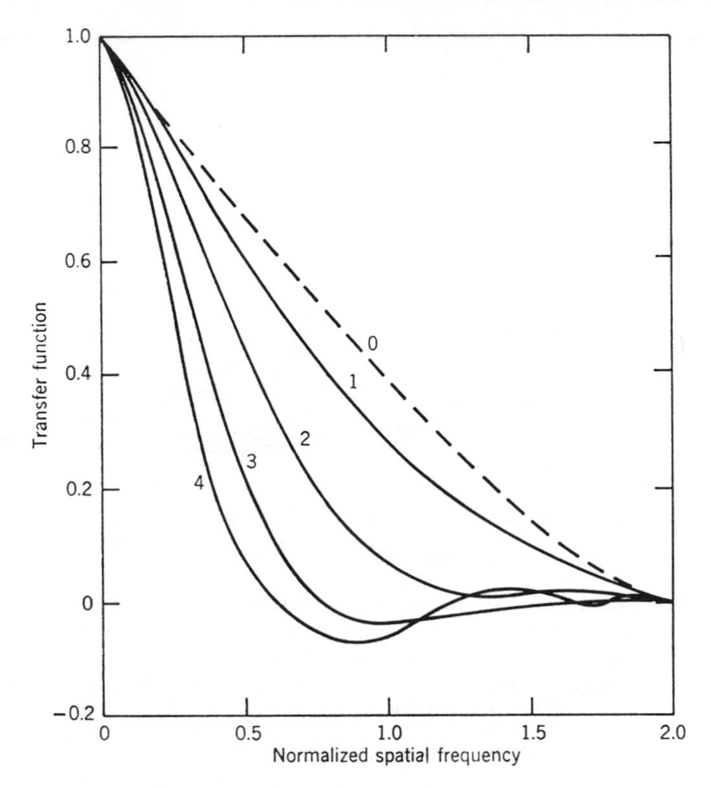

Figure 9.22 Modulation transfer function for a system having astigmatism. (The receiving plane is midway between the tangential and sagittal focal lines. The maximum wave distortion at the margin is n/π wavelengths where n is the number on the curve. The spatial frequency lines are oriented so as to be periodic along a line 45 deg to the tangential and sagittal planes) [Ref. 35].

does not fall below the required amplitude at κ_m. Some optical designers arbitrarily choose F^* so that κ_m is less than 20% of κ_c.

By comparing the MTFs shown in this chapter, one can conclude that the frequency response of an optical system is sensitive to even slight aberrations. As an aberration is increased from zero, an MTF will show a detectable degradation before the size of the central disk of a diffraction pattern shows a perceptible increase. Thus an MTF test is more sensitive in well corrected systems than is a resolution test.

A bar test chart, when considered as a one-dimensional function of radiance, is a square wave and is, therefore, rich in harmonics. The frequency $\kappa = 1/l$, defined earlier, is only the fundamental. The contrast for the harmonics declines rapidly with increasing frequency and reaches zero

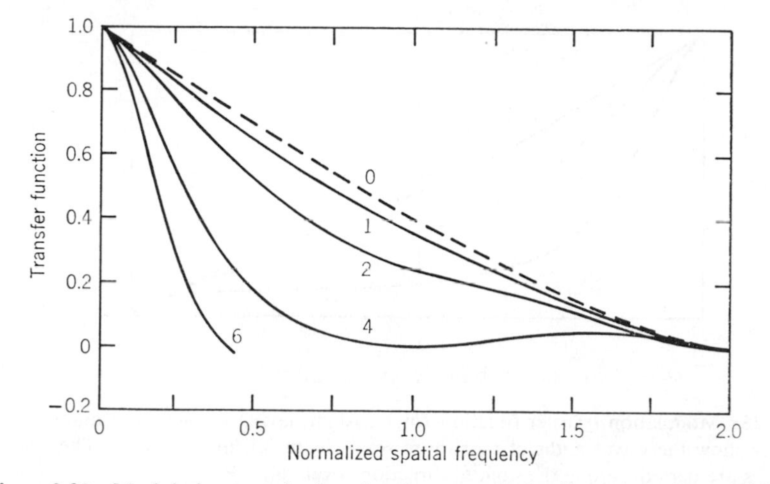

Figure 9.23 Modulation transfer function for a system having astigmatism. (This optical system and aberration is the same as those of Fig. 9.22. The spatial frequencies are now periodic along a line in the tangential or sagittal plane) [Ref. 35].

for harmonics greater than κ_c. Edges of the bar must therefore fade out, because the loss of harmonics has the effect of "rounding off" a square wave. Some writers have tried to define a square wave response function for optical systems, but since such a definition does not take into account the loss of harmonics and the consequent effects, their conclusions must be examined carefully.

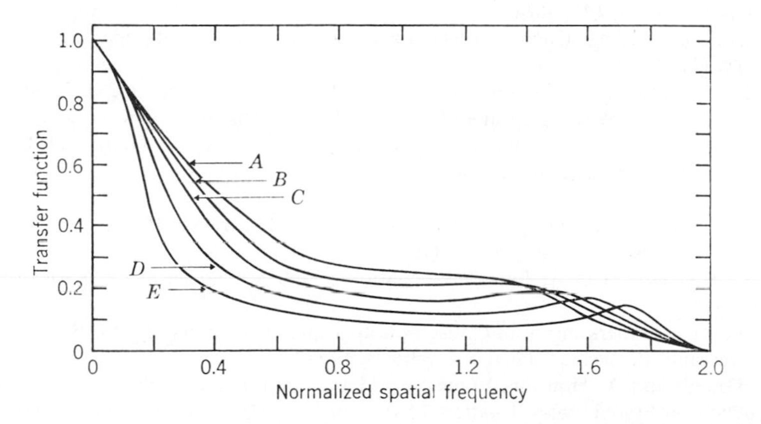

Figure 9.24 Modulation transfer function for a system with central circular obscuration and with one wavelength of wave aberration at the margin caused by fourth power spherical aberration. (The different curves are for different fractions of obscuration. Curve A is for 0.4 of the aperture obscured. The other curves are for increasing obscurations up to 0.8 going from Curve B to Curve E) [Ref. 37].

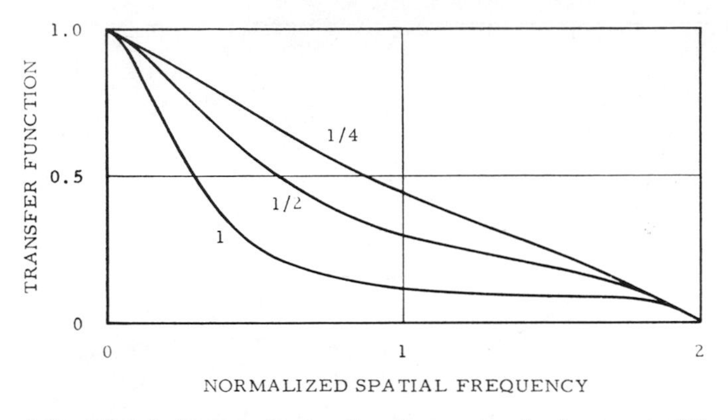

Figure 9.25 Modulation transfer function for a system having coma. (The numbers on the curves show the wavelengths of maximum wave distortion in each case. The spatial frequencies are periodic along the sagittal direction) [Ref. 34].

Any appreciable coherence in the beam also tends to make bar edges indistinct. Also, curve (b) of Fig. 9.17 even indicates that a bar chart test of this system would show 450 lines/mm as the resolution limit without revealing the serious phase reversal at 200 lines/mm.

REFERENCES

1. T. Orhaug, "On the Resolution of Imaging Systems," *Optica Acta*, **16,** 75 (1969).
2. P. B. Fellgett and E. H. Linfoot, "On the Assessment of Optical Images," *Trans. Roy. Soc. London*, **247,** 369 (1955).
3. J. H. Altman, "Image Quality Criteria for Data Recording and Storage," *J. SMPTE*, **76,** 629 (1967).
4. A. E. Conrady, *Applied Optics and Optical Design*, Vol. 1, pp. 132–135. Oxford University Press, 1929. (Reprinted by Dover Publications, Inc., New York, 1957.)
5. C. M. Sparrow, "Spectroscopic Resolving Power," *Astrophys. J.*, **44,** 76 (1916).
6. F. H. Perrin, "Methods of Appraising Photographic Systems," *J. SMPTE*, **69,** 151 (1960).
7. P. G. Roetling, E. A. Trabka, and R. E. Kinzley, "Theoretical Prediction of Image Quality," *J. Opt. Soc. Am.*, **58,** 342 (1968).
8. G. C. Higgins and L. A. Jones, "Evaluation of Image Sharpness," *J. SMPTE*, **58,** 277 (1952).
9. H. Molitz, "Bermerkungen sur Optisch-photographischen Abbildung von Strichrasten und vervandt Probleme," *Phot. Korr.*, **95,** 3, 19 (1959).
10. R. Barakat and A. Houston, "Line Spread Function and Cumulative Line Spread Function for Systems with Rotational Symmetry," *J. Opt. Soc. Am.*, **54,** 768 (1964).
11. N. W. McLachlan, *Bessel Functions for Engineers*, p. 76. Oxford University Press, Oxford, 1955.
12. F. H. Perrin, "The Structure of the Developed Image," in *The Theory of the Photographic Process*, Third Edition. C. E. K. Mees and T. H. James. Macmillan, New York, 1966.

13. D. N. Grimes and B. J. Thompson, "Two Point Resolution with Partially Coherent Light," *J. Opt. Soc. Am.*, **57,** 1330 (1967).
14. S. H. Rowe, "Light Distribution in the Defocused Image of a Coherently Illuminated Edge," *J. Opt. Soc. Am.*, **59,** 711 (1969).
15. B. J. Thompson, "Image Formation with Partially Coherent Light," in *Progress in Optics*, Vol. VII. Edited by E. Wolf. North-Holland Publishing Company, Amsterdam, 1969.
16. C. E. Thomas, "Coherent Optical Noise Suppression," *Appl. Optics*, **7,** 517 (1968).
17. E. H. Linfoot, *Fourier Methods in Optical Image Evaluation*. The Focal Press, London, 1964.
18. A. Papoulis, *Systems and Transforms with Applications in Optics*. McGraw-Hill Book Company, Inc., New York, 1968.
19. P. Jacquinot and B. Roizen–Dossier, "Apodization," in *Progress in Optics*, Vol. III. Edited by E. Wolf. North-Holland Publishing Company, Amsterdam, 1964.
20. J. W. Goodman, *Introduction to Fourier Optics*. McGraw-Hill Book Company, New York, 1968.
21. J. Tsujiuchi, "Correction of Optical Images by Compensation of Aberrations and by Spatial Frequency Filtering," in *Progress in Optics*, Vol. II. Edited by E. Wolf. North-Holland Publishing Company, Amsterdam, 1963.
22. K. Rosenhauer and K. J. Rosenbruch, "The Measurement of the Optical Transfer Function of Lenses," in *Reports on Progress in Physics*, Vol. XXX, Part I. Edited by A. C. Stickland. Institute of Physics and the Physical Society, London, 1967.
23. M. D. Waterworth, "A New Method for the Measurement of Optical Transfer Function," *J. Sci. Instrum.*, **44,** 195 (1967).
24. A. J. Montgomery, "Two Methods of Measuring Optical Transfer Functions with an Interferometer," *J. Opt. Soc. Am.*, **56,** 624 (1966).
25. K. Murata, "Instruments for the Measuring of Optical Transfer Functions," in *Progress in Optics*, Vol. V. Edited by E. Wolf. North-Holland Publishing Company, 1966.
26. P. Kuttner, "An Instrument for Determining the Transfer Function of Optical Systems," *Appl. Optics*, **7,** 1029 (1968).
27. P. Elias, D. S. Grey, and D. Z. Robinson, "Fourier Treatment of Optical Processes," *J. Opt. Soc. Am.*, **42,** 127 (1952).
28. P. Elias, "Optics and Communication Theory," *J. Opt. Soc. Am.*, **43,** 229 (1953).
29. R. Barakat, "Computation of the Transfer Function of an Optical System from the Design Data for Rotationally Symmetrical Aberrations," Part I, Theory, *J. Opt. Soc. Am.*, **52,** 985 (1962); Part II, Programming and Numerical Results (with M. V. Morello), *J. Opt. Soc. Am.*, **52,** 992 (1962).
30. H. H. Hopkins, "The Frequency Response of Defocused Optical Systems," *Proc. Roy. Soc.*, **A231,** 91 (1955).
31. L. Levi and R. H. Austig, "Tables of the Modulation Transfer Function of a Defocused Perfect System," *Appl. Optics*, **7,** 967 (1968).
32. G. Black and E. H. Linfoot, "Spherical Aberration and the Information Content of Optical Images," *Proc. Roy. Soc.*, **A239,** 522 (1957).
33. A. M. Goodbody, "The Influence of Spherical Aberration on the Response Function of an Optical System," *Proc. Phys. Soc.* London, **72,** 411 (1958).
34. K. Miyamoto, "Wave Optics and Geometrical Optics in Optics Design," in *Progress in Optics*, Vol. I. Edited by E. Wolf. North-Holland Publishing Company, Amsterdam, 1961.
35. M. De, "The Influence of Astigmatism on the Response Function of an Optical System," *Proc. Roy. Soc.*, **A233,** 91 (1955).

36. K. Rosenhauer, K.-J. Rosenbruch, and F.-A. Sunder–Plassman, "The Relations Between the Axial Aberrations of Photographic Lenses and Their Optical Transfer Functions," *Appl. Optics*, **5**, 415 (1966).
37. R. Barakat and A. Houston, "Transfer Function of an Annular Aperture in the Presence of Spherical Aberration," *J. Opt. Soc. Am.*, **55**, 538 (1965).
38. R. E. Hufnagel, "Significance of Phase of Optical Transfer Functions," *J. Opt. Soc. Am.*, **58**, 1505 (1968).
39. H. H. Hopkins, "The Aberration Permissible in Optical Systems," *Proc. Roy. Soc.* London, **B70**, 449 (1957).
40. R. Barakat, "Rayleigh Wavefront Criterion," *J. Opt. Soc. Am.*, **55**, 527 (1965).
41. R. Barakat and A. Houston, "Transfer Function of an Optical System in the Presence of Off-Axis Aberrations," *J. Opt. Soc. Am.*, **55**, 1142 (1965).
42. H. H. Hopkins and M. J. Yzuel, "The Computation of Diffraction Patterns in the Presence of Aberrations," *Optica Acta*, **17**, 157 (1970).
43. H. H. Hopkins and H. J. Tiziana, "A Theoretical and Experimental Study of Lens Centering Errors and Their Influence on Optical Image Quality," *Brit. J. Appl. Phys.*, **17**, 33 (1966).

10

The Elements of Spectroscopy

Our study of spectroscopy begins with the following expressions of Bouguer's and Beer's laws:

$$d\Phi = -\Phi \alpha_c \, dl. \tag{10-1}$$

$$\tau = \Phi/\Phi_0 = \exp\,[-\alpha_c(\lambda) l]. \tag{10-2}$$

$$\alpha_c = \epsilon_1 c_1 + \epsilon_2 c_2 + \epsilon_3 c_3 + \cdots. \tag{10-3}$$

These relationships were discussed in Chapter 2 in connection with Equations (2-24), (2-25), and (2-26). A related expression, recommended by the Society for Applied Spectroscopy [Refs. 1–3], defines the term *absorbance A*:

$$A = -\log_{10} \tau = abc, \tag{10-4}$$

in which τ is the internal transmittance, a is absorptivity, b is optical distance, and c is concentration. From Eq. (10-4), absorptivity is absorbance per unit concentration per unit distance.

As indicated by the defining expression, Eq. (10-2), internal transmittance is a bulk property of the optical material; thus, experimental determinations must eliminate losses at material interfaces and in other parts of the measuring system. This is usually done by first observing the radiant power transmission through an empty standard cell and then repeating the observation with the optical material to be measured in the standard cell. Because

$$(\Phi_{\text{material}})/(\Phi_{\text{empty}}) = (k\Phi)/(k\Phi_0) = \Phi/\Phi_0 = \tau, \tag{10-5}$$

where the factor k includes all system losses outside of the standard cell, the ratio of observed powers is equivalent to the defining ratio for internal transmittance. Throughout the remainder of this chapter "transmittance" will mean "internal transmittance" as defined above and explained in Chapter 2 following Eq. (2-24a).

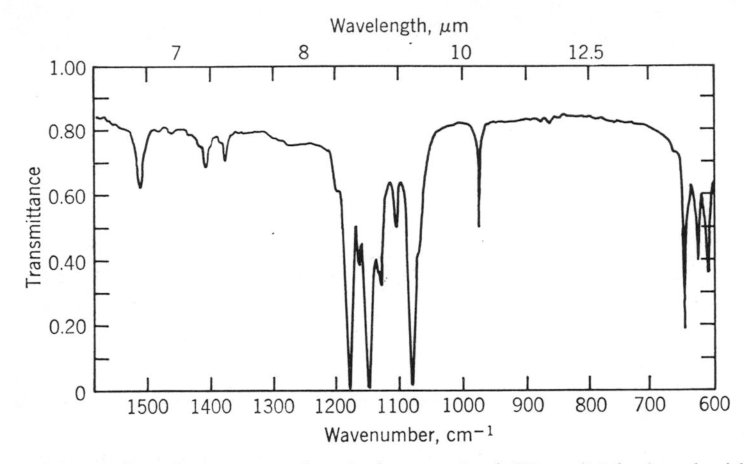

Figure 10.1 Infrared spectrum of a single crystal of KBr. (Melt doped with $CaSO_4$) [Ref. 21].

Both τ and a in Eq. (10-4) are wavelength dependent, even though this has not been explicitly indicated. Absorbance is sometimes called optical density, and the symbol d is then usually used instead of A.

The concentration c (Eq. (10-4)) may be given in several ways; however, it is always the quantity of an absorbing substance contained in a unit quantity of the optical material. Fairly common kinds of units for c are those of mass, volume, moles, and, in a gas, partial pressure. The product bc represents the amount of absorbing material per centimeter^2. For example, if c is in moles per centimeter^3 and b is in centimeters, the product is moles per centimeter^2 and the absorbtivity is centimeters^2 per mole.

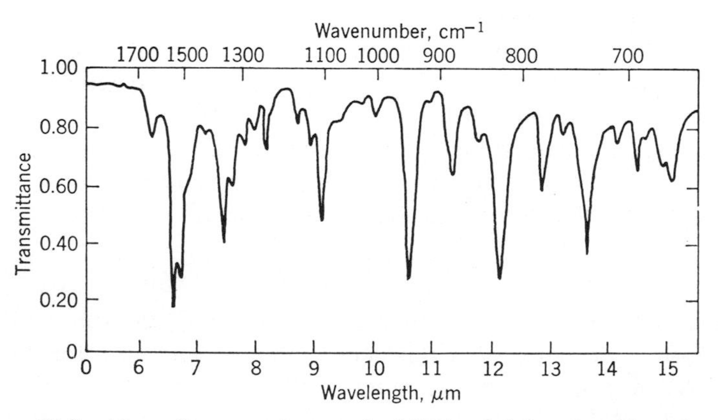

Figure 10.2 Absorption spectrum of 4,5-dimethyl-2-*p*-nitrochlorophenyl-*v*-triazole [Ref. 22].

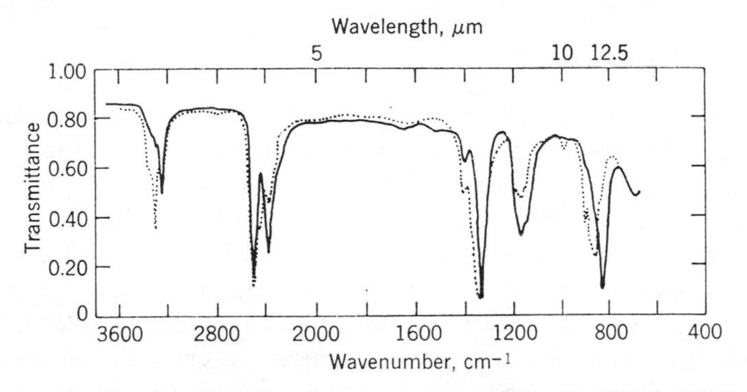

Figure 10.3 Polarized infrared spectrum of deuterated sulfamide (solid line: electric vector parallel to direction of crystal growth; broken line: electric vector perpendicular to direction of crystal growth) [Ref. 23].

Spectra may be plotted with transmittance as a function of wavenumber, wavelength, or frequency. Other units for the dependent variable are also found in the literature—sometimes even the deflection of a galvanometer is used. Other recommended units are absorptance $\alpha = (1 - \tau)$, absorption coefficient α_c, and absorbance A.

Figures 10.1–10.4 are examples of absorption spectra; each shows the measured transmittance as a function of wavelength for a given material. The narrow regions of low transmittance where relatively high absorption occurs are called absorption bands. Sometimes, particularly if the bands are quite narrow, they are called absorption lines. Absorption bands exist, as is explained briefly in Chapter 1, because each kind of molecule interacts, by absorption or emission, with only certain photons corresponding to specific wavelengths (energies).

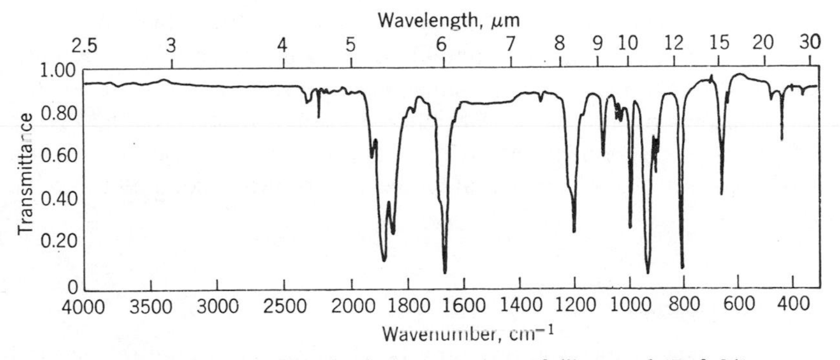

Figure 10.4 Infrared solution spectrum of diketene-d_4 [Ref. 24].

The Origins of Spectra

Although the kinetic (translational) energy of a free molecule can vary continuously, the energy associated with periodic motions—of the molecule as a whole (rotations), of the constituent atoms relative to one another (vibrations), and of the electronic structures that hold a molecule together—can have only certain discrete values, the quantized energy levels characterizing the stationary states of the molecule. Any simple process that changes an atom or molecule from one stationary state to another is called a transition; each transition requires the absorption or emission of energy to account for the energy change. The frequency associated with an absorbed or emitted photon is related to the energy change by the Bohr equation, Eq. (1-9). The total amount of energy, at a given frequency, absorbed by a sample of material is, therefore, proportional to the energy of an absorbed photon and to the number of molecules absorbing photons. The number of absorbing molecules depends in turn upon the power of the incident beam (the rate of photon incidence), the molecular concentration, and the distance traveled by the beam through the material.

Since, according to quantum theory, only certain stationary states are possible for each atom or molecule, each kind of molecule has a unique set of possible energy levels and allowed transitions between these levels; therefore, the specific photons that can be absorbed and emitted are characteristic of a given material, that is, a given absorption spectrum is produced by only one material. Although two chemically related substances may absorb identically in small regions of the spectrum, there will always be some region where great and obvious differences exist. Each material is also characterized by an emission spectrum with emission bands at the same wavelengths as the absorption bands.

The work of the spectroscopist is to measure the relative amounts of radiant energy absorbed or emitted at each frequency. From these measurements he gets information about whether or not a given material is present, the relative amounts of the various substances present in a mixture, or the structure of atoms or molecules. Because of the significance of the results gained from the measurement and study of spectra, the absorption spectrum is one of the first characteristics of a newly produced compound to be studied.

Types of Spectroscopy

The different fields of spectroscopy may be classified according to the kind of information sought or by the techniques used: interference spectroscopy, flash spectroscopy, reflectance spectroscopy, grill spectroscopy, etc.; but, more generally, the fields are distinguished according to the wavelength range in which measurements are made. For example, every element or compound, with a few exceptions, has strong absorption and emission bands somewhere in each of the ultraviolet-visible, infrared, and microwave regions.

Transitions involving relatively large amounts of energy, corresponding to extremely high frequencies, originate in the nucleus of an atom and are independent of the environment of the nucleus. If a sample is irradiated with photons of slightly less energy, electrons in the inner levels may be separated from the atoms. These photons correspond to frequencies in the x-ray region: the frequencies are related only to the structure of the atom, not to its state of chemical association with other atoms in compounds.

Further reduction in photon energy leads to frequencies in the ultraviolet and visible regions. The associated atomic spectra originate with displacement of valence electrons in the outer shells. These spectra are nearly independent of the chemical state.

Excitation—that is, a transition raising an atom or molecule to a higher energy state—can be caused by heating or ion bombardment in an electric arc, and emission spectra are then produced by return to the normal state (Fig. 10.5). In many instances, electronic energy absorbed by heating or ion bombardment is quickly converted to vibrational, rotational, and translational energy of the molecules; in other instances, emission from the electronic levels occurs, either immediately or after a short delay. The latter process is called fluorescence or phosphorescence.

To a first approximation, the energy of a molecule, not counting translational energy, is contained in states associated with: (1) rotation of the whole molecule, (2) vibrations of constituent atoms, and (3) intraelectronic motions. The energy levels for the vibrational states are much closer to each other in energy than the electronic states. Photons associated with vibrational transitions are in the infrared. The rotational levels are still closer, and the associated transition photons are in the far infrared and microwave regions.

The infrared spectroscopist is mostly concerned with the vibrational, rotational, and combination vibrational–rotational energy transitions. The problems of band absorption are most troublesome in the infrared because spectrum measuring instruments are quite limited in their ability to determine the bandwidth and profile shape. Appreciable overlapping of bands—particularly rotational bands—adds to the difficulty. These problems preclude any simple and direct test of band shape. Conditions are more favorable in the microwave region because a much narrower wavelength range, smaller than the typical bandwidth, can be detected.

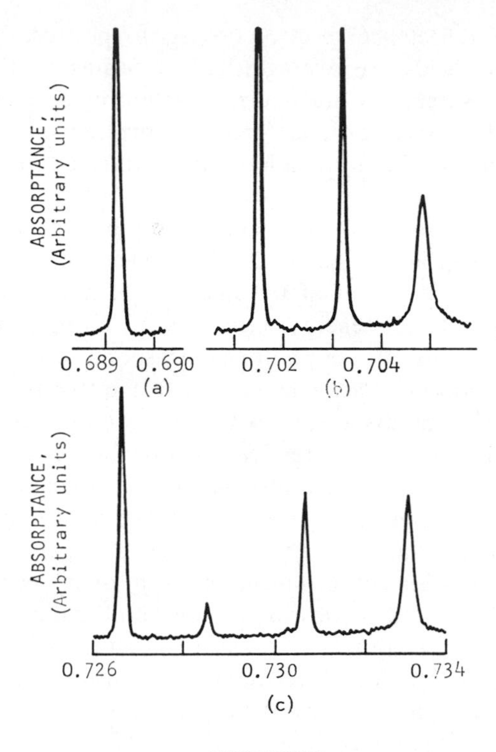

Figure 10.5 First three sets of sharp emission lines of Sm^{2+}:KCl at 10 K [Ref. 25].

Microwave and radio frequency radiant energy can be used in studies of very low-energy transitions arising from the reorientation of nuclear and electron spins in the presence of an applied magnetic field. These are called nuclear magnetic resonance and electron magnetic resonance, respectively; the latter being subdivided into paramagnetic and ferromagnetic effects.

Term Level Diagrams

Allowed transitions of a molecule are often shown graphically on a term (energy) level diagram in which the energy levels of the various stationary states are plotted along an energy scale. Figure 10.6 is an example. Each horizontal line represents a possible state for the molecule, and the line position indicates the molecular energy for that state. Energy increases upward in the figure. The rotational levels are represented by closely spaced,

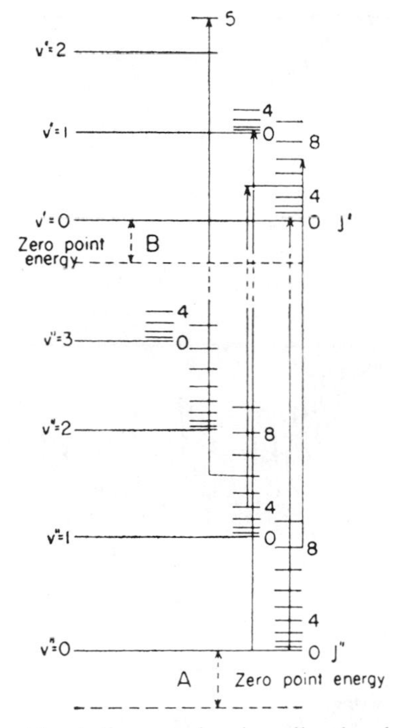

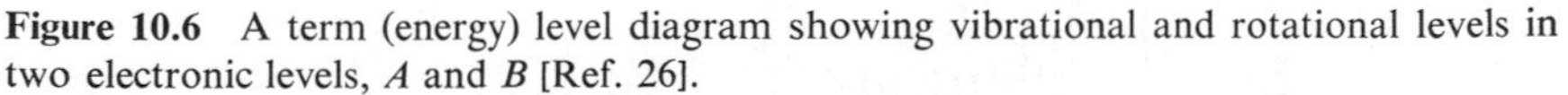

Figure 10.6 A term (energy) level diagram showing vibrational and rotational levels in two electronic levels, *A* and *B* [Ref. 26].

short horizontal lines and, in each series, are numbered 0, 1, 2, ... These numbers are the rotational quantum numbers *J*. The vibrational levels are represented by more widely spaced and longer lines and are numbered using vibrational quantum numbers *v*. The dotted horizontal lines at *A* and *B* are associated with electronic energy levels and are not drawn to the same scale used for the other levels.

The vertical lines, with arrows pointing upward, each joining a rotational level in one part of the diagram to a rotational level in another, represent allowed excitation transitions. A change in rotational, vibrational, and electronic energy may generally be involved in a single transition. Certain transitions between one vibration–rotation level to another in the same electronic level and also between one rotation level to another within one vibration level in certain instances can be allowed. Allowed transitions are determined by rigid quantum mechanical selection rules which are beyond the scope of the present discussion. An "allowed" transition is one whose occurrence is highly probable; a corresponding strong line or band will, therefore, appear in the spectrum. "Forbidden" transitions, on the other hand, have low probability; and if they show up at all in a spectrum, they will appear as very weak bands.

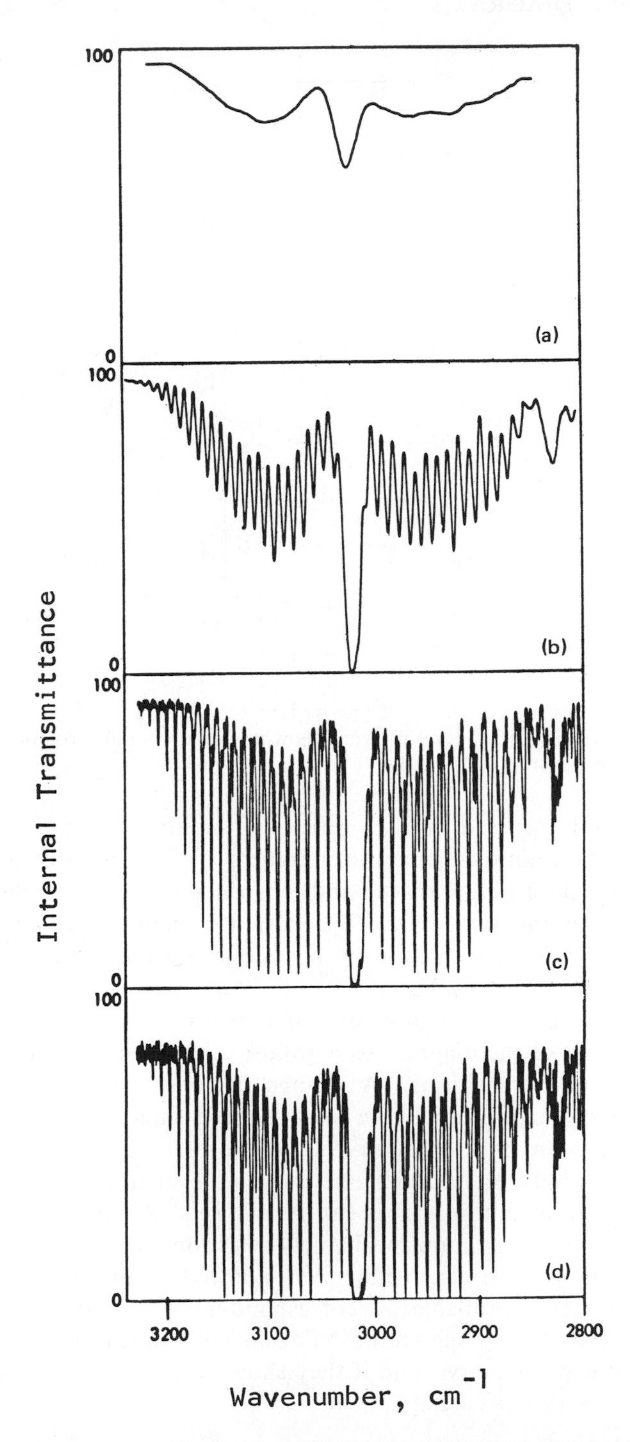

Figure 10.7 The H–C stretching band of methane recorded with spectrometers of different resolving powers: from top to bottom 0.5, 1, 5, and 20 cm^{-1} [Ref. 28].

Infrared Spectra

Infrared spectra can be obtained from all three states of matter: gases, liquids, and solids. In gases, the molecules have enough separation to vibrate and rotate at the same time, so they exhibit complex vibrational–rotational bands. In liquids the molecules can still vibrate quite freely, but their rotations are hampered so that their spectra are simpler. In solids the constraints are further tightened. Even in molecular crystals, which are only weakly bonded, the vibrations can take place only when the whole crystal lattice vibrates.

The method best suited for observing the spectra of gases and liquids is to observe absorption. The absorption in solids is often so strong that it is necessary to measure them as fine particles mixed with a transparent binder. Emission is almost never used in infrared spectroscopy of organic compounds because the high temperatures required to produce detectable amounts of radiant energy tend to destroy the groups being studied. Gases, however, can be studied by emission in the infrared (in flames and electric discharges).

When the vibrational bands of diatomic molecules in gases were first examined with spectrometers of low resolution, they consistently showed a double-humped appearance (Fig. 10.7a). As the technique improved, a rather striking fine structure of the humps emerged (Fig. 10.7c) and eventually became recognized as the result of molecular rotation. In a simple-minded way we may picture the diatomic molecule to be a tiny dumbbell rotating about an axis perpendicular to the handle. The handle is not necessarily perfectly rigid because the molecule can be vibrating and rotating at the same time.

A molecule consisting of more than two atoms bound together with quasi-elastic forces exhibits, when disturbed, a quite complicated vibration and a correspondingly complicated spectrum. However, even a motion of this kind can be resolved into a set of normal vibrations each of which is simple in the sense that all atoms contributing to one of a set vibrate with the same frequency and maintain constant phase relationships. Moreover, individual normal vibrations are independent of each other; that is, they can be excited one at a time without coupling with any other. Also, certain groups of atoms called chromophores, each in a particular structure, occur quite commonly in various different molecules. For example, a hydrogen atom is often chemically bonded to a carbon atom, the carbon in turn being bonded to other atoms to form the complete, complex molecule. The carbon–hydrogen group, as with a number of other simple atomic groups, produces a unique relatively simple spectrum that generally represents the group whenever it occurs in different molecules.

Band Shape and Band Broadening

Although each packet of energy emitted or absorbed by an atom or molecule is a discrete amount related by the Bohr equation [Eq. (1-9)] to a specific wavelength, the energy radiated from a group of atoms or molecules can never be recorded as a perfectly sharp line. The spectra of Figures 10.1–10.5 show all the bands to have appreciable width, although many seem to be quite narrow. The general band shape is shown in Fig. 10.8. The width of such a band, called the half width $\Delta\nu$, is the width between two points where the absorption value is half the peak absorption. The width of a recorded band is influenced by the nature of radiant energy, by the emitting or absorbing properties of the molecules, and by imperfections of the instrument making the recording. Spectral bands have values of $\nu_0/\Delta\nu$, where ν_0 is the center frequency, that range from 10^2 to 10^{12}. This great range of widths indicates that this characteristic is influenced by a correspondingly great variety of physical causes. Although a particular broadening effect does not always produce a typical width, line shapes can be classified according to their causes, and magnitudes can be assigned to them that are characteristic for the particular conditions under which they occur. The several causes contributing in a complex way are discussed in Refs. 4–12; but, as our discussion suggests, the simultaneous presence of several broadening effects makes the analysis of band shapes difficult. Figure 10.9, for example, shows an asymmetrical band shape that could be caused by overlapping triplets as

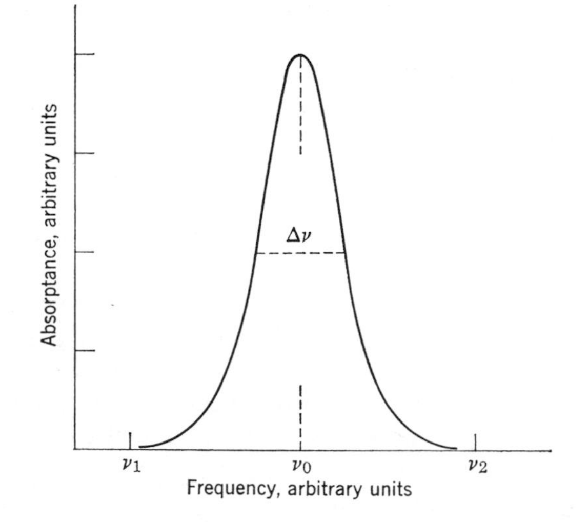

Figure 10.8 General band shape for an isolated absorption line.

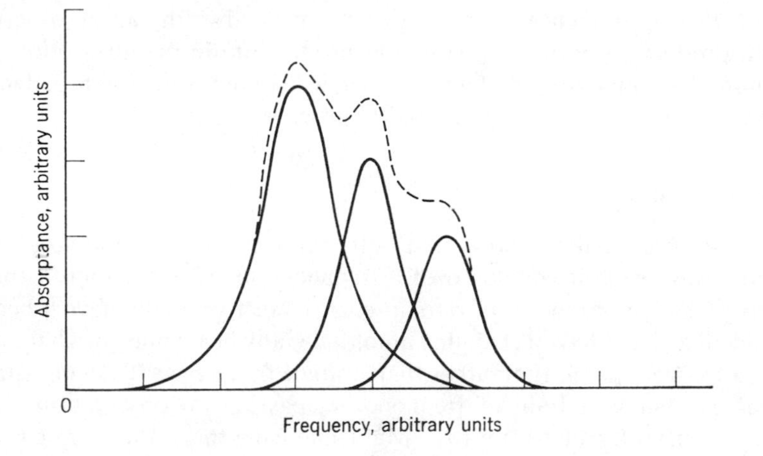

Figure 10.9 An asymmetrical band produced by three closely spaced absorption lines.

shown, but an accurate shape could hardly be determined from the overall band shape alone. The determination is further complicated because the shape of the true band profile may be distorted by the measuring instrument. Two principal problems occur in connection with the recorded shape of spectral bands: one is finding the correct mathematical expression for the actual band profile, and the other is correcting the experimentally observed profile for instrumental effects. Instrument limitations are discussed later in this chapter.

Spectroscopic Information

The spectroscopist generally wishes to know the position (central wavelength or wavenumber) of a band, its width, shape, and its "strength." The strength of an absorption band is its integrated absorptance

$$\alpha = \int_{\lambda_1}^{\lambda_2} \alpha(\lambda)\, d\lambda, \tag{10-6}$$

where $\alpha(\lambda)$ is negligible at wavelengths outside the range λ_1 to λ_2.

The two principal limitations on the accuracy of wavenumber measurements made by spectrometers of moderate resolution (0.2–1 cm^{-1}) are the reproducibility of reading the position of an absorption-line maximum and the reproducibility of setting the instrument to a specific wavenumber. Standard lines occurring in a standard sample are used for calibrating the wavelength setting.

Although fine differences in band shape and bandwidth can be observed between production during emission and production during absorption, we will ignore these subtleties. In fact, we assume here that emission broadening and absorption broadening are reciprocal processes.

Frequency Analysis

A simple Fourier analysis shows that both a finite wavetrain and a damped harmonic wave each is composed of a frequency band rather than a single frequency; so, since a beam of radiant energy must generally have a beginning and an end and have a changing amplitude, any beam must be characterized by a frequency spectrum rather than a single frequency. To demonstrate this, let us assume a train of frequency ν_0 passing an observation point during the time interval $-t_1$ to $+t_1$. Mathematically this is the infinite wave,

$$f(t) = \xi_0 \cos 2\pi\nu_0 t, \tag{10-7}$$

multiplied by the rectangular function,

$$\begin{aligned} R(t) &= 1 \qquad \text{when } -t_1 \le t \le t_1, \\ &= 0 \qquad \text{when } |t| > t_1. \end{aligned} \tag{10-8}$$

The Fourier transform of the product is the convolution of the transforms of the individual functions as follows [Ref. 13, p. 65]:

$$\begin{aligned} F(\nu) &= \xi_0\pi[\delta(\nu - \nu_0) + \delta(\nu + \nu_0)] * 2t_1 \operatorname{sinc} 2\pi\nu t_1 \\ &= 2\pi\xi_0 t_1[\operatorname{sinc} 2\pi(\nu - \nu_0)t_1 + \operatorname{sinc} 2\pi(\nu + \nu_0)t_1], \end{aligned} \tag{10-9}$$

which is plotted in Fig. 10.10. The band at $-\nu_0$ is part of the mathematical result but has no physical reality. The power in the beam is proportional to the square of $F(\nu)$ since $F(\nu)$ represents the amplitude of a wave at the

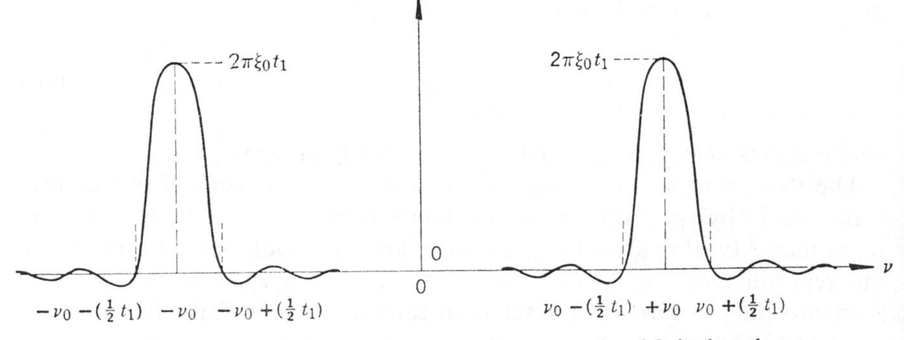

Figure 10.10 Frequency spectrum of a wave train of finite length.

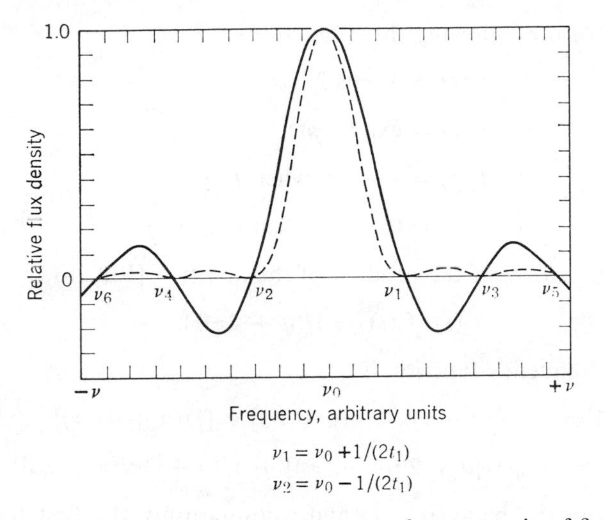

Figure 10.11 Normalized profile of the spectrum of a wave train of finite length (solid line: amplitude; broken line: radiant flux density).

frequency ν. Therefore the band (power) profile is

$$g(\nu) = 4\pi^2\xi_0^2 t_1^2 \operatorname{sinc}^2 2\pi(\nu - \nu_0)t_1, \tag{10-10}$$

which appears in Fig. 10.11 as a broken curve. The band height, maximum radiant flux density, is $4\pi^2\xi_0^2t_1^2$. The first zeros occur when $2\pi(\nu - \nu_0)t_1 = \pm\pi$, that is, at

$$\nu = \nu_0 \pm 1/(2t_1). \tag{10-11}$$

The width of the band between the first zeros is $1/t_1$. Half of this width, which is only approximately the half width, is often called the half width:

$$\Delta\nu = 1/(2t_1). \tag{10-12}$$

In the infrared where $\nu_0 \approx 10^{13}$ Hz, a wavetrain 0.1 sec in duration, as from a highly coherent laser, would have a relatively high Q:

$$Q_r \equiv \nu_0/\Delta\nu = 2\nu_0 t_1 = 2 \times 10^{12}. \tag{10-13}$$

A damped harmonic wave,

$$\begin{aligned} f(t) &= \xi_0 \exp(-at) \cos 2\pi\nu_0 t \qquad && \text{when } t \geq 0, \\ &= 0 && \text{when } t < 0, \end{aligned} \tag{10-14}$$

is the product of the following three functions:

$$f_1(t) = \xi_0 \cos 2\pi\nu_0 t, \tag{10-15}$$

$$f_2(t) = \exp(-at), \tag{10-16}$$

$$\begin{aligned} f_3(t) &= 1 \qquad \text{when } t \geq 0 \\ &= 0 \qquad \text{when } t < 0. \end{aligned} \tag{10-17}$$

The Fourier transform of the product of the last two [Ref. 13, p. 67] is

$$f_2(t) \cdot f_3(t) \leftrightarrow 1/(a + 2\pi i\nu). \tag{10-18}$$

Therefore the frequency function is

$$\begin{aligned} \hat{F}(\nu) &= \pi\xi_0[\delta(\nu - \nu_0) + \delta(\nu + \nu_0)] * [1/(a + 2\pi i\nu)] \\ &= \pi\xi_0\{1/[a + 2\pi i(\nu - \nu_0)] + 1/[a + 2\pi i(\nu + \nu_0)]\}. \end{aligned} \tag{10-19}$$

Again we ignore the band at $-\nu_0$ and examine only the first term in the brace. When the denominator is rationalized, the complex quantity can be written as an exponential:

$$\hat{F}(\nu) = \pi\xi_0[a^2 + 4\pi^2(\nu - \nu_0)^2]^{-1/2} \exp i\varphi(\nu), \tag{10-20}$$

where

$$\varphi(\nu) = \arctan\,[4\pi^2(\nu - \nu_0)^2/a^2]. \tag{10-21}$$

The power is proportional to $F(\nu)$ times its complex conjugate, so the band profile is

$$g(\nu) = \pi^2\xi_0^{\,2}[a^2 + 4\pi^2(\nu - \nu_0)]^{-1}. \tag{10-22}$$

Natural Line Width

A molecule or atom remains in an excited state until spontaneous decay to a lower energy state takes place or until a downward transition is induced by interaction with other molecules, atoms, or photons. The lifetime Δt_1, the most probable time that the molecule is in the excited state, leads to an uncertainty ΔU in the energy of the state, a consequence of the Heisenberg uncertainty principle of quantum theory, according to the relationship

$$\Delta U \cdot \Delta t_i \geq h/2\pi, \tag{10-23}$$

where h is Planck's constant. This leads in turn to an uncertainty in the frequency by

$$\Delta\nu_0 = \Delta U/h \geq 1/(2\pi\Delta t_i). \tag{10-24}$$

This uncertainty in frequency makes desirable the definition of a "natural" spectral line width as at least as wide as the magnitude of the uncertainty.

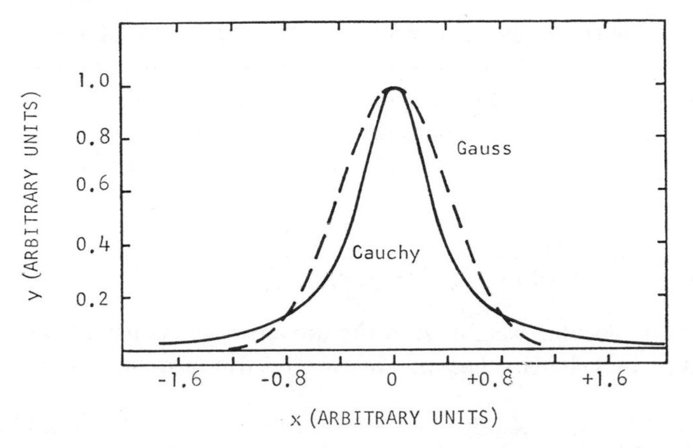

Figure 10.12 Comparison of the profiles of Gauss and Cauchy curves of the same peak height and total area.

Quantum theory yields an expression for the probability of energy between U and $U + dU$ for an atom or a molecule. The profile of a natural line shape can be derived from this expression. For the simpler atomic and molecular configurations, the profile is close to Eq. (10-22). Because this equation was first derived by analyzing a damped harmonic wave, the term "radiation damping" is still used in connection with natural line width. The line profile of a naturally broadened line is

$$g(\nu) = [1/(4\pi^2 \Delta t_i)]\{(\nu - \nu_0)^2 + [1/(4\pi \Delta t_i)]^2\}^{1/2}, \tag{10-25}$$

whose maximum value, when $\nu = \nu_0$, is

$$g_m = 4 \Delta t_i. \tag{10-26}$$

Then,

$$g(\nu) = (1/\pi^2 g_m)[(\nu - \nu_0)^2 + (1/\pi g_m)^2]^{-1}. \tag{10-27}$$

If $g(\nu)$ is set equal to $(\frac{1}{2})g_m$,

$$\Delta\nu = 2/(\pi g_m) = 1/(2\pi \Delta t_i). \tag{10-28}$$

This half width turns out to be the same as the uncertainty in frequency given by Eq. (10-24). The shape of the line is shown in Fig. 10.12 as the "Cauchy" curve.

Doppler Broadening

When a molecule has a velocity component parallel to the line of sight, an observer will note a shift in frequency. The apparent frequency is given by the classical Doppler formula,

$$\nu = \nu_0[1 \pm (v/c)]. \tag{10-29}$$

Here v is the velocity component, and ν_0 is the frequency of the radiant energy as it would be observed from the radiating molecule. Because of a Maxwellian distribution in the amplitudes of the velocity components, caused by thermal motion, the apparent frequency will have a corresponding probability distribution curve. The radiant power profile can be shown to have a Gauss shape:

$$g(\nu) = \left[\frac{Mc^2}{2\pi RT\nu_0^{\,2}}\right] \exp\left\{-\left[\frac{Mc^2}{2RT\nu_0^{\,2}}\right](\nu - \nu_0)^2\right\}, \tag{10-30}$$

where M is the molecular weight, R is the universal gas constant, and T is the temperature in kelvins. The maximum spectral power is seen to be

$$g_m = Mc^2/(2\pi RT\nu_0^{\,2}), \tag{10-31}$$

and the "half width" is

$$\Delta\nu = 2.64(\nu_0/c)(RT/M)^{1/2}. \tag{10-32}$$

Perturbation Broadening

In the infrared range, encounters between molecules—so-called collisions—in a gas under normal conditions of temperature and pressure are the principal influence on the width and shape of bands. Broadening results from the perturbation of energy levels in the absorbing or radiating molecules during the close approach of molecules of either similar or different kinds. An atom or molecule absorbs or emits at a relatively discrete frequency during the time between encounters; however, the radiation process stops abruptly or changes abruptly to a slightly different frequency when the collision occurs. A number of mechanisms are involved in this type of broadening, all of which are more effective at higher pressures. The process becomes especially complicated when the two molecules are alike, so that a certain kind of resonance occurs between them during the interaction. Detailed calculations give a band shape similar to that of Eq. (10-27) but with the term $(1/\pi g_m)$ replaced by a term proportional to the pressure. The value of this term varies appreciably from one gas to another.

Conclusions About True Band Shape

Still other causes of band broadening occur, some of which are asymmetric about the band center. In the three kinds that we have discussed, radiation damping and collision broadening have been shown to give profiles of the same form. Mathematically, these are Cauchy functions of the form

$$y_c = a/(b^2 + x^2). \tag{10-33}$$

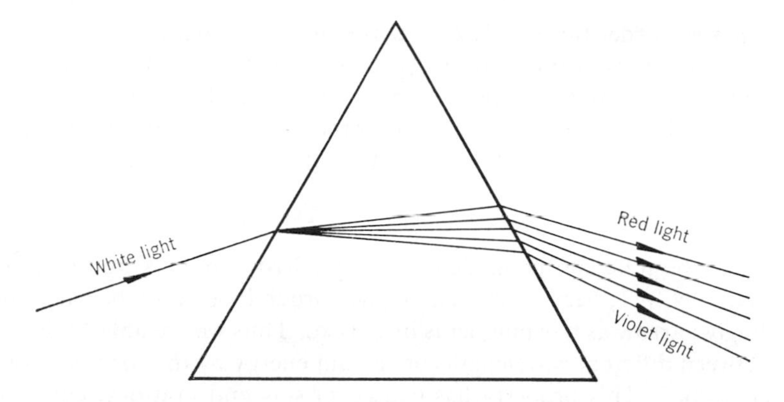

Figure 10.13 Deviation and dispersion of a light beam by a prism.

Doppler broadening has the Gauss form

$$y_g = a' \exp(-x^2/\sigma^2). \tag{10-34}$$

If these two profiles are normalized to both unit peak height and unit area, the following unique solutions result [see Ref. 12]:

$$y_c = (1/\pi^2)\{1/[(1/\pi^2) + x^2]\}, \tag{10-35}$$

$$y_g = \exp(-\pi x^2). \tag{10-36}$$

These two functions are shown graphically in Fig. 10.12.

The curves of Fig. 10.12 show that the measuring instrument should so accurately record a spectrum that one could determine if the recorded band shape fits the Gauss curve or the Cauchy curve, the latter of which is more narrow at the half-intensity points and broader at the base. Then one could tell whether the broadening was due to pressure broadening or to Doppler broadening.

Classical Devices for Wavelength Separation

A study of instruments for measuring spectra and of related practical problems can begin by noting in Fig. 8.3 that a shaped piece of glass separates the colors of a beam of white light into differently directed beams; the optimum shape for this purpose is the prism proportioned so that a right section is an isosceles triangle as in Fig. 10.13. Another means for separating the colors of a white light beam is suggested by Fig. 5.7 and Eq. (5-48). The main peaks in the diffraction pattern of the grating occur where $\pi k_x d = \pi x' d/\lambda z_0 = m\pi$, that is, at

$$x_p' = m\lambda z_0/d, \tag{10-37}$$

where m is the order ($m = 0, 1, 2, 3 \ldots$) designating a particular peak. (The symbol d is used here for the grating spacing instead of x_1.) Equation (10-37) shows that the position of a peak, except the zero order peak, is dependent on wavelength. The lobe of the zero order peak terminates at the axis (goes to zero) where $(2N + 1)\pi x'd/\lambda z_0 = \pi$, that is, where

$$x_w' = \lambda z_0/[(2N + 1)d]. \tag{10-38}$$

Therefore, the lobe can be made extremely narrow by making the number of "slits," $2N + 1$, very large. The higher order lobes also become correspondingly narrow as this number is increased. Thus we are able to discriminate between different wavelengths of radiant energy by the space separation of their peaks. This property has caused prisms and gratings, collectively called dispersors, to become the fundamental devices in spectrum-measuring instruments.

Figure 10.14 is a schematic of a classical spectrum-measuring instrument. It consists of an entrance slit of width a_1 located at the focal point of a collimating lens L_1, a second collimating lens L_2 called the exit collimator, whose focal plane is the plane of the exit slit, and a dispersor between the two collimators. The long dimension of the slit is perpendicular to the plane of the figure.

A diverging, ideally monochromatic beam is shown emerging from the entrance slit. After the beam becomes collimated by passing through the lens, it is incident on the dispersor, whose function is to change the direction of the wavefronts an angular amount depending on the wavelength. The second lens then forms an image of the entrance slit on the exit slit plane. A separate image is formed for each wavelength in a heterogeneous beam, and the several images are distributed along the exit slit plane. We assume that the plane wavefronts leaving the dispersor are rectangular of width a (the long dimension is perpendicular to the figure) and that the entrance slit constitutes a line source that can be represented by a delta function $\delta(x)$.

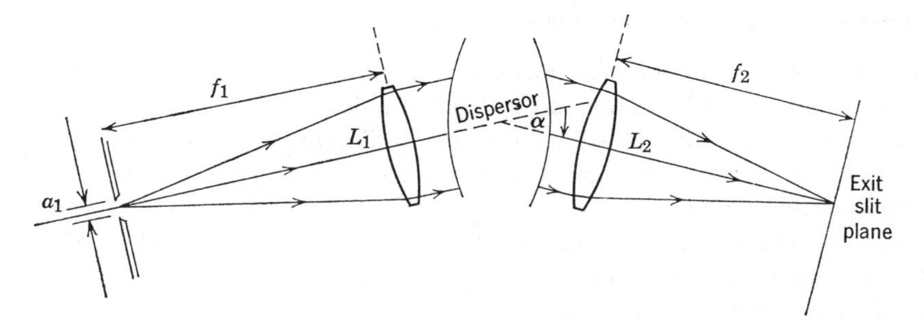

Figure 10.14 Schematic of a general type of classical spectrometer.

Then, if the optical system is free of aberrations, the diffraction pattern is the line-spread function given by Eq. (5-24), except for changing z_0 to f (the focal length of L_2) and changing $2a$ to a (the beam width):

$$W_s(x') = W_0 \operatorname{sinc}^2 (\pi a x'/\lambda f). \tag{10-39}$$

This equation describes the image of the ideal entrance slit since the convolution of a function with the delta function gives the function itself, positioned according to the position of the delta function [see Eq. (5-25)]. Because the angular half width of the main lobe of the pattern [see Fig. 5.4a and Eq. (10-39)] at its base is π radians,

$$x' = f\lambda/a \tag{10-40}$$

at the "edge" of the lobe. Equation (10-40) and Fig. 5.4 show that even a "perfect" instrument broadens an assumed zero-width spectral line.

A number of generic terms are applied to the several forms of spectrum measuring instruments. A spectrograph employs a photographic plate, placed in the exit slit plane (Fig. 10.14) to record all lines in a spectral range simultaneously. The radiant power passing through the optical system is integrated over time, so the quantity recorded by the instrument is a function of radiant energy.

A spectrometer is an instrument with an entrance slit and one or more exit slits. Measurements are made of the radiant power passing through a slit either by scanning the spectral range point by point, or by simultaneously measuring at several spectral positions with different slits. The quantity measured, therefore, is a function of radiant power.

A spectrophotometer is a spectrometer equipped with associated equipment to read out the ratio, or a function of the ratio, of the radiant powers of two beams as a function of spectral position. The two beams may be separated in time, space, or both. We will use spectrometer as a general term when there is no reason to be specific about the kind of instrument.

Definitions of Spectroscopy Terms

Ideally a spectrometer should allow us to examine the details of the true spectrum—no matter how fine. Figure 10.7 shows how differently the fine structure may appear when measurements are made by different instruments. Terms used to characterize instruments and dispersors are defined in the following paragraphs.

To separate two closely spaced lines, the two band centers must be displaced enough that the two images, the broadened recorded lines, do not overlap too much. For the ideal situation discussed in connection with Eq. (10-39), where broadening by the instrument is caused only by diffraction, the two bands are just resolvable, according to the Rayleigh criterion, when the first zero of one line coincides with the peak of the other. However, other effects—aberrations, finite aperture widths, time lag of the recorder, scattering in light sensitive emulsions, imperfect adjustments, and others—cause further instrumental broadening. Generally, the "resolution limit" is the smallest interval $\delta\lambda$ between two just distinguishable wavelengths. This value can rarely be determined with precision.

The angular change of direction of the beam as a function of wavelength is called the angular dispersion,

$$D_i = \delta\alpha/\delta\lambda, \tag{10-41}$$

of the dispersor. A practical parameter for the complete instrument is the linear dispersion X, which depends on the focal length:

$$X = \delta x'/\delta\lambda = f\,(d\alpha/d\lambda). \tag{10-42}$$

Resolving power is the ratio of the mean wavelength of two closely spaced lines, a "doublet," to their wavelength separation when they are just resolved:

$$\mathcal{R} = \lambda/\delta\lambda. \tag{10-43}$$

Broadening of an ideally monochromatic line is a characteristic common to all spectrum measuring instruments. The amount of broadening and profile distortion varies considerably among the several types of instruments and largely decides how well one can determine the fine details or structure of a given spectrum. The general expression that takes into account all the instrumentally induced distortions of the true band shape is called the instrument function (also called instrumental line shape profile or slit function) and is designated by $a(\lambda)$. Figure 10.15 shows the effect of

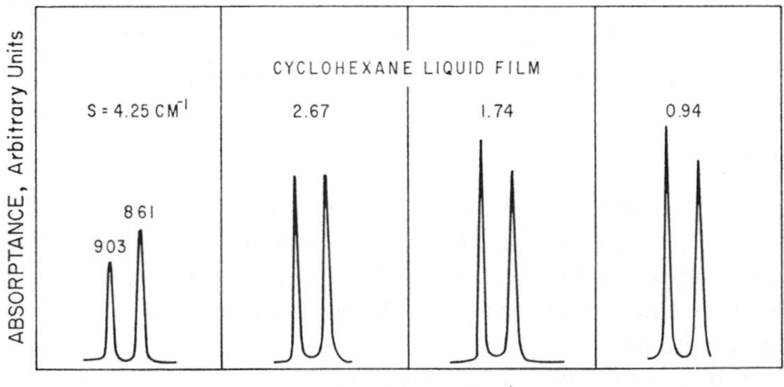

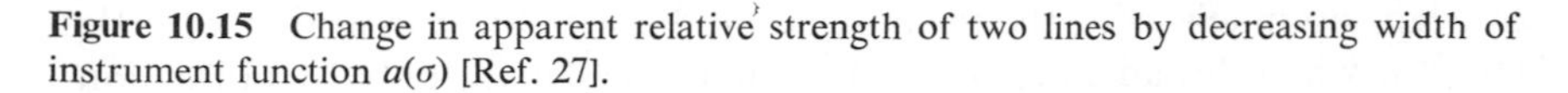

Figure 10.15 Change in apparent relative strength of two lines by decreasing width of instrument function $a(\sigma)$ [Ref. 27].

improving the resolution by reducing the width of the instrument function. It is interesting to note that the relative heights of the two bands of cyclohexine at 903 and 861 cm^{-1} actually reverse with a narrowing of the instrument function.

Instead of being a line, an actual entrance slit has an appreciable width x_0 so that its image function $f_i(x')$ has to be found by the convolution of the line spread function, Eq. (10-39), with the rectangular function $\mathscr{R}_0(x')$ that describes the slit:

$$f_i(x') = W_s(x') * \mathscr{R}_0(x'), \tag{10-44}$$

where $\mathscr{R}_0(x')$ is defined by

$$\begin{aligned} \mathscr{R}_0(x') &= 1 \qquad \text{when } |x'| \leq (x_0/2), \\ &= 0 \qquad \text{when } |x'| > (x_0/2). \end{aligned} \tag{10-45}$$

If the width of the exit slit is x_i, the slit can be described by the rectangular function

$$\begin{aligned} \mathscr{R}_i(x') &= 1 \qquad \text{when } |x'| \leq (x_i/2), \\ &= 0 \qquad \text{when } |x'| > (x_i/2); \end{aligned} \tag{10-46}$$

and the theoretical instrument function for an aberration-free, perfectly adjusted spectrometer is

$$\begin{aligned} f_s(x') &= f_i(x') * \mathscr{R}_i(x') \\ &= W_s(x') * \mathscr{R}_0(x') * \mathscr{R}_i(x'). \end{aligned} \tag{10-47}$$

Since x' can in principle be related to wavelength λ by the linear dispersion, Eq. (10-42), the instrument function can be expressed as a function of wavelength. Thus,

$$f_s(x') = a(\lambda). \tag{10-48}$$

Prisms as Dispersors

Figure 10.16 shows a right section of a dispersing prism with a ray passing through it. The plane of the section is chosen so that it is the plane of incidence of the ray at each surface. The angle β at A is called the refracting angle or prism angle. We find the deviation angle α of a ray as follows: The quadrilateral $APRQ$ is constructed with right angles APR and AQR so that $\beta + PRQ = 180$ deg. Also, in the triangle PRQ, $\theta_1' + \theta_2 + PRQ =$ 180 deg; therefore,

$$\beta = \theta_1' + \theta_2. \tag{10-49}$$

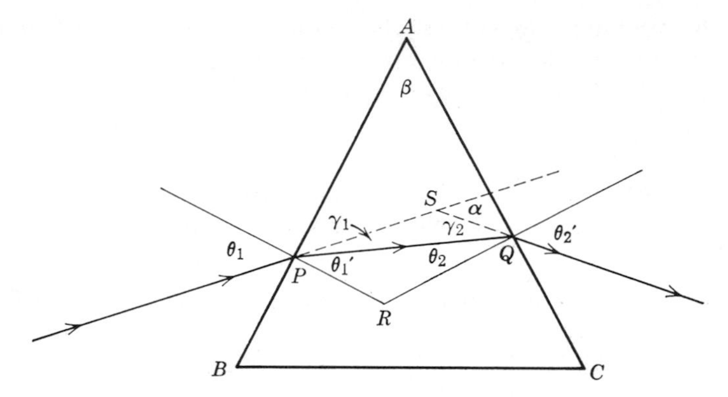

Figure 10.16 Deviation of a ray by a prism.

In the triangle PQS, because α is the exterior angle,

$$\begin{aligned}\alpha &= \gamma_1 + \gamma_2 \\ &= \theta_1 - \theta_1' + \theta_2' - \theta_2.\end{aligned} \tag{10-50}$$

For instrument efficiency, it appears desirable to accommodate as wide a collimated beam as possible through the prism; α has to be restricted to prevent vignetting by the edges of the prism. To reach some reasonable mode of operation, we explore the relationships between the parameters involved. From Equations (10-49) and (10-50),

$$\beta + \alpha = \theta_1 + \theta_2'. \tag{10-51}$$

If n is the refractive index of the prism material for the monochromatic light ray under consideration, Snell's law gives

$$\sin \theta_1 = n \sin \theta_1', \tag{10-52}$$

$$\sin \theta_2' = n \sin \theta_2. \tag{10-53}$$

By adding Equations (10-52) and (10-53) and then applying a trigonometric identity,

$$\sin \tfrac{1}{2}(\theta_1 + \theta_2') \cos \tfrac{1}{2}(\theta_1 - \theta_2') = n[\sin \tfrac{1}{2}(\theta_1' + \theta_2) \cos \tfrac{1}{2}(\theta_1' - \theta_2)]. \tag{10-54}$$

By rearranging and substituting from Equations (10-49) and (10-51),

$$\sin \tfrac{1}{2}(\alpha + \beta) = n \sin \left(\frac{\beta}{2}\right) \frac{\cos \frac{1}{2}(\theta_1' - \theta_2)}{\cos \frac{1}{2}(\theta_1 - \theta_2')}. \tag{10-55}$$

It can be shown that $\sin \frac{1}{2}(\alpha + \beta)$ is a minimum when $\theta_2' = \theta_1$; and for a given β, α is a minimum when $\sin \frac{1}{2}(\alpha + \beta)$ is a minimum. So, for minimum deviation, α_m, the required condition is that the ray must pass symmetrically

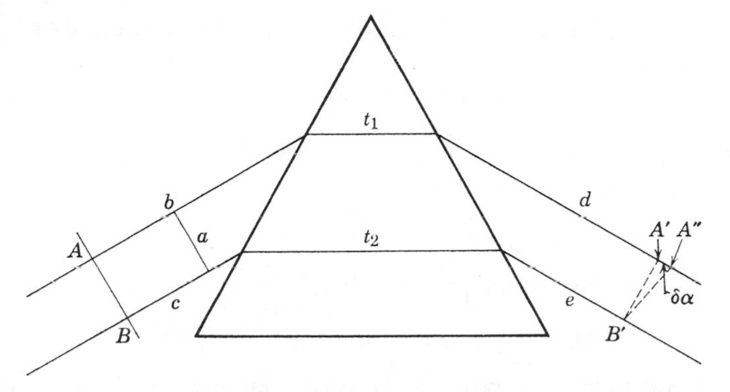

Figure 10.17 Dispersion by a prism.

through the prism. Then the relationship between α_m and β (since, by symmetry, θ_1' will equal θ_2) is

$$\sin \tfrac{1}{2}(\alpha_m + \beta) = n \sin \beta/2. \tag{10-56}$$

To find the dispersion of the prism, we begin by assuming a collimated beam of width a incident on the prism in the direction for minimum deviation at a median wavelength (Fig. 10.17). The optical distances traveled by rays on each edge of the beam are represented by AA' and BB'. If the beam consists of radiant energy at two slightly different wavelengths, the corresponding wavefronts will have slightly different directions, differing by $\delta\alpha$, at $A'B'$ where the second wavefront is represented by $A''B'$. To a first approximation, the rays at the two wavelengths travel along the same paths, one at each wavelength along AA' and a second pair along BB'. If $\delta\lambda$ represents the small difference in wavelength and δn the consequent small change of refractive index of the prism material, equating the optical distances traveled by the edge rays at each wavelength gives the following two equations

$$b + nt_1 + d = c + nt_2 + e, \tag{10-57}$$

$$b + (n + \delta n)t_1 + d + a\delta\alpha = c + (n + \delta n)t_2 + e, \tag{10-58}$$

where $a\delta\alpha$ approximates the extra distance, A' to A'', traveled by the upper ray at the shorter wavelength. Subtraction of Eq. (10-57) from Eq. (10-58) gives

$$\delta n t_1 + a\delta\alpha = \delta n t_2, \tag{10-59}$$

$$(\delta\alpha/\delta n) = (t_2 - t_1)/a. \tag{10-60}$$

If the beam "fills the aperture," that is, the wavefronts are as large as the projected prism face, then t_1 is zero, and t_2 is the thickness t of the prism at its base so that

$$(\delta\alpha/\delta n) = t/a. \tag{10-61}$$

The refractive index variation for most glasses can be expressed as a power series as follows:

$$n = A_0 + A_1/\lambda^2 + A_2/\lambda^4 + \cdots, \tag{10-62}$$

where the coefficients are constants for a particular glass and are small enough that to a first approximation:

$$(\delta n/\delta\lambda) = -2A_1/\lambda^3. \tag{10-63}$$

The angular dispersion of the prism is therefore:

$$(\delta\alpha/\delta\lambda) = (\delta\alpha/\delta n)(\delta n/\delta\lambda) = -2tA_1/a\lambda^3. \tag{10-64}$$

The theoretical resolving power for a prism can be found by working with the relationships derived above. From Eq. (10-61),

$$(\delta\alpha/\delta\lambda)(\delta\lambda/\delta n) = (t/a), \tag{10-65}$$

$$(\delta\alpha/\delta\lambda) = (t/a)(\delta n/\delta\lambda). \tag{10-66}$$

If a ray passes symmetrically through the prism, there is no vignetting, and the beam width is unchanged by passage through the prism. The separation between the peak of a theoretical instrument function and the first minimum, the resolution limit separation, is given by Eq. (10-40). In that equation, x'/f is equivalent to $\delta\alpha$. Substituting for $\delta\alpha$ in Eq. (10-66) gives the resolving power:

$$\mathcal{R} = (\lambda/\delta\lambda) = t(\delta n/\delta\lambda). \tag{10-67}$$

Gratings as Dispersors

The simple grating described in Chapter 5 would hardly be a practical dispersor in a modern spectrometer because too many orders share the radiant energy and too much energy goes into the zero order where there is no wavelength discrimination. However, a properly designed and ruled grating is indeed a practical dispersor [Refs. 14–17]. Usually the grating is a reflecting type with a periodic structure consisting of grooves cut into an aluminum coating that has been vapor-deposited onto a glass surface. The grooves are cut, or rather ruled, by diamond cutting tools controlled by elaborate ruling machines. A section of the surface perpendicular to the groove length has the general profile shown in Fig. 10.18. The angle of the groove β is 100–120 deg. Such patterns are called blazed gratings; the blaze angle α is the angle between a large facet and the mean grating surface. The blaze angle and the groove angle are adjusted so that a large fraction of the diffracted energy occurs in one of the lower orders. The direction of this angle with respect to a mean-surface normal is the angle of diffraction. As much as 75–90% of the incident energy can be concentrated into a single

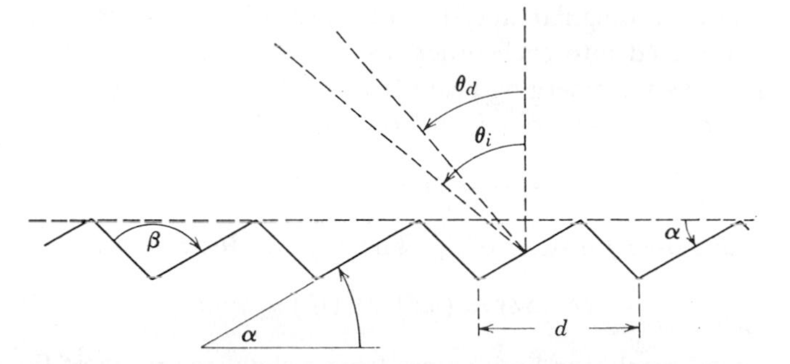

Figure 10.18 Surface profile of a blazed reflection grating.

order. Spacing constants, represented by d, corresponding to 300–600 lines/mm are common, and 2400 lines/mm are possible [Ref. 14]. The smallest usable spacing constant, however, is $\frac{1}{2}\lambda$ at the shortest wavelength to be applied to the grating.

The glass surface, before coating, is shaped and polished to optical flatness for plane gratings, or to a spherical shape for concave gratings. Plastic replicas of fine ruled gratings are of such high quality that the most commonly used gratings are replicas. The sizes of gratings vary, but sizes from 4 to 8 in. are common.

For highest resolution the grating is used in an "autocollimation" mode: the angle of incidence θ_i (with a normal to the mean grating surface) is approximately equal to the angle of diffraction θ_d; and both angles are large (64–75 deg) and both are on the same side of the normal.

A study of Eq. (5-45) shows that the beam diffracted into a given order consists of essentially plane wavefronts. If the value of x_w' given by Eq. (10-38) is substituted in Eq. (5-45),

$$W_g(x') = W_s(x')(\sin^2 \pi)/\sin^2 [\pi/(2N + 1)]. \qquad (10\text{-}68)$$

This equation shows that for large $(2N + 1)$ (i.e., many grooves) the argument of the sine in the denominator is so small in the vicinity of the main peak that the sine can be approximated by its argument. When this approximation is made and both numerator and denominator of Eq. (5-45) are multiplied by $(2N + 1)^2$, this equation is obtained:

$$W_g(x') \approx (2N + 1)^2 W_s(x') \frac{\sin^2 [(2N + 1)\pi k_x d]}{[(2N + 1)\pi k_x d]^2}$$

$$\approx (2N + 1)^2 W_s(x') \operatorname{sinc}^2 [(2N + 1)\pi k_x d] \qquad (10\text{-}69)$$

(As in an earlier development in this chapter, d is used instead of x_1 for grating spacing.) Since the sinc function describes the diffraction pattern produced by a rectangular aperture of width $(2N + 1)d$ [see Eq. (5-24)], the beam diffracted into each order, as it leaves the grating, has the same characteristics as the emerging beam from such an aperture.

The linear dispersion is obtained from Eq. (10-37):

$$(\delta x'/\delta\lambda) = mf/d, \tag{10-70}$$

where we have used f instead of z_0. The angular dispersion is

$$(\delta\alpha/\delta\lambda) = (\delta x'/\delta\lambda)(1/f) = m/d. \tag{10-71}$$

The theoretical resolution limit comes from a slight rewriting of Eq. (10-38):

$$\delta\alpha = \frac{x_w'}{f} = \frac{x_w'}{z_0} = \lambda/[(2N + 1)d]. \tag{10-72}$$

If this value of $\delta\alpha$ is substituted into Eq. (10-71), the resolving power is obtained:

$$\mathscr{R} = (\lambda/\delta\lambda) = m(2N + 1). \tag{10-73}$$

Hence, for a given grating width, a given order, and also (though not discussed in our development) a given angle of incidence, the theoretical resolving power is determined by the total number of lines (grooves) in the grating.

The product $(2N + 1)d$, appearing in Eq. (10-69), is the width of the grating. The width of the diffracted beam determines the line-spread function, Eq. (10-39). Therefore, whenever the diffraction angle θ_d is too large for its cosine to be approximated by unity, the width of the grating must be reduced by $\cos\theta_d$. Thus, the resolution limit, Eq. (10-72), will increase by the factor $(1/\cos\theta_d)$; and the resolving power, Eq. (10-73), will decrease by $\cos\theta_d$.

Scanning a Spectrum

No matter what kind of instrument is used to measure a continuous spectrum, the instrument can never "look" at an infinitesimally narrow region of the spectrum. If the radiant power of the beam to be studied is dispersed along the exit slit plane as shown in Fig. 10.14, a narrow window, a slit properly placed in the dispersed beam, allows a portion of the radiant energy in the vicinity of a mean wavelength λ_i to pass through to a radiant energy detector. When the window is moved relative to the dispersed beam, it scans all wavelengths of the spectrum; but at every instant energy from a band, rather than a single wavelength, passes through. This is shown graphically in Fig. 10.19. All the radiant power within the band about λ_i will be recorded

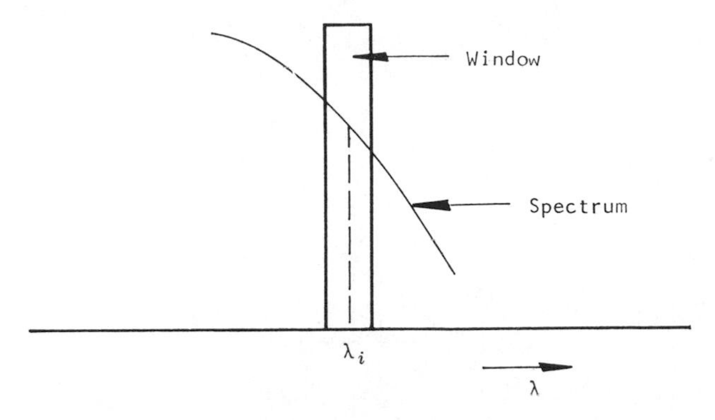

Figure 10.19 Scan of a spectrum with a "window."

by the instrument as the power received at the single wavelength λ_i. If we assume an infinitesimally narrow line and a rectangular window, the recorded band would be as wide as the window and rectangular in shape. This example is the physical equivalent of the mathematical convolution of a rectangular function with a delta function. In general, all spectrometers incorporate some form of a "spectral window," and its characteristics are expressed in an instrument function $a(\lambda)$.

Convolution of Instrument Function and True Spectrum

If the energy distribution over a spectrum is represented mathematically as a function of wavenumber by $\varphi(\sigma)$, the distribution recorded by a spectrometer can be calculated by a convolution process. Each quasi-monochromatic component of the true spectrum $\varphi(\sigma)\,d\sigma$, at σ, must be operated on by the instrument function centered at σ. In Fig. 10.20, some of the radiant power at σ will contribute to the observed power at an arbitrary point σ'; similarly, other components of the true spectrum also make contributions to the apparent radiant power at σ'. The total power at σ' can be evaluated as follows:

$$f(\sigma') = \int_{-\infty}^{+\infty} a(\sigma' - \sigma)\varphi(\sigma)\,d\sigma. \tag{10-74}$$

From our earlier chapters, Eq. (10-74) is recognized as the convolution of a with φ.

If the instrument function were an infinitesimally sharp line, it could be represented by the Dirac delta function $\delta(\sigma - \sigma')$. Then the convolution of the true spectrum with the delta function would produce the true spectrum:

$$\varphi(\sigma') = \int_{-\infty}^{+\infty} \delta(\sigma - \sigma')\varphi(\sigma)\,d\sigma. \tag{10-75}$$

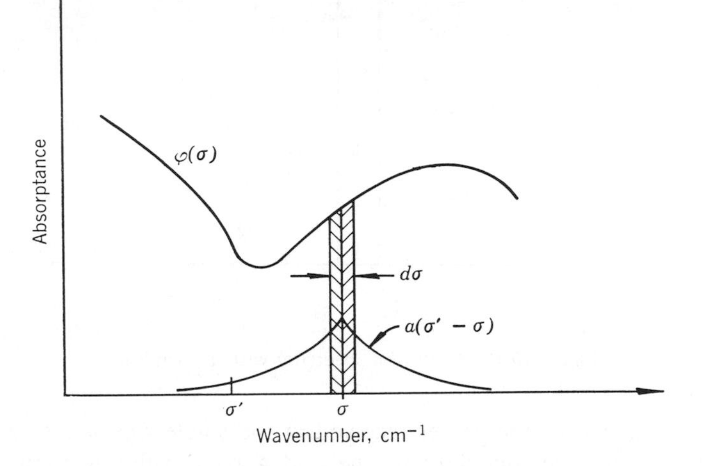

Figure 10.20 Application of the instrument function to a spectrum.

From the preceding discussion, it is obvious that the representation of the true spectrum improves with sharpening of the instrument function. This corresponds to narrowing the spectral window. Unfortunately, changing the characteristics of any spectrometer in this direction inevitably reduces the radiant power available for the measurement process (exciting a detector, exposing a film, etc.) until the signal-to-noise limit is reached. In the vicinity of this limit, the instrument designer or operator gains a better resolved spectrum at the expense of a noisy (less reliable) signal. Because the instrument is energy limited, a compromise has to be made at a resolution capability considerably less than ideal.

In order that the recorded spectrum may show a total power proportional to that in the true spectrum, the instrument function must be normalized to unity so that

$$\int_{-\infty}^{+\infty} a(\sigma)\, d\sigma = 1. \tag{10-76}$$

The proper function $a(\sigma)$ would therefore contain the necessary normalizing constant. Generally, the total power is not required, and a spectrometer measures only the relative power as a function of wavelength.

It is difficult to determine $a(\sigma)$ with precision experimentally [Refs. 18–20], chiefly because the widths of available emission or absorption bands are not infinitesimal compared with the width of the instrument function. Radiant energy produced by lasers does have a very narrow bandwidth, but its high coherence causes a new set of optical problems that may negate the bandwidth advantage.

Sometimes the causes of instrumental distortions can be separated and an instrument function estimated or determined theoretically for each cause. For example, there may be electronic causes such as a slow detector time constant or a limited electronic bandpass. Also, there may be aberrations in the optical system or diffraction effects that cause distortion of the spectrum. If $a_e(\sigma)$ is the instrument function due to the electronics and $a_0(\sigma)$ the instrument function due to optical effects, the overall instrument function is found by the convolution of the two:

$$a(\sigma') = \int_{-\infty}^{+\infty} a_e(\sigma' - \sigma) a_0(\sigma)\, d\sigma. \tag{10-77}$$

REFERENCES

1. H. K. Hughes, "Suggested Nomenclature in Applied Spectroscopy: Report of the Joint Committee on Nomenclature," *Anal. Chem.*, **24,** 1349 (1952).
2. E 131-65T, "Tentative Definitions of Terms and Symbols Relating to Molecular Spectroscopy," *1966 Book of ASTM Standards*, p. 651.
3. E 135-66, "Standard Definitions of Terms and Symbols Relating to Emission Spectroscopy," *1966 Book of ASTM Standards*, p. 655.
4. R. M. Goody, *Atmospheric Radiation*, Chapters 2 and 3 and References. Oxford University Press, London, 1964.
5. R. G. Breene, Jr., "Line Width," in *Handbuch der Physik*, Vol. 27. Edited by S. Flugge. Springer-Verlag, Berlin, 1964.
6. R. G. Breene, Jr., *The Shift and Shape of Spectral Lines*. Pergamon Press, New York, 1961.
7. A. Di Giacomo, "Impact Approximation for Line Shape in the Microwave Region," *Nuovo Cimento*, **34,** 473 (1964).
8. A. Di Giacomo, "On the Validity of the Semiclassical Treatment of Collisions in Problems of Pressure Shift and Broadening of Spectral Lines," *Nuovo Cimento*, **44,** 140 (1966).
9. C. L. Chaney, "Integrated Spectral Line Intensities," *Spectrochim. Acta*, **32A,** 1 (1967).
10. B. Bizzerides, "Radiation Absorption Phenomena in Gases—I. General Theory of Line Broadening," *J. Quant. Spectro. Radiative Transfer*, **7,** 353 (1967).
11. R. Norman Jones et al., "A Statistical Approach to the Analysis of Infrared Band Profiles," *Can. J. Chem.*, **41,** 750 (1963).
12. K. S. Seshadri and R. N. Jones, "The Shape and Intensities of Infrared Absorption Bands—A Review," *Spectrochim. Acta*, **19,** 1013 (1963).
13. A. Papoulis, *Systems and Transforms with Applications in Optics*. McGraw-Hill Book Company, Inc., New York, 1968.
14. G. W. Stroke, "Diffraction Gratings," in *Handbuch der Physik*, Vol. XXIX. Edited by S. Flugge. Springer-Verlag, Berlin, 1967.
15. G. W. Stroke, "Ruling, Testing and Use of Optical Gratings for High Resolution Spectroscopy," in *Progress in Optics*, Vol. II. Edited by E. Wolf. North-Holland Publishing Company, Amsterdam, 1963.
16. R. P. Madden and J. Strong, "Diffraction Gratings," in *Concepts of Classical Optics*. J. Strong, W. H. Freeman and Company, San Francisco, 1958.

17. D. Richardson, "Diffraction Gratings," in *Applied Optics and Optical Engineering*, Vol. V, Part II. Edited by R. Kingslake. Academic Press, New York, 1969.
18. A. Girard and P. Jacquinot, "Principles of Instrumental Methods in Spectroscopy," in *Advanced Optical Techniques*. Edited by A. C. S. Van Heel. North-Holland Publishing Company, Amsterdam, 1967.
19. S. G. Routain, "Real Spectral Apparatus," *Sov. Phys.—Usp.*, **66**, 245 (1958).
20. K. Freis and H. H. Gunthard, "Distortion of Spectral Line Shapes by Recording Instruments," *J. Opt. Soc. Am.*, **51**, 83 (1961).
21. J. C. Decius, E. H. Coker, and G. L. Brenna, "Vibrational Spectra of Sulfate Ions in Alkali Halide Crystals," *Spectrochim. Acta*, **19**, 2181 (1963).
22. E. Borello and A. Zicchina, "Infrared Spectra of *v*-Triazoles I. 2 Aryl-*v*-Triazoles," *Spectrochim. Acta*, **19**, 1703 (1963).
23. T. Uno, K. Machida, and K. Hanai, "Infrared Spectra of Sulfamide and Sulfamide-d4," *Spectrochim. Acta*, **22**, 2065 (1966).
24. J. R. Durig and J. N. Willis, Jr., "The Vibrational Spectra and Structure of Diketine and Diletane-d4," *Spectrochim. Acta*, **22**, 1299 (1966).
25. W. E. Bron and W. R. Heller, "Rare-Earth Ions in the Alkali Halides. I. Emission Spectra of Sm^{2+}-Vacancy Complex," *Phys. Rev.*, **136**, A1436 (1964).
26. W. West, *Chemical Applications of Spectroscopy*, Part I. (Vol. IX of *Techniques of Organic Chemistry*. Edited by A. Weissberger.) Second Edition. Interscience Publishers, New York, N.Y., 1968.
27. H. J. Sloane, "High Resolution Infrared Spectroscopy in the Analysis of Liquids and Solutions," *Appl. Spectrosc.*, **16**, 5 (1962).
28. C. D. Kennedy, "High Resolution Instrumentation: Advantages," in *Progress in Infrared Spectroscopy*, Vol. I. Edited by H. A. Szymanski. Plenum Press, New York, 1962.

11

Interference Spectroscopy

This chapter deals in some detail with the Fourier transform method of interference spectroscopy [Refs. 1–3], one method by which a spectrum, either in original or modified form, can be analyzed. Introductory topics, as follows, have been discussed in earlier chapters: how a radiant energy beam which has an inherent distribution of its energy over a range of wavelengths (a spectrum) may have its spectrum modified by transmission through an absorbing material; the significance of absorption and emission spectra; the terminology of spectroscopy; the reasons for an instrument function; the operation of classical, dispersing spectrometers; and how the convolution of the instrument function with the true spectrum gives the recorded spectrum. Interference spectroscopy consists of two processes: the first, experimental, is the recording of an autocorrelation function called the *interferogram*; and the second, mathematical, is the computation of the spectrum from the interferogram. An adequate signal-to-noise ratio requires that the interfering beams be coherent enough to record an interference pattern of high visibility. So the present discussion begins appropriately by extending the earlier treatment of coherence (see Chapter 1), followed by a demonstration of how spectral information is contained in the autocorrelation function.

Superposition of Beams

As described in Chapter 1, a beam of radiant energy may be divided and then the two parts superposed after traveling different paths. Let $\hat{V}_1(t)$ and $\hat{V}_2(t)$ be the analytic signal representations of the beams at the two pinholes (see Fig. 1.4). In the region of superposition Q the beams are represented by $\hat{k}_1\hat{V}_1(t - \theta_1)$ and $\hat{k}_2\hat{V}_2(t - \theta_2)$. The several θ are time delays, the times required for the energy to travel from the respective pinholes. The several $\hat{k}$

are geometrical factors that depend upon the pinhole sizes and the distances and directions to the region of superposition. In the theory of diffraction, these constants are considered pure imaginaries [Ref. 4, p. 500]. We shall not discuss them further here except to give the following properties for later use:

$$\hat{k}_1\hat{k}_1^* = |k_1|^2,$$
$$\hat{k}_2\hat{k}_2^* = |k_2|^2, \tag{11-1}$$
$$\hat{k}_1\hat{k}_2^* = \hat{k}_1^*\hat{k}_2 = |k_1k_2|.$$

The superposed beams can be represented by the sum of their analytic signals:

$$\hat{V}(t) = \hat{k}_1\hat{V}_1(t - \theta_1) + \hat{k}_2\hat{V}_2(t - \theta_2). \tag{11-2}$$

The radiant flux density is found by substituting in Eq. (1-45):

$$W = \langle[\hat{V}(t)\hat{V}^*(t)]\rangle, \tag{11-3}$$

$$\hat{V}(t)\hat{V}^*(t) = [\hat{k}_1\hat{V}_1(t - \theta_1) + \hat{k}_2\hat{V}_2(t - \theta_2)]$$
$$\times [\hat{k}_1^*\hat{V}_1^*(t - \theta_1) + \hat{k}_2^*\hat{V}_2^*(t - \theta_2)], \tag{11-4}$$

$$\hat{V}(t)\hat{V}^*(t) = \hat{k}_1\hat{k}_1^*\hat{V}_1(t - \theta_1)V_1^*(t - \theta_1)$$
$$+ \hat{k}_2\hat{k}_2^*\hat{V}_2(t - \theta_2)\hat{V}_2^*(t - \theta_2)$$
$$+ \hat{k}_1\hat{k}_2^*\hat{V}_1(t - \theta_1)\hat{V}_2^*(t - \theta_2)$$
$$+ \hat{k}_1^*\hat{k}_2\hat{V}_1^*(t - \theta_1)\hat{V}_2(t - \theta_2). \tag{11-5}$$

Since the analytic signal is stationary, a statistical quantity obtained from it does not depend upon our choice of zero for time. That is, the quantity we obtain from the time average of $\hat{V}_1(t - \theta_1)\hat{V}_1^*(t - \theta)$ is the same as from $\hat{V}_1(t)\hat{V}_1^*(t)$. Then the first and second terms of Eq. (11-3), as expanded in Eq. (11-5), are as follows:

$$\langle[\hat{k}_1\hat{k}_1^*\hat{V}_1(t - \theta_1)\hat{V}_1^*(t - \theta_1)]\rangle = |k_1|^2 \langle[\hat{V}_1(t)\hat{V}_1^*(t)]\rangle$$
$$= |k_1|^2 W_1, \tag{11-6}$$

$$\langle[\hat{k}_2\hat{k}_2^*V_2(t - \theta_2)V^*(t - \theta_2)]\rangle = |k_2|^2 W_2. \tag{11-7}$$

Here $|k_1|^2 W_1$ and $|k_2|^2 W_2$ are the respective radiant flux densities in the region of superposition that would be produced by the beams acting alone. If we set $\theta = \theta_2 - \theta_1$ and then shift the origin of time so that $(t' + \theta - \theta_2) = (t + \theta)$ and $(t' - \theta_2) = t$ (where the primes refer to the origin before the shift), the third and fourth terms of Eq. (11-5) become

$$|k_1k_2|\ \hat{V}_1(t+\theta)\hat{V}_1^*(t),$$

and

$$|k_1k_2|\ \hat{V}_1^*(t+\theta)\hat{V}_2(t).$$

Because each is the complex conjugate of the other, their sum is

$$2\ \mathscr{Re}\ |k_1k_2|\ \hat{V}_1(t+\theta)\hat{V}_2^*(t).$$

Then the radiant flux density of Eq. (11-3) is, by assembling the evaluations of the four terms in Eq. (11-5),

$$W = |k_1|^2\ W_1 + |k_2|^2\ W_2 + 2\ |k_1k_2|\ \mathscr{Re}\ \langle[\hat{V}_1(t+\theta)\hat{V}_2^*(t)]\rangle. \tag{11-8}$$

Complex Degree of Coherence

The last term of Eq. (11-8) is of interest in our discussion of coherence. The quantity $\langle[\hat{V}_1(t+\theta)\hat{V}_2^*(t)]\rangle$ is called the mutual coherence $\hat{\Gamma}_{12}(\theta)$ of the two beams, so, by definition,

$$\hat{\Gamma}_{12}(\theta) = \langle[\hat{V}_1(t+\theta)\hat{V}_2^*(\theta)]\rangle. \tag{11-9}$$

A more commonly used term, the complex degree of coherence $\gamma_{12}(\theta)$, is defined by normalizing the mutual coherence as follows:

$$\hat{\gamma}_{12}(\theta) = \frac{\hat{\Gamma}_{12}(\theta)}{(W_1W_2)^{1/2}} = \frac{\langle[\hat{V}_1(t+\theta)\hat{V}_2^*(t)]\rangle}{[(\langle\hat{V}_1\hat{V}_1^*\rangle)(\langle\hat{V}_2\hat{V}_2^*\rangle)]^{1/2}}. \tag{11-10}$$

A well-known relationship, Schwarz's inequality [Ref. 5, p. 63], shows that the magnitude of the numerator in Eq. (11-10) is always less than or equal to that of the denominator. Therefore,

$$|\gamma_{12}(\theta)| \leq 1. \tag{11-11}$$

Let $[\alpha(\theta) + 2\pi\nu_0\theta]$ be the argument of the complex degree of coherence where ν_0 is a mean frequency in the beams. Then Eq. (11-8) can be written

$$W = |k_1|^2\ W_1 + |k_2|^2\ W_2 + 2\ |k_1k_2|\ (W_1W_2)^{1/2}\ |\gamma_{12}(\theta)| \cos\ [\alpha(\theta) + 2\pi\nu_0\theta]. \tag{11-12}$$

Frequency Information

In an experiment (discussed in greater detail later) with a Michelson interferometer, one can set $|k_1| = |k_2| = 1$, $W_1 = W_2$, and $\alpha(\theta) = 0$. Under these conditions, Eq. (11-12) reduces to

$$W = 2W_1[1 + |\gamma_{12}(\theta)| \cos 2\pi\nu_0\theta]. \tag{11-13}$$

If the beams are now made completely coherent, that is, $|\gamma_{12}(\theta)| = 1$, the equation becomes the classical

$$\begin{aligned} W &= 2W_1[1 + \cos 2\pi\nu_0\theta] \\ &= 4W_1 \cos^2 \pi\nu_0\theta. \end{aligned} \tag{11-14}$$

By substituting for ν_0 and θ as follows:

$$\nu_0 = c/\lambda_0 = c\sigma_0, \tag{11-15}$$

$$\theta = \delta/c, \tag{11-16}$$

the equation takes the form,

$$W = 4W_1 \cos^2 \pi\sigma_0\delta, \tag{11-17}$$

where σ_0 is the wavenumber, and δ is the difference between the optical paths traveled by the two beams.

Frequency Spectrum From Correlation Functions

The simple form of Eq. (11-17) emphasizes that spectrum information is contained in the interference pattern just as it is in the more general mutual coherence $\hat{\Gamma}_{12}(\theta)$, a correlation function. From the definition of the notation we have been using,

$$\langle[\hat{V}_1(t + \theta)\hat{V}_2^*(t)]\rangle = \lim_{T\to\infty} \frac{1}{2T} \int_{-T}^{+T} \hat{V}_1(t + \theta)\hat{V}_2^*(t)\, dt, \tag{11-18}$$

the cross-correlation function. The cross-power spectrum is obtained by taking the Fourier transform [Ref. 6, p. 65, et seq.]. To conform with practice in designating correlation functions, let us use the notation

$$\hat{C}_{12}(\theta) = \hat{\Gamma}_{12}(\theta). \tag{11-19}$$

If the two beams came from the same source in such a way that $\hat{V}_1(t) = \hat{V}_2(t)$, the expression in Eq. (11-18) would become the autocorrelation function. Then the Fourier transform gives the power density spectrum $\hat{B}(\nu)$ as follows:

$$\hat{B}(\nu) = \int_{-\infty}^{+\infty} \hat{C}_{11}(\theta) \exp(-2\pi i\nu\theta)\, d\theta, \tag{11-20}$$

and by the inversion theorem,

$$\hat{C}_{11}(\theta) = \int_{-\infty}^{+\infty} \hat{B}(\nu) \exp(2\pi i\nu\theta)\, d\nu. \tag{11-21}$$

The following substitutions put these equations into notation more common to spectroscopy:

$$\begin{aligned} \theta &= \delta/c, \\ \nu &= \sigma c, \\ \hat{C}_{11}(\theta) &= \hat{I}(\delta), \\ \hat{B}(\nu) &= \hat{I}(\sigma). \end{aligned} \tag{11-22}$$

Then

$$\hat{I}(\delta) = \int_{-\infty}^{+\infty} \hat{I}(\sigma) \exp(2\pi i \sigma \delta)\, d\sigma, \tag{11-23}$$

$$\hat{I}(\sigma) = \int_{-\infty}^{+\infty} \hat{I}(\delta) \exp(-2\pi i \sigma \delta)\, d\delta. \tag{11-24}$$

Thus, if we can record $\hat{I}(\delta)$, a function of path difference which is proportional to the autocorrelation, we can obtain the spectrum.

Visibility of Fringes

The flux density represented by Eq. (11-12) has a maximum value when $\cos[\alpha(\theta) + 2\pi\nu_0\theta] = +1$ and a minimum when this function becomes -1. The fringe visibility as given by Eq. (1-46) can then be written

$$\mathscr{V} = \frac{2\,|k_1k_2|\,(W_1W_2)^{1/2}\,|\gamma_{12}(\theta)|}{|k_1|^2\,W_1 + |k_2|^2\,W_2}, \tag{11-25}$$

which shows that the visibility of fringes is proportional to the degree of coherence; the visibility is zero when the coherence is zero and is a maximum when the degree of coherence is unity. Equation (11-17), for example, gives a visibility of unity because the equation was derived for a completely coherent beam.

Two beams having essentially no coherence will still interfere in the sense that there will be a combined effect due to the presence of two beams, but the instantaneous effects will be unobservable (see Chapter 1).

The Michelson Interferometer

We begin the study of interferometry by considering coherent beams, that is, by considering ideal wave trains. To show how a wave train can be divided by amplitude division into two beams and caused to interfere, we use the

Michelson interferometer shown schematically in Fig. 11.1. The incident wave train originating at Q is represented by Eq. (1-14):

$$\xi = \xi_0 \cos(2\pi\nu_0 t - \varphi). \tag{11-26}$$

At the beamsplitter the wave train is divided, a portion being reflected and the remainder transmitted. The reflected portion is again reflected at mirror M_1 so that it travels back toward the beamsplitter. At the beamsplitter the wave will be divided again, part being reflected and the rest transmitted toward Q'. Similarly, the portion transmitted at the first encounter will be reflected from mirror M_2 and then a part reflected from the beamsplitter toward Q'. With perfect alignment of the beamsplitter and mirrors, the two parts of the wave train traveling toward Q' are exactly parallel.

If φ_2 is the phase of one of the waves at x_2, the other wave will have a phase $\varphi_2 - (2\pi\delta/\lambda)$ to take into account the difference δ in the two paths. The expressions for the two waves are then

$$\begin{aligned} \xi_1 &= (\xi_0/2)\cos(2\pi\nu_0 t - \varphi_2), \\ \xi_2 &= (\xi_0/2)\cos[2\pi\nu_0 t - \varphi_2 - (2\pi\delta/\lambda)]. \end{aligned} \tag{11-27}$$

The resultant field quantity is

$$\xi = \xi_1 + \xi_2. \tag{11-28}$$

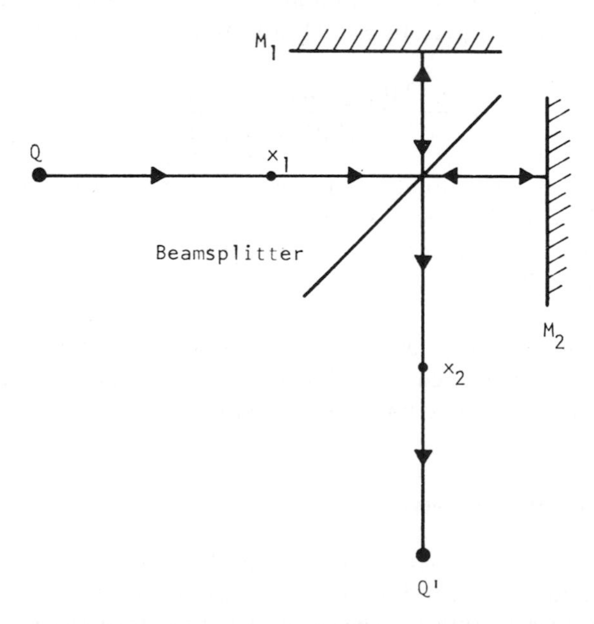

Figure 11.1 Schematic of a Michelson interferometer.

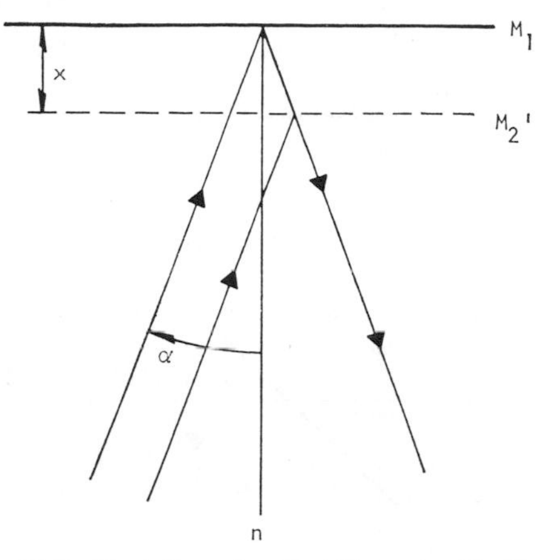

Figure 11.2 Variation of path difference with direction.

When the coherent ξ_1 and ξ_2 are combined, the amplitude and phase of ξ are found to be $[\xi_0 \cos(\pi\delta/\lambda)]$ and $(2\pi\nu_0 - \varphi_2)$, respectively. The average of the square of ξ over a large number of cycles, with a convenient choice of units, gives

$$W = (\xi_0^2/4)[1 + \cos(2\pi\delta/\lambda)] \tag{11-29}$$

for the radiant flux density. As indicated by Eq. (11-29), whenever

$$(2\pi\delta)/\lambda = n\pi \quad \text{or} \quad \delta = n\lambda/2 \qquad n = 1,3,5,7,\ldots, \tag{11-30}$$

the flux density becomes zero. Whenever

$$(2\pi\delta)/\lambda = m\pi \quad \text{or} \quad \delta = m\lambda/2, \qquad m = 0,2,4,6,\ldots, \tag{11-31}$$

the flux density is a maximum. In words: When the path difference is an odd number of half wavelengths, the two waves cancel; when the path difference is an even (including zero) number of wavelengths, the flux density is a maximum.

Formation of Fringes

If we assume that many wave trains arriving at the interferometer of Fig. 11.1 have the same frequency ν_0, but are arriving from all directions within a solid angle whose half-cone angle is α_m (which is assumed small, only a few degrees), a significant variation in path difference will result. These can be evaluated by looking toward the beamsplitter from Q', seeing through to M_1, and also seeing the image of M_2 formed by reflection at the beam splitter. A schematic diagram constructed from this point of view, Fig. 11.2, shows that

$$\delta = 2x \cos \alpha. \tag{11-32}$$

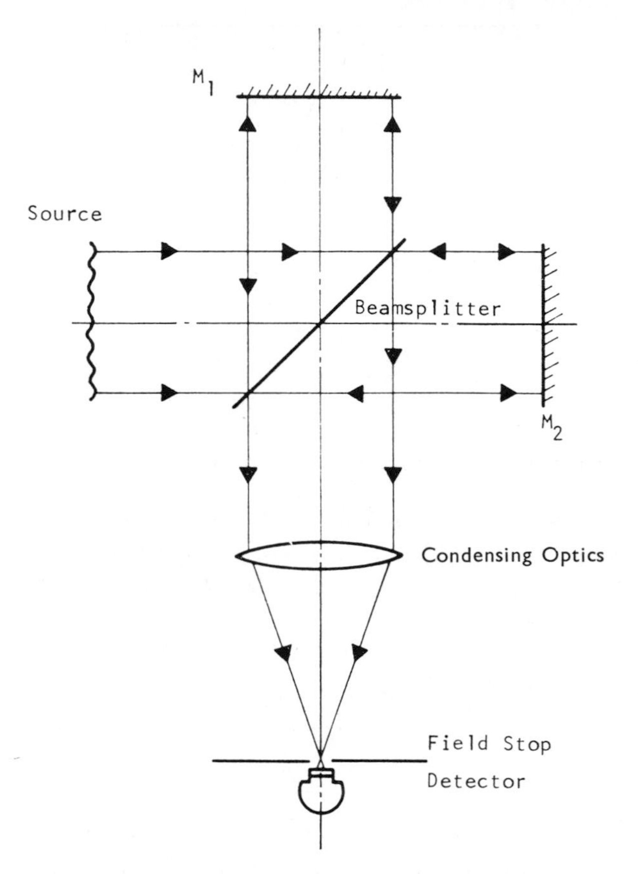

Figure 11.3 Localizing fringes at a field stop.

Destructive interference occurs when

$$\delta = 2x \cos \alpha = n\lambda/2, \tag{11-33}$$

and constructive interference when

$$\delta = 2x \cos \alpha = m\lambda/2. \tag{11-34}$$

For given values of x, λ, and n or m, the angle α determines the direction for either destructive or constructive interference. Because of this characteristic and the symmetry about a normal to the mirrors, circular fringes are seen from Q' with the eye focused at infinity; so it is said that fringes are formed at infinity. One sees a central disk surrounded by alternating bright and dark rings. If, while the fringe pattern is observed, the distance between mirrors x is reduced gradually, each fringe will pass through a sequence of alternations, the bright fringes becoming dark and the dark ones becoming bright. The diameters of the disk and rings will also gradually increase until the central disk completely fills the field of view. At this position, provided reflections have produced the same phase changes in the two beams, the path difference and phase difference are zero, and the central fringe will be a bright fringe.

By imaging the fringes, normally formed at infinity, in the focal plane of an optical system, their apparent location can be controlled. A radiant energy detector can be placed at the image (Fig. 11.3). If the detector is masked by a circular aperture or field stop no larger than the central fringe and centered on the fringe, the detector will produce an alternating electrical signal as x (Fig. 11.2) is increased (or decreased) and the fringe alternates from bright to dark.

The Interferogram

If one of the mirrors is moved with a constant velocity v, the path difference will be, from Eq. (11-32), a function of time:

$$\delta = 2vt \cos \alpha. \tag{11-35}$$

The phase difference is

$$2\pi\delta/\lambda = (4\pi vt \cos \alpha)/\lambda,$$

which can be written in terms of wavenumber σ:

$$2\pi\delta/\lambda = 4\pi\sigma vt \cos \alpha.$$

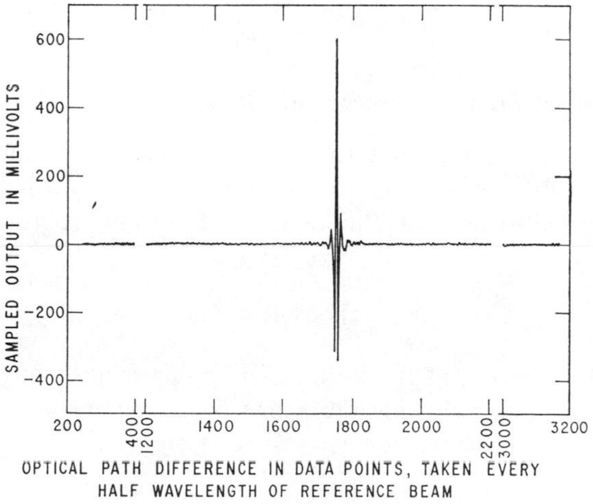

Figure 11.4 Example of a recorded interferogram.

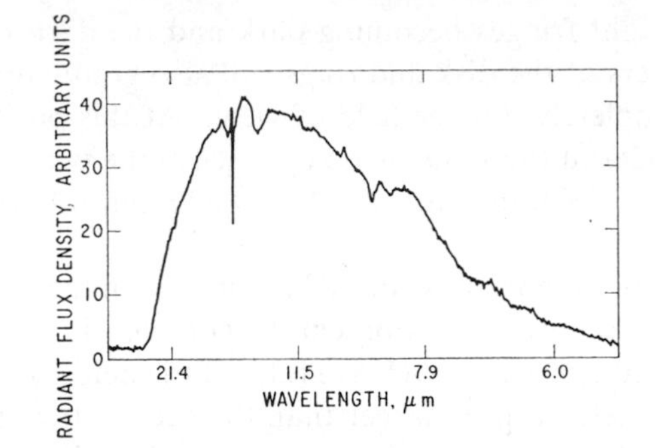

Figure 11.5 The spectrum computed from the interferogram of Fig. 11.4.

The equation for the signal produced by the detector [see Eq. (11-29)], except for a scale constant, is

$$I(t) = (\xi_0^2/4)[1 + \cos(4\pi\sigma v t \cos\alpha)]. \qquad (11\text{-}36)$$

If α is so small for all those directions accepted by the detector aperture that $\cos\alpha \approx 1$ and if the detector and amplifiers pass only alternating current, the signal is

$$I(t) = (\xi_0^2/4)\cos 4\pi\sigma v t. \qquad (11\text{-}37)$$

This equation shows that $2\sigma v$ is the frequency of the electrical signal produced; the signal contains information about the wavenumber σ of the radiant energy.

The record of the electrical signal $I(t)$ is called an interferogram. Figure 11.4 is an example of a recorded interferogram, and Fig. 11.5 is the spectrum computed from it.

Interferogram Signal for a Heterochromic Beam

If the interferometer is subjected to a general incoherent radiant energy beam, the radiant flux densities of the several wave trains will add and the detector will respond to the total flux density. The interferogram signal will be

$$I'(t) = \int_{\sigma_1}^{\sigma_2} I(\sigma)\cos 4\pi\sigma v t \, d\sigma, \qquad (11\text{-}38)$$

where $I(\sigma)$ is the spectrum of the radiant energy passing through the interferometer. If $I(\sigma)$ is zero outside the range of wavenumbers σ_1 to σ_2, the limits of integration can be written as $-\infty$ to $+\infty$:

$$I(\delta) = \int_{-\infty}^{+\infty} I(\sigma)\cos 2\pi\sigma\delta \, d\sigma, \qquad (11\text{-}39)$$

where $\delta/2v$ has been substituted for t. Equation (11-39) is recognized as the Fourier cosine transform of $I(\sigma)$, which is a real function and can be made an even function by mathematically defining a negative wavenumber spectrum $I(-\sigma) = I(\sigma)$. Then the cosine transform is a complete Fourier transform and $I(\delta)$ is itself a real and even function, being symmetrical about the point of zero path difference. The spectrum can be found by taking the inverse cosine transform

$$I(\sigma) = \int_{-\infty}^{+\infty} I(\delta) \cos 2\pi\sigma\delta \, d\delta. \tag{11-40}$$

So long as the path difference δ is small compared with the average length of the wave trains, sometimes called the coherence length, the input beam is sufficiently coherent to record an interferogram.

The Instrument Function

The recorded spectrum (we might now say recorded and computed spectrum), as pointed out in Chapter 10, can never be a perfect representation of the true spectrum. Departures from the ideal arise from the limitations of the interferometer instrument function. Two major limitations, the finite range of path difference and the finite field of view, which are discussed below, are straightforward.

One can never experimentally record the complete interferogram from $-\infty$ to $+\infty$ that is needed for Eq. (11-40). The function $I(\delta)$ will be recorded only between some minimum path difference δ_1 and some maximum path difference δ_2. For convenience we shall let $\delta_1 = -L$ and $\delta_2 = +L$. The ideal interferogram is in effect multiplied by a rectangular function $D(\delta)$ defined as

$$\begin{aligned} D(\delta) &= 1 \qquad -L \leq \delta \leq +L, \\ &= 0 \qquad |\delta| > L. \end{aligned} \tag{11-41}$$

The computed spectrum will be

$$I_1(\sigma) = \int_{-\infty}^{+\infty} D(\delta) I(\delta) \cos 2\pi\sigma\delta \, d\delta, \tag{11-42}$$

instead of the true spectrum given by Eq. (11-40). As already demonstrated in earlier chapters, the Fourier transform of the product of two functions is the convolution of the inverse transforms of the functions:

$$I_1(\sigma') = \int_{-\infty}^{+\infty} I(\sigma' - \sigma) a(\sigma) \, d\sigma, \tag{11-43}$$

where $a(\sigma)$ is the Fourier transform of $D(\delta)$, and $I(\sigma)$ is the transform of the complete function from $-\infty$ to $+\infty$, $I(\delta)$. $I(\sigma)$ is the true spectrum, and $a(\sigma)$ is the instrument function.

The Fourier transform of $D(\delta)$ is

$$a(\sigma) = \int_{-\infty}^{+\infty} D(\delta) \cos 2\pi\sigma\delta \, d\delta$$

$$= \int_{-L}^{+L} \cos 2\pi\sigma\delta \, d\delta$$

$$= 2L(\sin 2\pi\sigma L)/(2\pi\sigma L). \tag{11-44}$$

As is indicated in earlier chapters, $(\sin x)/x$ is a function called sinc x. The shape of this function is demonstrated by comparing Eq. (5-24) and Fig. 5.4. In the present discussion, the first zero occurs where $\sigma = 1/2L$. If $1/2L$ is taken as the half width of the instrument function, the assumed path difference $2L$ (equal to $\delta_2 - \delta_1$) alone determines the theoretical resolution limit of the interferometer.

The sinc function is usually not a suitable instrument function because of its shape. In particular, a "satellite"—a band of small strength very near a band of large strength—could hardly be distinguished from a "foot" (subsidiary loop) of the sinc function. To avoid this ambiguity, the experimenter will remove the feet of the instrument function, that is, he will "apodize." To accomplish this, the rectangular function $D(\delta)$ has to be multiplied by an apodizing function $A(\delta)$. The resolution, however, can never be improved by apodizing; in fact, the resolution is usually appreciably degraded. A number of apodizing functions have been discussed in the literature [Refs. 1 and 7–10].

The Finite Field of View

The size of the solid angle Ω from which wave trains are accepted by the interferometer affects the modulation index, that is, the amplitude of the a-c signal produced, and determines the usable maximum path difference. To investigate these relationships, we assume a centered system with a circular field stop placed in front of the detector at the focal plane. From Eq. (11-29), with $\delta = 2x \cos \alpha$ and $a = 2\pi x/\lambda$, we write

$$W = (\xi_0^2/4)[1 + \cos (2a \cos \alpha)]. \tag{11-45}$$

By multiplying this flux density by area, we can get an expression for power, to which the interferometer detector responds. A useful element of area for this purpose is the annulus dA with radius r centered on the optic axis in the field stop:

$$dA = 2\pi r\, dr. \tag{11-46}$$

This area subtends a solid angle $d\omega$ [see Eq. (3-43)]:

$$d\omega = 2\pi \sin \alpha\, d\alpha, \tag{11-47}$$

where α is the angle determined by r/f, f being the focal length. This angle is the direction from which radiant power received by the system passes through and falls on the area dA. At first let us consider only that radiant power received in the solid angle $d\omega$, and treat W given by Eq. (11-45) as the flux density over dA.

The power passing through the annulus dA is

$$dP = [(\pi \xi_0^2 r)/2][1 + \cos (2a \cos \alpha)]\, dr. \tag{11-48}$$

The total power passing through the field stop, if ξ_0 is constant with respect to r, is

$$P = (\pi\xi_0^2/2) \int_0^{r_m} r[1 + \cos (2a \cos \alpha)]\, dr. \tag{11-49}$$

The maximum radius is related to the solid angle Ω determined by the field stop

$$\Omega \approx \pi r_m^2/f^2 = \pi\alpha_m^2. \tag{11-50}$$

Since α is small, usually less than 0.1 rad, we can approximate $\cos \alpha$ by only two terms of its expansion:

$$P = (\pi\xi_0^2/2)\left\{\int_0^{r_m} r\, dr + \int_0^{r_m} r \cos [2a(1 - r^2/2f^2)]\, dr\right\}. \tag{11-51}$$

By changing the variable, the second integral can be easily evaluated:

$$u = 2a[1 - (r^2/2f^2)],$$

$$du = -(2a/f^2) r\, dr.$$

When $r = 0$, $u = u_1 = 2a$. When

$$r = r_m, \qquad u = u_2 = 2a[1 - (r_m^2/2f^2)]. \tag{11-52}$$

So

$$P = (\pi\xi_0^2/2)\left\{(r_m^2/2) + (f^2/2a)\int_{u_1}^{u_2} \cos u\, du\right\}. \tag{11-53}$$

Upon integration, the expression for the power can be put in the form,

$$P = [(\pi \xi_0^2 f^2 \Omega)/4]\{1 + \operatorname{sinc} (a\Omega/2\pi) \cos a[2 - (\Omega/2\pi)]\}, \tag{11-54}$$

by applying an identity, $\sin x - \sin y = 2[\sin (x - y)/2][\cos (x + y)/2]$, to the integral and by using the definition, $\operatorname{sinc} x = (\sin x)/x$.

When $a = 0$, which corresponds to zero path difference because $a = 2\pi x/\lambda$ (where x is the path difference), the sinc factor and the cosine factor are both unity and the power has its first maximum:

$$P_{m1} = \xi_0^2 f^2 \Omega/2. \tag{11-55}$$

Because $\Omega/2\pi \ll [2 - \Omega/2\pi]$, the cosine factor will go through many cycles before the sinc factor reaches its first zero as a increases. The power, as a function of $a\Omega/2\pi$, is plotted in Fig. 11.6. The modulation index is defined as

$$M_i = \frac{P_{\max} - P_{\min}}{P_{\max} + P_{\min}}. \tag{11-56}$$

The maxima occur very close to where the cosine factor is $+1$ and the minima occur very close to where the cosine factor is -1.

For the assumed range of values for Ω (from 0 to about 0.03), the following approximations can be written:

$$\begin{aligned} P_{\max} &= (P_{m1}/2)[1 + \operatorname{sinc}(a\Omega/2\pi)], \\ P_{\min} &= (P_{m1}/2)[1 - \operatorname{sinc}(a\Omega/2\pi)], \end{aligned} \tag{11-57}$$

$$\begin{aligned} M_i &= \operatorname{sinc}(a\Omega/2\pi) \\ &= \operatorname{sinc}(\pi x \alpha_m^2/\lambda). \end{aligned} \tag{11-58}$$

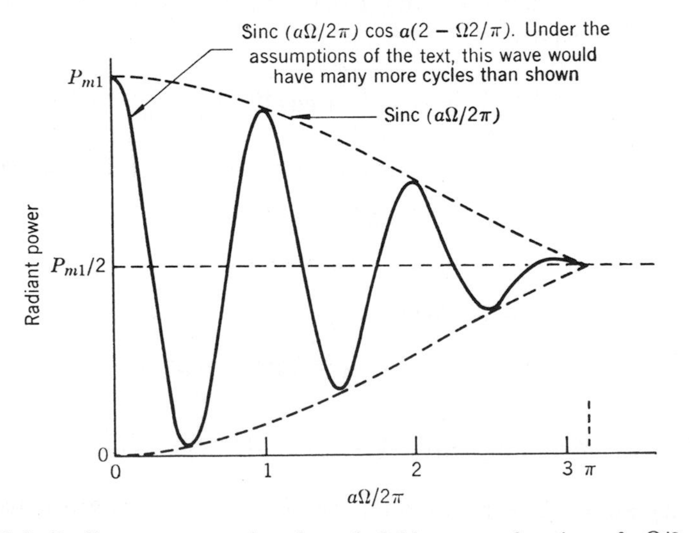

Figure 11.6 Radiant power passing through field stop as function of $a\Omega/2\pi$ showing decrease in modulation index.

Equation (11-58) indicates that the electrical signal from the detector, which is proportional to the modulation index, will go to zero when

$$x\alpha_m^{\ 2} = \lambda. \tag{11-59}$$

The significance of this equation is that the size of the solid angle from which wave trains are accepted restricts the resolution at a given wavelength by setting a limit to the practical maximum excursion of the mirror (the modulation index is discussed further in Ref. 11).

Instrument Defects

Various imperfections and misalignments of the optical components cause a broadening of the instrument function. For example, the mirrors, instead of being "flat," may be slightly spherical, or they may have pits due to imperfect polishing. The effects of these and other imperfections, as well as of noise in the system, are treated in a number of references [Refs. 1 and 12–14]. To show one approach to handling instrument defects, we discuss the mirror misalignment problem in the following paragraphs. This example will show how nearly perfect an alignment must be achieved for either near-maximum fringe visibility or near-maximum modulation index.

The geometry is illustrated in Fig. 11.7, showing a cross section of a circular aperture, and in Fig. 11.8, showing schematically the relative positions of one mirror M_1 and the image of the other M_2'. A hypothetical plane M_1' parallel to M_1, intersecting M_2' at the center of the aperture, is shown for reference. The angle of misalignment α is assumed small, of the order of 100 μrad or less, so that the change in ray direction at M_2 is not of significance. At any given x, the total variation in optical path difference $2\alpha x$ must be examined, however. Only wave trains traveling perpendicular to M_1

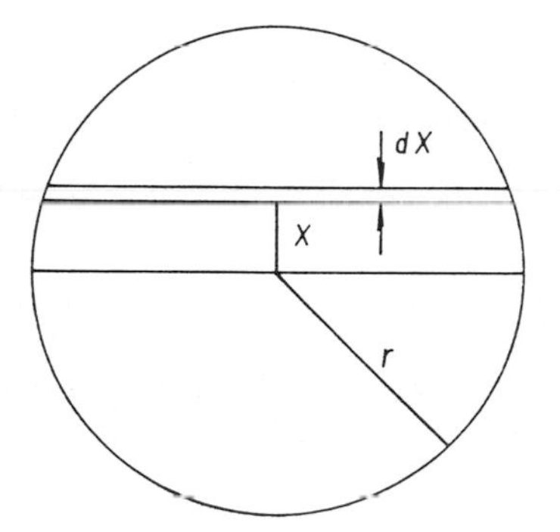

Figure 11.7 Element of area in aperture of interferometer.

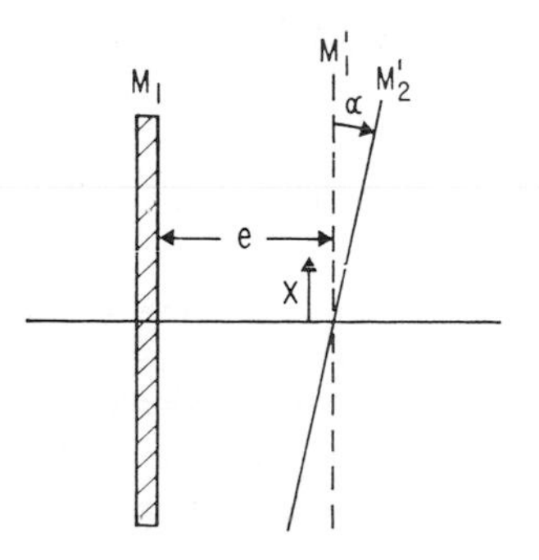

Figure 11.8 Schematic showing geometry of misalignment of mirrors.

are to be considered. The contribution to the power passing through to the field stop by an element of area dA of the aperture is, from Eq. (11-29):

$$dP = (\xi_0^2/4)[1 + \cos(2\pi\sigma\delta)]\,dA. \tag{11-60}$$

The misalignment gives a path difference, varying over the aperture, of $(\delta + 2\alpha x)$ instead of the ideal δ given in Eq. (11-60). If the element of area is $dA = 2(r^2 - x^2)^{1/2}\,dx$ (see Fig. 11.7), the total power passing through the field stop is

$$P = (\xi_0^2/4)\int_{-r}^{+r} 2(r^2 - x^2)^{1/2}[1 + \cos 2\pi\sigma(\delta + 2\alpha x)]\,dx, \tag{11-61}$$

$$P = (\xi_0^2/4)\int_{-r}^{+r} 2r[1 - (x/r)^2]^{1/2}$$

$$\times\,[1 + \cos 2\pi\sigma\delta \cos 4\pi\sigma\alpha x - \sin 2\pi\sigma\delta \sin 4\pi\sigma\alpha x]\,dr,$$

$$P = \xi_0^2/4)\Bigg\{\int_{-r}^{+r} 2r[1 - (x/r)^2]^{1/2}\,dx$$

$$+ 2r\cos 2\pi\sigma\delta\int_{-r}^{+r}[1 - (x/r)^2]^{1/2}\cos 4\pi\sigma\alpha x\,dx$$

$$- 2r\sin 2\pi\sigma\delta\int_{-r}^{+r}[1 - (x/r)^2]^{1/2}\sin 4\pi\sigma\alpha x\,dx\Bigg\}. \tag{11-62}$$

The value of the first integral in the brackets is the area of the aperture, πr^2. The third term contains a symmetrical integral of an odd function $[f(x) = -f(-x)]$ and is, therefore, zero. The middle term turns out to be the first order Bessel's function $J_1(a)$ [Ref. 15]:

$$J_1(a) = (a/\pi)\int_{-1}^{+1}(1 - t^2)^{1/2}\cos at\,dt, \tag{11-63}$$

if $a = 4\pi\sigma\alpha r$ and $t = x/r$; and

$$P = (\pi r^2\xi_0^2)\{1 - [2J_1(a)/a]\cos 2\pi\sigma\delta\}. \tag{11-64}$$

The interferogram, a function of δ, is proportional to the variable term in Eq. (11-64):

$$I(\delta) = (\pi r^2\xi_0^2)[2J_1(a)/a]\cos 2\pi\sigma\delta. \tag{11-65}$$

Because the term in brackets is constant (not a function of δ), it will be part of the constant coefficient when the Fourier transform of $I(\delta)$ is taken. In the Fourier transform,

$$I(\sigma) = (\pi r^2\xi_0^2)[2J_1(a)/a]\int_{-\infty}^{+\infty}\cos 2\pi\sigma_0\delta\cos 2\pi\sigma\delta\,d\delta, \tag{11-66}$$

the factor $[2J_1(a)/a]$ will, therefore, appear as a distortion in amplitude on the reconstructed spectrum. The amplitude will fall to zero, for instance, when the wavenumber makes $a = 3.83$; and the amplitude will be 0.99 of its value at perfect alignment when a is reduced to 0.283.

Sampling the Interferogram

The mathematical processing of the interferogram to accomplish the Fourier transformation is usually done by digital computers. Data for this kind of computation are obtained by sampling at regular intervals δ_n along the interferogram as shown graphically in Fig. 11.9. Nonuniform spacing of data points is a significant cause of distortions in the spectrum. For high accuracies the necessary constancy in the velocity of the moving mirror and regularity in the "clock," which triggers the commands to sample, are hard to achieve. These problems are avoided by using an auxiliary interferometer to mark off the δ intervals.

A nearly monochromatic source of radiant energy—a neon or mercury glow lamp with a narrow bandpass optical filter to isolate a single band—is used to illuminate a portion of the aperture. The central fringe produced by this source produces a nearly perfect cosine wave in an auxiliary detector. Thus, the reference beam and the beam to be tested pass through the same interferometer—fixed mirror, beamsplitter, and moving mirror—and points along the cosine reference signal correspond precisely to positions of the moving mirror independent of the mirror velocity and a "clock." For example, the passage of the reference signal through zero could be used to

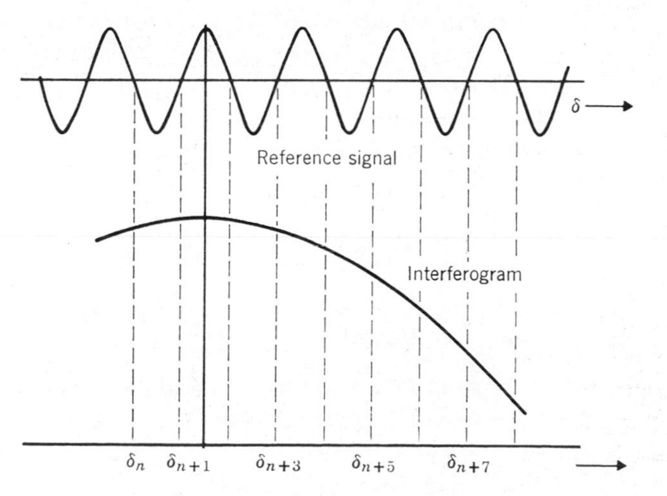

Figure 11.9 Sampling the interferogram.

initiate a command to sample the interferogram and thus cause two data samples to be taken for each cycle of the reference signal.

Computing the Spectrum

The Fourier transforms of $\hat{I}(\sigma)$ and $\hat{I}(\delta)$ are given by Equations (11-23) and (11-24). However, since these functions are real functions and are (or can be made to be) even functions, that is,

$$\hat{I}(\delta) = \hat{I}(-\delta) \qquad \text{and} \qquad \hat{I}(\sigma) = \hat{I}(-\sigma), \tag{11-67}$$

Equations (11-39) and (11-40) turn out to be the complete transforms, so we can treat these functions by using the notation of real numbers. The function $I(\sigma)$ is a power density spectrum, and $I(\delta)$ may be considered to be its autocorrelation function by defining the square of an amplitude function:

$$|\Phi(\delta)|^2 = I(\delta),$$

and writing Eq. (11-24) as

$$I(\sigma) = \int_{-\infty}^{+\infty} |\Phi|^2 \exp(-2\pi i \sigma \delta)\, d\delta. \tag{11-68}$$

Since $|\Phi|^2$ is the product of a function times itself, the Fourier transform is the convolution of the Fourier transform of the function with itself:

$$I(\sigma) = \int_{-\infty}^{+\infty} f_1(\sigma') f_1(\sigma - \sigma')\, d\sigma', \tag{11-69}$$

where $f_1(\sigma')$ is the Fourier transform of $\Phi(\delta)$. From its definition, $|\Phi(\delta)|^2$ is proportional to the square of the electric field quantity in the interfering beam. Equation (11-69) is usually called the autocorrelation function.

Calculation of the spectrum from a sampled interferogram can be done by a Fourier series approximation. A convenient form is given in Stratton [Ref. 16, p. 287]:

$$I(\sigma) = \sum_{n=-\infty}^{\infty} c_n \exp(in\pi\sigma/l), \tag{11-70}$$

$$c_n = \frac{1}{2l} \int_{-l}^{+l} I(\sigma) \exp(-in\pi\sigma/l)\, d\sigma. \tag{11-71}$$

The value of the function $I(\sigma)$ is assumed negligible everywhere outside the region $-l \le \sigma \le +l$. These equations can be written in the symbols and in the forms used in this chapter by letting

$$-l = \sigma_1 \tag{11-72}$$

and

$$c_n = a_n/2\sigma_1. \tag{11-73}$$

Then

$$I(\sigma) = \frac{1}{2\sigma_1} \sum_{n=-\infty}^{+\infty} a_n \exp(-in\pi\sigma/\sigma_1),$$

$$a_n = -\int_{+\sigma_1}^{-\sigma_1} I(\sigma) \exp(in\pi\sigma/\sigma_1)\, d\sigma. \tag{11-74}$$

The sign of the integral can be changed by changing the signs of the limits:

$$a_n = \int_{-\sigma_1}^{+\sigma_1} I(\sigma) \exp(in\pi\sigma/\sigma_1)\, d\sigma. \tag{11-75}$$

Now if $a_n = a(\delta_n)$ and $\delta_n = (n/2\sigma_1)$, Eq. (11-75) can be written as

$$a(\delta_n) = \int_{-\sigma}^{+\sigma} I(\sigma) \exp(2\pi i\sigma\delta_n)\, d\sigma,$$

and, since the power density spectrum is zero outside of $-\sigma_1 \leq \sigma \leq +\sigma_2$,

$$a(\delta_n) = \int_{-\infty}^{+\infty} I(\sigma) \exp(2\pi i\sigma\delta_n)\, d\sigma. \tag{11-76}$$

The right-hand sides of Eq. (11-76) and Eq. (11-23) are the same except that Eq. (11-76) has $\delta = \delta_n = n/2\sigma_1$. Thus $a(\delta_n) = I(\delta_n)$, or

$$I(\delta_n) = \int_{\infty}^{+\infty} I(\sigma) \exp 2\pi i\sigma\delta_n\, d\sigma. \tag{11-77}$$

Then, by applying Eq. (11-74),

$$I(\sigma) = (1/2\sigma_1) \sum_{n=-\infty}^{+\infty} I(\delta_n) \exp(-in\pi\sigma/\sigma_1). \tag{11-78}$$

Thus we need only to know $I(\delta_n)$, the amplitude of the interferogram at each δ_n where $\delta_n = (n/2\sigma_1)$. Since we are still talking about real and even functions,

$$I(\sigma) = (1/2\sigma_1) \sum_{n=-\infty}^{+\infty} I(\delta_n) \cos(n\pi\sigma/\sigma_1). \tag{11-79}$$

Of course, $I(\delta_n)$ is available not to $n = \infty$ but only up to a maximum $n = N_1$ called the maximum lag. With this limit in Eq. (11-79), an approximate value of $I(\sigma)$ is

$$I_1(\sigma) = (1/2\sigma_1) \sum_{n=-N}^{+N} I(\delta_n) \cos(n\pi\sigma/\sigma_1). \tag{11-80}$$

The maximum lag, corresponding to the finite maximum mirror excursion, produces the instrument function and the need to apodize as was discussed earlier in this chapter.

One practical problem occurs when the measured values of $I(\delta_n)$ are not symmetrical, that is, when the δ_n are not symmetrical about the point corresponding to zero path difference. Another problem is caused by the sampling points being time-shifted by a small amount k so that instead of being $(n/2\sigma_1)$ they are $[(n/2\sigma_1) + k]$. Both of these problems can be easily overcome if one is willing to calculate the complete Fourier transform:

$$I_1(\sigma) = (1/2\sigma_1)\left\{\left[\sum_{n=-N}^{N} I(\delta_n)\cos(\pi\sigma n/\sigma_1)\right]^2 + \left[\sum_{n=-N}^{N} I(\delta_n)\sin(\pi\sigma n/\sigma_1)\right]^2\right\}^{1/2}. \tag{11-81}$$

Many computer programs are being written to perform the computation of the spectrum, and the process is being discussed widely in the literature [Refs. 10, 11, and 17–28]. Several of these papers show that much can be done to improve the fidelity of the computed spectrum.

REFERENCES

1. J. Connes, "Recherches sur la Spectroscopie par Transformation de Fourier," *Rev. d'Opt.*, **40,** 45, 116, 171, 231 (1961). English translation: "Spectroscopic Studies Using Fourier Transforms" by Mme. Janine Connes, NAVWEPS Report 8099, NOTS TP 3157, 1963.
2. G. A. Vanasse and H. Sakai, "Fourier Spectroscopy," in *Progress in Optics*, Vol. VI. Edited by E. Wolf. North-Holland Publishing Co., Amsterdam, 1967.
3. H. N. Rundle, "Construction of a Michelson Interferometer for Fourier Spectroscopy," *J. Res. NBS*, **69C,** 5 (1965).
4. M. Born and E. Wolf, *Principles of Optics*. Pergamon Press, Oxford, 1965.
5. A. Papoulis, *The Fourier Transform Integral and Its Applications*. McGraw-Hill Book Company, Inc., New York, 1962.
6. J. S. Bendat, *Principles and Applications of Random Noise Theory*. John Wiley & Sons, Inc., New York, 1958.
7. A. S. Filler, "Apodization and Interpolation in Fourier Transform Spectroscopy," *J. Opt. Soc. Am.*, **54,** 762 (1964).
8. F. P. Parshin, "Apodization in Fourier Spectroscopy," *Opt. Spectrosc.*, **13,** 418 (1962).
9. S. G. Rautian, "Real Spectral Apparatus," *Sov. Phys.—Usp.*, **66,** 245 (1958).
10. P. Jacquinot, "Apodization," *Progress in Optics*, Vol. III. Edited by E. Wolf. North-Holland Publishing Company, Amsterdam, 1964.
11. E. R. Peck, "Integrated Flux from a Michelson or Corner Cube Interferometer," *J. Opt. Soc. Am.*, **45,** 931 (1955).
12. B. A. Kiselev and P. F. Parshin, "Some Distortions in Fourier Spectroscopy," *Opt. Spectrosc.*, **12,** 169 (1962).
13. E. V. Loewenstein, "On the Correction of Phase Errors in Interferograms," *Appl. Optics*, **2,** 491 (1963).

14. R. A. Williams and W. S. C. Chang, "Resolution and Noise in Fourier-Transform Spectroscopy," *J. Opt. Soc. Am.*, **56,** 167 (1966).
15. G. N. Watson, *A Treatise on the Theory of Bessel Functions*, p. 48. Cambridge University Press, Cambridge, 1958.
16. J. A. Stratton, *Electromagnetic Theory*. McGraw-Hill Book Company, Inc., N.Y., 1941.
17. R. B. Blackman and J. W. Tukey, "The Measurement of Power Spectra from the Point of View of Engineering," *Bell Syst. Tech. J.*, **37,** 185, 485 (1958).
18. J. W. Cooley and J. W. Tukey, "An Algorithm for Machine Calculation of Complex Fourier Series," *Math. Comp.*, **19,** 297 (1965).
19. M. L. Forman, "Fast Fourier-Transform Technique and Its Application to Fourier Spectroscopy," *J. Opt. Soc. Am.*, **56,** 978 (1966).
20. M. L. Forman, W. H. Steel, and G. A. Vanasse, "Correction of Asymmetric Interferograms Obtained in Fourier Spectroscopy," *J. Opt. Soc. Am.*, **56,** 59 (1966).
21. G. A. Vanasse, "Feger Series Approximation to Spectral Functions Obtained by Interferometric Techniques," *J. Opt. Soc. Am.*, **52,** 472 (1962).
22. H. Sakai and G. A. Vanasse, "Accuracy of Zero Path Difference Determination in Fourier Spectroscopy for Narrow-Band Spectra," *J. Opt. Soc. Am.*, **57,** 844 (1967).
23. S. T. Fisher, "On the Fourier Transformation of Sampled Interferograms Having a Small Number of Samples," *Appl. Optics*, **4,** 256 (1965).
24. H. Sakai and G. A. Vanasse, "Hilbert Transform in Fourier Spectroscopy," *J. Opt. Soc. Am.*, **56,** 131 (1966).
25. W. H. Steel and M. L. Forman, "Examples of Errors Occurring in Fourier Spectroscopy Due to Hilbert Transform Effects," *J. Opt. Soc. Am.*, **55,** 982 (1965).
26. *IEEE Trans Audio and Electroacoustics*, Vol. AU-15, June issue, 1967.
27. W. T. Cochran, "What Is the Fast Fourier Transform?" *Proc. IEEE*, **55,** 1664 (1967).
28. C. M. Randall, "Fast Fourier Transform for Unequal Number of Input and Output Points," *Appl. Optics*, **6,** 1432 (1967).
29. N. Jacobi, "Resolution and Noise in Fourier Spectroscopy," *J. Opt. Soc. Am.*, **58,** 495 (1968).
30. J. M. Dowling, "Signal and Noise in Two-Beam Interferometry," *Appl. Optics*, **6,** 1581 (1967).

12

Radiant Energy Detectors

Any device that reveals the presence of incident radiant energy can be called a detector. A transformation to another form of energy or an interaction of the radiant energy with matter is required to produce an effect that can be observed either directly or indirectly. When the transformation is to thermal energy—either directly or through an intermediate process—a rise in temperature is manifested by a change in the electrical conductivity of a material, a change in the voltage produced by a thermocouple, a change of pressure in a gas cell (the Golay cell), or a sensation of warmth or pain in the skin. When the incident radiant energy causes electronic excitation, that is, atoms experience transitions to higher energy levels, electrons may be emitted from certain surfaces or electron-hole pairs may be generated in a semiconductor. The latter effect either causes a change of conductivity or produces a voltage across a junction diode.

How we think about the radiant energy spectrum is influenced by the way the energy's presence is revealed to us. The "visible" band of wavelengths is so named, for instance, because the human eye is an efficient detector in this range. In fact, our designations of the neighboring invisible bands come from the eye's interpretation of the boundary wavelengths—*ultraviolet* for the wavelengths shorter than the visible and *infrared* for the wavelengths longer than the visible. Our discussion of radiant energy detectors in this chapter will be confined to the ultraviolet, visible, and infrared bands, with emphasis on the infrared band.

Classification

Like the eye, a number of detectors consist of an extensive sensing surface on which an image is focused. Among these are photographic film, the various television camera tubes, large arrays of tiny discrete detectors, and the infrared evaporograph; the last-named produces a visible image by the differential evaporation of an oil film exposed to the infrared "image." In contrast are the *point* detectors, which typically average the effects of radiant energy incident on their sensitive surfaces. The two classes merge when point detectors are assembled in arrays—as in the eye—so that jointly they can resolve an image.

Detectors can also be classified according to the transformation mechanism. On this basis, useful devices tend to be either *thermal detectors* or *photon detectors*. The latter are known also as *quantum detectors*. Here we confine ourselves to detectors that exhibit some kind of electrical response.

In a thermal detector, the incident radiant energy raises the temperature of the detector material and, in turn, affects some temperature-dependent electrical property of the material. In a photon detector, on the other hand, the incident photons transfer their energy *directly* to the electrons in the device material, and heating plays no significant part in the detection process.

The two modes of operation suggest some of the differences between thermal detectors and photon detectors. Because energy has to be accumulated by the thermal detector in the form of heat, this device is slow compared with the photon detector, in which no "storing" process is required. Actually, time constants of thermal detectors are typically in milliseconds, whereas those of photon detectors are in microseconds.

A good part of Chapter 10 emphasizes the limited wavelength range over which certain photon-to-particle energy exchanges can be made. This limitation applies in general to photon detectors. However, the transformation efficiency of photon energy to heat in a thermal detector is relatively independent of wavelength; thus, thermal detectors characteristically have relatively uniform response over a wider range of wavelengths.

Detector Parameters

Detectors are chosen on the basis of their performance parameters, their required operating environment, and their associated circuits. The commonly used parameters are defined and discussed in the following paragraphs [Refs. 1–4].

Responsivity $\mathscr{R}$ is the ratio of the detector output increment to the increment of incident radiant power causing the output increment. If the incident power is chopped (see the section on response time), the detector output increment is measured by the rms value of the fundamental component of the electrical output, and the incident power increment is measured by the rms value of the fundamental component of the chopped radiant power. Responsivity is one of the parameters indicating the sensitivity of detectors. For phototubes its units are commonly amperes per lumen; for infrared detectors, its units are usually volts or amperes per watt. Spectral responsivity $\mathscr{R}_\lambda$ applies to the condition where the incident power is within a narrow wavelength band about the wavelength λ; it is blackbody responsivity $\mathscr{R}_T$ where the incident power has the spectral distribution of a blackbody at temperature T.

Where threshold levels of radiant power are to be detected, the signal-to-noise ratio becomes important. A parameter called detectivity expressing "sensitivity" to weak signals is defined as the ratio of the responsivity to the rms noise:

$$D = \mathscr{R}/V_{\text{rms}}, \tag{12-1}$$

when $\mathscr{R}$ is in volts per watt, or

$$D = \mathscr{R}/I_{\text{rms}}, \tag{12-2}$$

when $\mathscr{R}$ is in amperes per watt. Detectivity, therefore, is in reciprocal watts.

In an important group of detectors, including most semiconductor photon detectors, the detectivity varies inversely with the square root of detector area A_c and inversely with the square root of the electrical bandwidth Δf. For these detectors, D^*, *Dee Star*, is defined as the detectivity normalized to unit area and unit bandwidth by the relation

$$D^* = (A_c\,\Delta f)^{1/2} D. \tag{12-3}$$

The units of D^* are cm $\text{Hz}^{1/2}$ watt^{-1} and can apply to spectral $D^*(\lambda, f)$ or blackbody $D^*(T, f)$. The chopping frequency f and either the wavelength λ or the blackbody temperature T are specified. For detectors limited by photon noise, the detector solid angle and the nature of the background seen by the detector are usually specified. In the notation $D^{**}(\lambda, f)$, the second star signifies that the detector solid angle is less than 2π.

Noise Equivalent Power (NEP), the reciprocal of detectivity, which may be considered a minimum detectable power, is the signal power required to produce a signal-to-noise ratio of unity when the detector is operated under specified conditions. The term NEP stands for either spectral noise equivalent power or blackbody noise equivalent power. The latter requires that the radiant power signal have the appropriate spectral characteristic. The detector area, electrical bandwidth, and chopping frequency should be given to make the specification complete.

Detector solid angle Ω is the solid angle from which the detector, in the form of a small wafer, receives radiant energy. Of course, a detector in this configuration receives radiant energy from all directions, that is, from a complete imaginary hemisphere about the wafer. However, if a cold shield is interposed between the wafer and part of the hemisphere, only the "uncovered" portion contributes to the solid angle. The shield is assumed so cold that the radiant energy received from it is negligible.

Response Time

In a typical test setup, the source power is mechanically chopped so that the incident radiant power $\Delta\Phi$ is a sequence of rectangular pulses of uniform width and separation with pulse width Δt equal to the pulse separation. When the chopping rate is slow so that the response time of the detector τ is very short compared with Δt, that is, $\tau \ll \Delta t$, the voltage pulses from the detector are essentially rectangular of width Δt. As the chopping rate is gradually increased, causing a decrease in Δt, the output, for any detector, will ultimately fall off (Fig. 12.1). The response time, also called the time constant, of a detector may be defined in several ways. A commonly used definition relates the time constant τ to detector cut-off frequency f_c by

$$\tau = 1/(2\pi f_c), \tag{12-4}$$

which assumes that the responsivity varies with frequency according to the relation:

$$\mathscr{R} = \mathscr{R}_0\{[1 + (2\pi f\tau)^2]^{1/2}\}^{-1}, \tag{12-5}$$

where $\mathscr{R}_0$ denotes the responsivity at zero frequency. The cut-off frequency is, therefore, the half-power point or the frequency at which the responsivity is 0.707 the d-c responsivity.

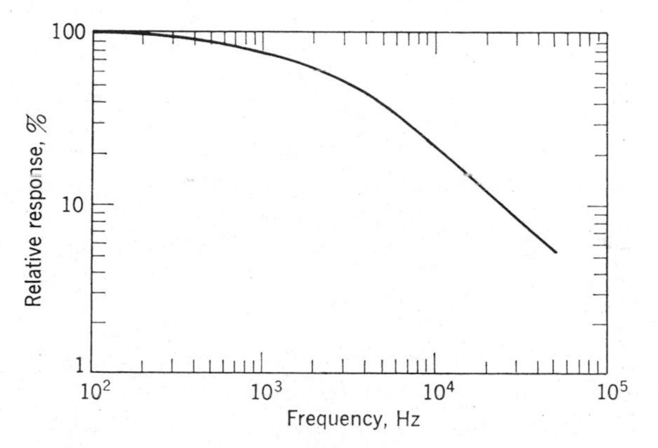

Figure 12.1 A detector frequency response function.

Background-Limited Condition

An optical system with properly positioned stops, as is discussed in Chapter 7, can limit the radiant power reaching the detector to power that originates from a specified target. However, these very stops, baffles, and other structures surrounding the detector form an enclosure that is in effect a background of the target, and the detector and background radiate to each other. Because the arrival or the emission of a photon is a random event, even though the average net incident power is constant, the fluctuation in the photon arrival or emission rate is a statistical quantity distributed according to the Bose–Einstein formula (see Chapter 3). If the background functions as a blackbody, the mean photon arrival rate is given by Eq. (3-7). The statistical variance and standard deviation can be found by the usual methods; the standard deviation turns out to be the rms background photon noise [Ref. 4, pp. 381–394]. This is expressed by the following equation (not derived here) in which the symbols have the same meanings as in Eq. (3-7) and $\overline{(\Delta\mathscr{M}^2)}_\lambda$ is the variance or the mean square of the fluctuations about the mean of the arrival rate at the wavelength λ:

$$\overline{(\Delta\mathscr{M}^2)}_\lambda = \overline{\mathscr{M}}_\lambda[\exp(hc/k\lambda T)][\exp(hc/k\lambda T) - 1]^{-1}. \tag{12-6}$$

The bar over $\mathscr{M}_\lambda$ signifies that $\mathscr{M}_\lambda(T)$ of Eq. (3-7) is now a *mean* value. The square root of $\overline{(\Delta\mathscr{M}^2)}_\lambda$ is the standard deviation referred to above. An approximation of this value can be made, as in the Wien law, when

$$\exp(hc/k\lambda T) \gg 1$$

as follows:

$$[\overline{(\Delta\mathscr{M}^2)}_\lambda]^{1/2} \approx (\overline{\mathscr{M}}_\lambda)^{1/2}. \tag{12-7}$$

This approximation is acceptable for backgrounds near 300 K and wavelengths near 10 μm, that is, when $\lambda T < 0.31$ cm K.

Photon noise becomes particularly significant in the infrared range. When all other noises in a system have been reduced so that photon noise from the background predominates, the system is said to be *background limited*.

Thermal Detectors

In order that incident radiant energy can be absorbed efficiently, the sensing element of a thermal detector must be in thermal contact with a radiant-energy-receiving surface that is approximately a blackbody—usually a blackened metal foil. Surface blackening techniques include depositing

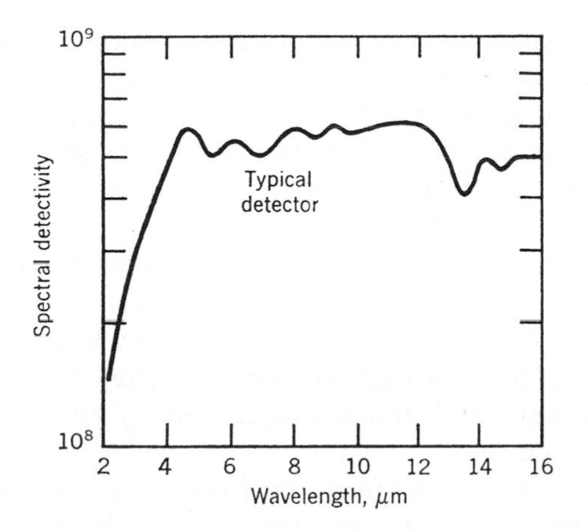

Figure 12.2 Spectral detectivity for a thermister bolometer.

soot from flames, evaporating gold blacks onto the surface, and electrolytically forming special coats. Sometimes black paints and lacquers are applied. None of these, unfortunately, gives a flat absorption characteristic over all wavelengths (Fig. 12.2).

The two most often used of the thermal detectors [Refs. 1–5] are the *bolometer* and the *thermocouple*.

The electrical resistance of a bolometer element changes with temperature [Refs. 6–8]. Two matched elements are usually used and are arms of a Wheatstone bridge. One is exposed; the other is shielded and mounted in such a way that it remains at ambient temperature. The unbalance current (or voltage) caused by heating of the exposed element indicates the absorption of radiant energy. Many refinements, depending on the application, can be made to improve this basic detecting system [see Ref. 1, Chapter III]. Among the characteristics sought in the element itself are adequate strength in extremely thin sheet form, a high temperature coefficient of resistance, and high absorptance. Details of mounting the element are important; in some instances, surrounding the element with a vacuum greatly improves performance.

A thin cross section is desirable to minimize heat capacity and, therefore, the time constant. Early bolometers were of platinum foil because of its strength. Sputtering and other techniques are currently employed to achieve the thin films that function as elements. Metals forming bolometer elements have temperature coefficients of resistance of about 0.5% per degree Celsius. This figure can be boosted to over 4% by substituting certain semiconductor materials for the metals.

Another type of bolometer, the thermistor bolometer, is produced by a specially developed technology which includes sintering mixtures of metallic oxides, mounting the resulting flakes on electrically insulating substrates, and mounting the substrates on metallic heat sinks [see Ref. 2, p. 273]. Minimum detectable powers for bolometers are in the neighborhood of a few times 10^{-10} watt for a 1-Hz bandwidth [see Ref. 1, p. 99].

The predominant noise in well-constructed bolometers is usually Johnson noise. However, since there are reasons for making the bias voltage large, and because the cut-off frequency is usually relatively low (less than a kilohertz), another kind of noise—called excess noise, current noise, or $1/f$ noise—may become significant. Both Johnson noise and $1/f$ noise are discussed in later sections of this chapter.

A thermocouple in its simplest form consists of two electrical conductors, of different materials, connected in series to form an electrical loop with two bimetallic junctions. Interposed in this loop, usually at one of the junctions, is some means for measuring the net electromotive force developed by *thermoelectric power* [Ref. 9]. The electromotive force results from a temperature difference between the two junctions. In thermometer applications, one junction is usually held at a fixed reference temperature (for instance, that of melting ice), and the other is put in thermal contact with the material whose temperature is being measured.

A thermocouple applied to radiant energy detection is equipped with an intercepting surface, called a *receiver*, which is in close thermal contact with one junction. The reference junction, besides being shielded from the energy to be detected, is attached to a body of large thermal capacity so that its thermal time constant is very long compared to any important dynamic properties of the detected energy signal. When ambient fluctuations are still disturbing even with this precaution, a second thermocouple is connected to the electrical detector in opposition to the first. The second pair of junctions is identical to the first except that neither junction is exposed to the signal energy. Then any disturbing ambient condition, other than the signal, will tend to be cancelled out.

Early workers with thermocouple detectors (say, before 1885) were limited by imperfect means to measure the thermocouple electromotive force. To overcome this problem, they usually connected 10 or more thermocouples in series-aiding with all the "hot" junctions intercepting the radiant energy and all the reference or "cold" junctions thermally connected to the body of large thermal capacity. This combination is called a *thermopile*.

The combinations of materials that can be used to form thermocouples is almost limitless. Antimony-bismuth was an early favorite with its thermoelectric power of about 100 μV/°C. At considerable loss in thermoelectric power, silver was often substituted for antimony because silver could more easily be drawn into the fine wire that the typical design required. The inevitable compromise between several thermal and electrical properties produces a more durable, less microphonic design with modern semiconductor materials than with the early single-element thermocouple combinations.

Many of the thermal considerations that apply to bolometers apply also to thermocouples. The thermocouple receiver, like the bolometer element, must have a small cross section to keep the thermal time constant short. Blackened gold foil as thin as 0.3 μm has been used. Again, mounting this part of the detector in vacuum is often found good practice.

Thermocouple detectors have about the same, or possibly slightly better, responsivities compared with bolometers. Comparisons, however, are difficult to make because of the many variables in specifications. For instance, in a given type of thermocouple, sensitivity can be traded for time constant in almost exact inverse proportion. The nature of thermal detectors limits their frequency response whenever they must also have high responsivity: bolometers to a few hundred hertz and thermocouples to less than 10 Hz. Thermocouples are limited by Johnson noise in the junction resistance although there is another noise, called temperature noise (fluctuations in the element temperature), that is also important.

Photon Detectors

Photon detectors can be conveniently divided into *photoelectric detectors* and *semiconductor detectors*. Although there is considerable overlap in the vicinity of the visible–infrared boundary, photoelectric detectors are sensitive principally in the visible and ultraviolet wavelengths, and semiconductor detectors are most useful in the infrared.

Photoelectric Detectors

In Chapter 1, the photoelectric effect is discussed under the subtitle, *Photon Nature of Light*. The significant optical event in a photoelectric detector is the conversion of a photon to kinetic electron energy in sufficient amount to free the electron from the illuminated surface. Then, when some means are provided to count the electrons so affected, the detector function is complete. The emitting surface is usually contained in an evacuated transparent envelope with provisions for applying a voltage between the emitting surface and another electrode, the anode [Ref. 10]. Straightforward electric circuit techniques are used to amplify the signal.

Three types of phototubes are the vacuum phototube, the gas-filled phototube, and the (vacuum) multiplier phototube. The vacuum phototube is the simplest, comprising a photocathode, an anode, an envelope, and a base for

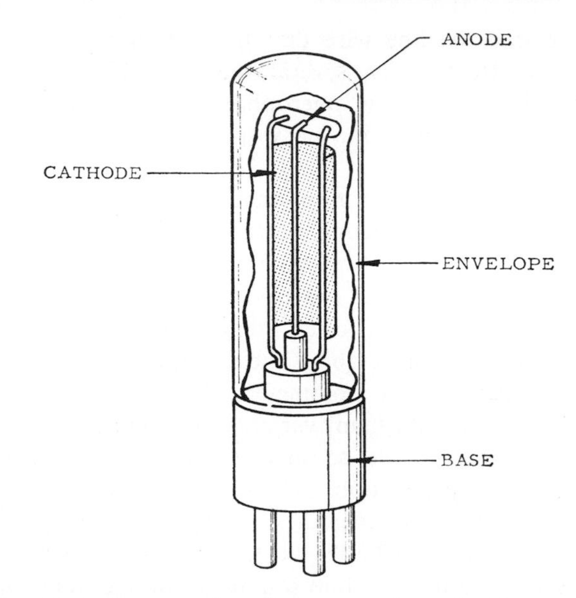

Figure 12.3 Typical construction of a vacuum phototube (Courtesy of RCA).

electrical connections. The photocathode is made of a material having a low work function and has a shape compatible with certain optical requirements to receive the radiant energy and to emit electrons efficiently to the anode. A typical construction is shown in Fig. 12.3.

Sometimes the phototube is designed so that gains of a million or more can be achieved inside the envelope by a succession of secondary electron emission steps. In these *multiplier phototubes* (often called "photomultipliers"), the initial electrons strike the first positive electrode forcefully enough to splash out an increased number of electrons, and this process is repeated on a sequence of electrodes, called *dynodes*, of increasing potential

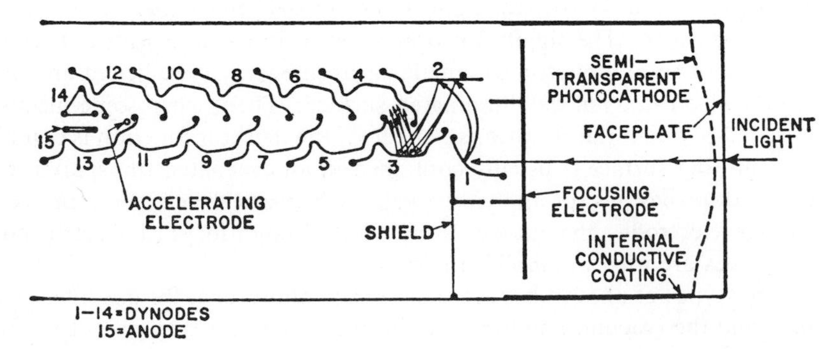

Figure 12.4 Schematic of a linear-type dynode construction (Courtesy of RCA).

until the final collecting electrode, the anode, is reached. Figure 12.4 shows schematically the arrangement of the electrodes in a 14-dynode multiplier phototube.

Vacuum phototubes and multiplier phototubes, when operated in proper electrical, optical, and environmental conditions, have a photocurrent that is linear with the incident radiant power over a wide range. The incident-flux–current relationship is usually given as a sensitivity in microamperes per lumen. The dark current, when the tube is operated in specified conditions, including complete darkness, is also given with the tube data. The sensitivity (which is responsivity) divided by the dark current gives an NEP in lumens.

The inherent response time of vacuum phototubes, including multiplier phototubes, is exceedingly short. The time delay between the incident light and the resulting emission of electrons has been too short to measure. The electron transit time from cathode to anode does introduce a time delay. However, it is rather the *variation* in transit time, not the time delay as such, that limits the frequency response. A large number of electrons make up each current pulse. Thus, there will be variations in the initial velocity with which electrons in a given pulse are emitted, and, among the many electrons, a variation in distance from the point of emission on one electrode to the point of incidence on the next electrode. The limiting time constant usually is a resistance–capacitance time constant resulting from the load resistance and the anode capacitance to other electrodes of the tube and to other components of the electric circuit. Vacuum phototubes have exhibited uniform responses up to about 100 MHz, depending principally upon transit time.

Another way of realizing appreciable gain inside the photoelectric detector envelope, say up to about 100 times, is to introduce a small amount of inert gas into the evacuated envelope. The emitted electrons tend to multiply the current flow between electrodes by ionizing the gas. The price of gain, however, is a combination of increased time constant, accelerated fatiguing, loss of linearity, and certain instabilities that deny use of the detector for photometry, particularly for low-level light measurements.

The characteristics of the photoemissive surface are critical to the operation of the photoelectric detector. First in importance is the least amount of energy, called the work function, required to emit an electron from the surface. Because of the relationship between photon energy and wavelength [see Eq. (1-9)], the work function sets the longest wavelength to which the detector will be sensitive. The lowest work functions are found among the alkali metals. Cesium has the lowest, 1.9 eV, which establishes 0.65 μm as its cutoff wavelength. However, combinations of materials can do even better. The best found so far is a silver–oxygen–cesium surface: 0.98 eV and 1.25 μm. Figure 12.5 shows the relative spectral response of commonly used photocathodes.

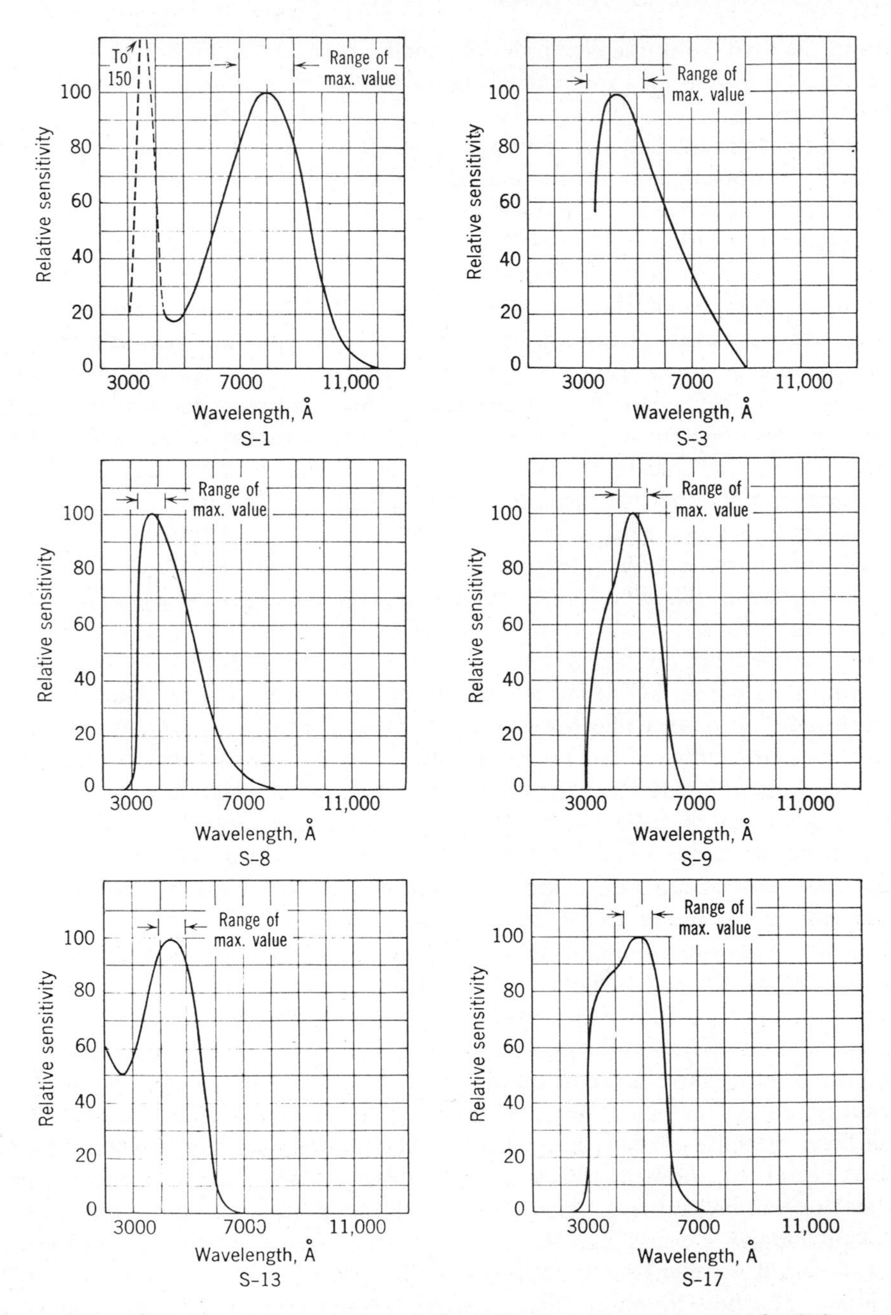

Figure 12.5 Spectral response curves for RCA phototubes (Courtesy of RCA).

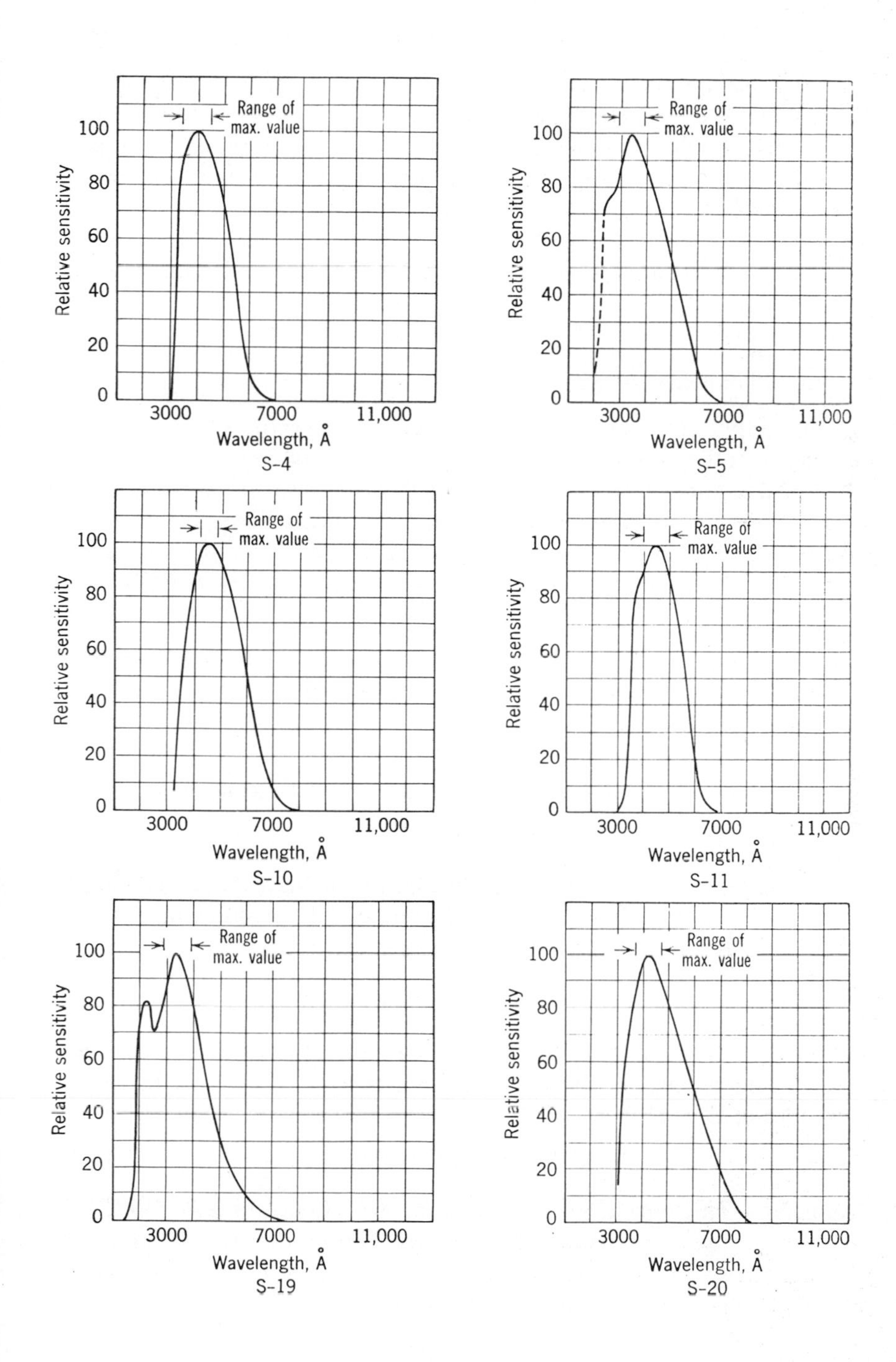
Range of max. value
100
80
60
40
20
0
Relative sensitivity
3000
7000
11,000
Wavelength, Å
S-4
Range of max. value
Relative sensitivity
Wavelength, Å
S-5
Range of max. value
Relative sensitivity
Wavelength, Å
S-10
Range of max. value
Relative sensitivity
Wavelength, Å
S-11
Range of max. value
Relative sensitivity
Wavelength, Å
S-19
Range of max. value
Relative sensitivity
Wavelength, Å
S-20

Though the spectral response of a phototube depends primarily upon the cathode material, the envelope, or window, by its spectral transmittance, controls the wavelength range of photons that can reach the cathode, so the spectral response of a phototube also depends on the envelope material.

Multiplier phototubes at room temperature have detected as few as 3×10^4 photons/sec; under special laboratory conditions, including cooling by liquid air, the detection level has been reduced to about 1 photon/sec [see Ref. 1].

Noise in Phototubes

In many applications of radiant energy detection, extremely small radiant powers are to be detected. As the signal power $\Delta\Phi$ becomes smaller and smaller, it becomes increasingly difficult to distinguish the signal-produced current (or voltage) from unwanted, random fluctuations in current (or voltage), which constitute the noise. Two significant types of noise occur in phototubes: Johnson noise (also called Nyquist and thermal noise) and shot noise.

Johnson-Nyquist noise. This type of noise [Refs. 11 and 12] is generated by the random motion of charged carriers moving about at thermal velocities. It is *white noise*, that is, the noise per unit bandwidth is constant with frequency. This type of noise is the most basic, being exhibited by all resistive materials. The thermal noise as an electrical power P_T is given by

$$P_T = 4kT\,\Delta f, \tag{12-8}$$

where Δf is the noise bandwidth in hertz, and k and T have the same definitions as in Eq. (12-6). Equation (12-8) can be modified by multiplying both sides by the resistance R in which P_T is being generated:

$$\overline{V_T{}^2} = 4kTR\,\Delta f, \tag{12-9}$$

where $V_T{}^2$ is the average of the squared Johnson–Nyquist open-circuit noise voltage. The equivalent short-circuit noise current is obtained by dividing both sides of Eq. (12-9) by R^2:

$$\overline{I_T{}^2} = (4kT\,\Delta f)/R. \tag{12-10}$$

Shot noise. This type of noise [Ref. 2, p. 308; Ref. 4, p. 376; Ref. 13] is caused by the quantum nature of the electric current through electron tubes: the "bunching" of electrons because of variations in the electron emission rate. The rms noise i_s as a fluctuating current is

$$i_s = (2qI\,\Delta f)^{1/2}, \tag{12-11}$$

where q is the electronic charge (1.6×10^{-19} coulomb) and I is the mean cathode emission current. Because of the gain μ in multiplier phototubes, the noise current is

$$i_m = \mu(2qI\,\Delta f)^{1/2}. \tag{12-12}$$

The shot noise, expressed as a voltage produced across a load resistor R, is

$$V_m = \mu R(2qI\,\Delta f)^{1/2}. \tag{12-13}$$

If Johnson noise and shot noise are plotted as functions of load resistance R, a crossover point occurs because one increases as the square root of R and the other as R. By equating the two noise voltages, the crossover load resistance can be solved for as follows:

$$R_c = 2kT/(\mu^2 qI). \tag{12-14}$$

If the load resistor is larger than R_c, the predominant noise is shot noise; but if it is less than R_c, the predominant noise is Johnson noise. Both the noise voltage and the signal voltage increase linearly with R when $R > R_0$, so the signal-to-noise ratio does not change significantly. The value R_c is approximately an optimum load resistance since in the Johnson noise range the signal-to-noise ratio increases as the square root of R.

Values of μ^2 can easily be as high as 10^{11} or 10^{12}, so typical values of R_c for simple diode phototubes and multiplier phototubes differ considerably. For the simple diode, R_c would be several thousand megohms, which is somewhat larger than practical values of load resistance. For the multiplier phototube, R_c could be less than 100 ohms. Consequently the R–C time constant in phototube circuits can be much less with multiplier tubes than with simple diodes.

Semiconductor Detectors

Because semiconductor detectors, being the most recent family to be developed, are the least likely to be discussed in available optical texts, we treat them more thoroughly than the other detectors.

The Nature of Semiconductor Materials

The characteristics of semiconductor materials that make them useful in radiant energy detectors are most easily discussed in terms of electronic energy level diagrams such as Fig. 10.6, which was explained in Chapter 10. The diagrams that apply particularly to semiconductors are shown in Fig. 12.6.

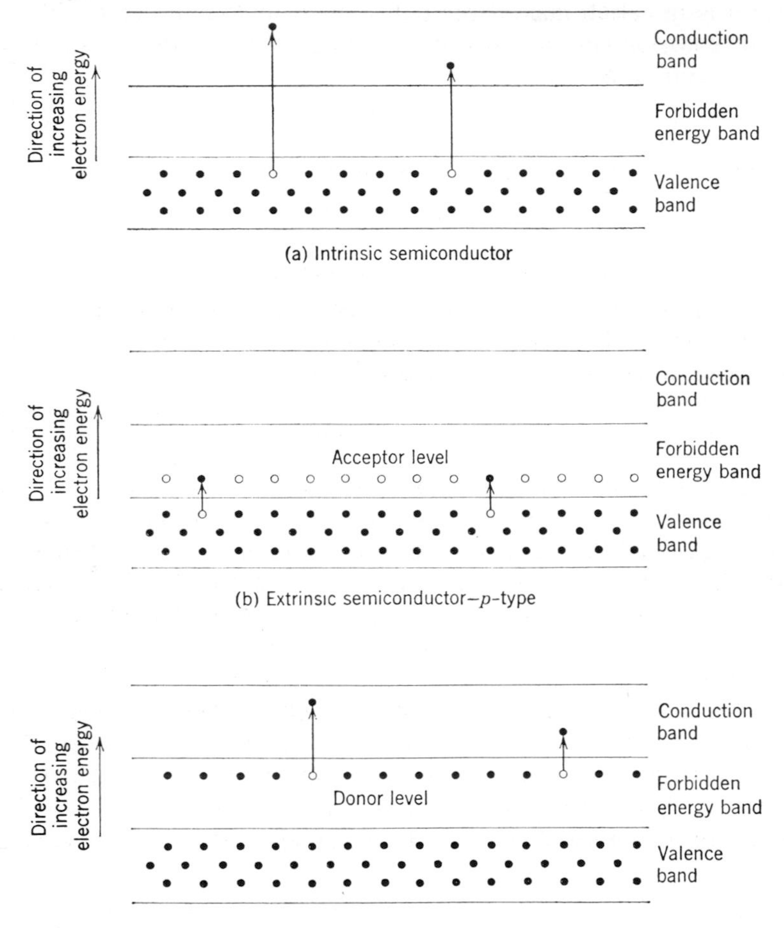

Figure 12.6 Energy bands in semiconductors: (a) intrinsic semiconductor, (b) extrinsic semiconductor—*p*-type, (c) extrinsic semiconductor—*n*-type.

In *intrinsic* semiconductors, the discrete energy levels that electrons may occupy are so closely consecutive, where they occur at all, that only bands (each of which consists of many discrete levels), rather than the levels themselves, are shown. Thus, in Fig. 12.6a, the intrinsic semiconductor is shown to have a valence band (lowest energy band), a forbidden energy band, and a conduction band (highest energy band). The existence of the forbidden energy band is the peculiarity of semiconductors that makes them useful in radiant energy detectors as well as in transistors and many other electronic applications.

As indicated by the diagram, transitions from the valence band must involve energy greater than the width of the forbidden energy band to occur at all. When electrons, indicated by black dots in the diagram, move from their valence band levels to the conduction band, their vacated levels are called *holes* and are indicated by open circles in the diagram.

Besides the conventional current in the conduction band, consisting of negative electrons, a current associated with the holes can be set up in the valence band. Although the valence electrons are relatively fixed with respect to their atoms in the crystalline semiconductor material, an electron adjacent to a hole can be induced by, say, an electric field to occupy the hole position—thereby creating a new hole in the location from whence it came. As this process is repeated, one can see that the hole progresses in a direction opposite to the electron movement and, in effect, has a positive charge. Thus, we can think of the hole movement in the valence band as a current flow of positive charges that adds to the conventional current of electrons in the conduction band.

With deliberate introduction of impurities or through inadvertent imperfections, intrinsic semiconductor material acquires additional properties indicated in the diagrams shown as Figures 12.6b and 12.6c. The material then becomes extrinsic semiconductor material—*p*-type if the majority carrier is positive, that is, consisting of holes; *n*-type if the majority carrier is negative, consisting of the electrons in the conduction band.

The predominant mechanism in *p*-type material is shown in Fig. 12.6b. Impurities or crystal defects have established an *acceptor* level in the forbidden energy band. The nature of the sites accounting for the acceptor level is to receive electrons and to hold them for an appreciable length of time. If the acceptor level is close enough to the valence band to allow photon or thermally excited electrons to be captured in sufficient number, the holes in the valence band will function as charge carriers, and the material will be known as *p*-type.

In *n*-type material, one or more *donor* levels near the upper edge of the forbidden energy band have electrons that, with photon or thermal excitation, spend a significant proportion of their time in the conduction band functioning as negative charge carriers (Fig. 12.6c).

In actual semiconductor material, all three kinds of behavior, indicated by the three parts of Fig. 12.6, can occur simultaneously. Designation as intrinsic or extrinsic depends upon which effect is predominant under the specified conditions. In fact, the photovoltaic detector, which is discussed in detail later in this chapter, depends upon a junction between *p*-type and *n*-type *intrinsic* material for its operation. Under conditions of just thermal excitation, the acceptor and donor levels hold electrons or contribute electrons, respectively, to establish the remaining holes in the *p* region or the excited electrons in the n region as the majority carriers. However, to be sensed, a photon has to create an electron–hole pair; that is, the photon has to excite an electron with energy equal to at least the width of the forbidden energy gap to produce both an electron in the conduction band and a hole in the valence band. Because this latter behavior is associated with *intrinsic* material, the detector is so designated, even though the conditions under just thermal excitation suggest the two diagrams of Fig. 12.6 labeled *extrinsic*.

Having sorted all semiconductor materials into three subgroups according to energy band diagrams, we may have some interest in how the more familiar classifications of *electrical conductor* and *electrical insulator* look in terms of these diagrams. In an electrical conductor, the forbidden energy band is negligible; thus, there is an abundance of thermally excited electrons to function as carriers in the conduction band. On the other hand, the forbidden energy band is so great in an electrical insulator that hardly any carriers can be developed at all.

The theory of semiconductors and the application of these materials to radiant energy detection—particularly in the infrared—are extensively discussed in the literature [Refs. 1–3, 9, and 14–16].

Semiconductor Detector Noise

Before the functioning of specific kinds of semiconductor detectors is discussed, a general review of electronic noise in semiconductors is desirable to provide a background for performance characteristics.

Noise in semiconductor photodetectors (I_n) is composed of the Johnson–Nyquist (thermal) noise (I_T) [see Refs. 11 and 12], generation–recombination noise (I_{G-R}), shot noise (in photodiodes) (I_S), and excess or low frequency noise (I_f). Since the noises from these several sources are uncorrelated, their powers add. By assuming that these powers are dissipated in a common load resistor, we can write:

$$I_n^2 = I_T^2 + I_{G-R}^2 + I_S^2 + I_f^2. \tag{12-15}$$

Johnson–Nyquist noise has already been discussed in connection with phototubes.

The detector element and the load resistor of the detector–amplifier combination are often cooled to reduce their thermal noise contributions. Then the Johnson–Nyquist noise of each element must be considered separately. In a hypothetical example of a detector element having a resistance of 10^5 ohms at 30 K, a bias resistor of 10^5 ohms at 77 K, and a parametric amplifier with an input capacitor having a 10^9 ohm resistive component at

300 K and at the operating frequency band Δf [see Eq. (12-10)], the squared thermal noise is

$$I_T^2 = 4k\,\Delta f(30/10^5 + 77/10^5 + 300/10^9)$$
$$= (\Delta f)(5.9 \times 10^{-26})\ \text{amperes}^2, \tag{12-16}$$

by assuming that the three noise sources are in parallel. Then

$$V_T^2/\Delta f = \frac{I_T^2 R^2_{\text{parallel}}}{\Delta f} \cong [(5.9 \times 10^{-26})/\Delta f][10^5/2]^2$$
$$= 1.5 \times 10^{-16}\ \text{volts}^2\ \text{Hz}^{-1}. \tag{12-17}$$

Generations and recombinations of carriers in a semiconductor are statistical in nature; that is, individual generations and recombinations are independent of one another. At steady state the total number of generations must equal the total number of recombinations over a long interval of time, but instantaneous fluctuation in the total number of carriers available in a given volume of a semiconductor can occur. Fluctuation in the carrier concentration directly affects, for instance, the filament resistance in that volume of the semi-conductor. Under an applied bias current, the drop in potential across that element of the semiconductor will fluctuate about some average value. The generation–recombination current noise per unit bandwidth has been derived by van Vliet [Ref. 17]:

$$I^2_{G-R} = I_{dc}^2 4t_c\,\Delta f(1 - P_i)/[N(1 + (\omega t_c)^2)], \tag{12-18}$$

where I_{dc} is the bias current, t_c is the carrier lifetime, N is the total number of free carriers, ω is $2\pi f$, and P_i is the probability that the donor and acceptor levels are ionized. The value P_i is very nearly zero for any photoconductor in a background-limited condition.

The generation–recombination noise is flat, or white, with frequency until $(\omega t_c)^2$ grows to the vicinity of unity where the noise current begins to roll off with frequency. Decreasing the background radiation on a background-limited photoconductor will decrease the applied electric field required for optimum operation, and this in turn will require less bias current. Generation–recombination noise reduction is best approached by writing Eq. (12-18) in terms of the electronic field E across the background-limited detector:

$$I^2_{G-R} = 4E^2 w\mu^2 q^2 J_r \mathcal{N} t_c^2\,\Delta f/\{l[1 + (\omega t_c)^2]\}, \tag{12-19}$$

where w is the width of the detector element, μ is the carrier mobility, q is the charge on the electron, J_r is the photon flux density for background photons within the bandgap for the detector, $\mathcal{N}$ is the quantum efficiency, and l is the length of the detector element. One cannot, unfortunately, reduce the noise by simply reducing all the parameters in the numerator while increasing the length l of the detector element. Analysis of the signal shows that these same parameters are important in the expression for signal strength.

Shot noise is generally associated with vacuum tubes and phototubes, but photovoltaic and other semiconductor diodes also exhibit a shot-type noise which is generated by the charged carriers as they pass through the space charge region between the p- and n-type layers. As in the vacuum tube, the shot noise current per unit bandwidth is given by [Ref. 18]:

$$I_S^2/\Delta f = 2qI_{dc}. \tag{12-20}$$

Shot noise in semiconductor diodes is white and is the dominant noise in background-limited photovoltaic detectors.

Excess noise, also known as low frequency noise and as $1/f$ noise, is the least understood of all the noise mechanisms but has been shown to follow the general empirical formula:

$$I_f^2/\Delta f = K' I_{dc}^2/(fAd). \tag{12-21}$$

Here K' is a constant, I_{dc} is the total current through the detector element, f is the frequency, A is the detector area, and d is the detector thickness. The exponents for both I_{dc} and f sometimes take on values different from those shown in Eq. (12-21). Also, the noise in photovoltaic detectors has been observed to vary inversely as the peripheral area rather than the total area, probably because of the existence of a surface shunt. In many semiconductor photodetectors, excess noise has been associated with particular surface treatments; however, even when these surface treatments are optimized, other sources for low frequency noise predominate over the background noise. Other sources of excess noise include electrical contacts, bulk and surface traps in the material, and minority carrier fluctuations.

A hypothetical noise current spectrum for a photoconductor at medium and extreme bias is shown in Fig. 12.7. At low frequency the excess noise predominates for both bias currents. The change in slope at low frequency is common for most photoconductors. Since the hypothetical detector is background-limited, generation–recombination (G–R) noise predominates in the central portion of the noise spectrum, rolling off with frequency as ωt_c approaches unity. The generation–recombination continues to roll off until Johnson–Nyquist noise becomes the limit. Since the signal decays with the same function as the G–R noise, the signal-to-noise ratios usually remain constant until the G–R noise approaches the Johnson–Nyquist noise level. The photoconductor noise has the form,

$$I_n^2/\Delta f = 4kT/R + 4I^2 t_c/[N(1 + (\omega t_c)^2)] + KI^2/f, \tag{12-22}$$

and the G–R noise (second term) predominates for a background-limited photoconductor; the shot noise term is usually not observed.

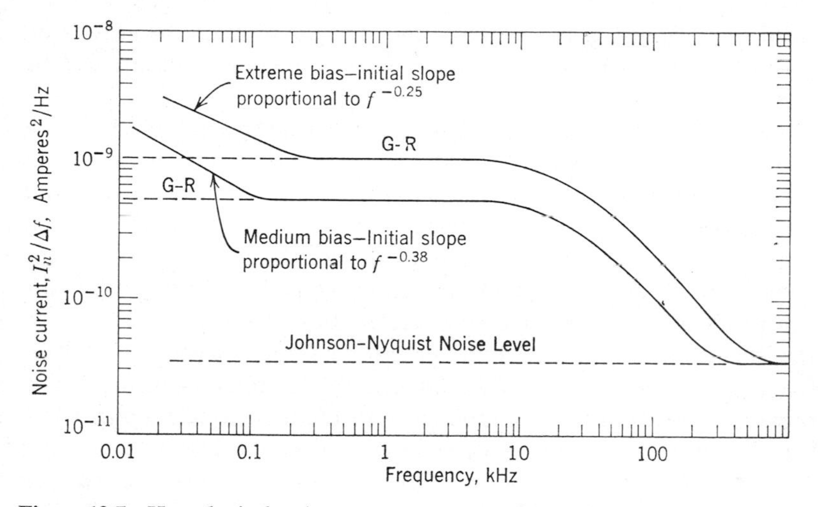

Figure 12.7 Hypothetical noise current spectrum for a photoconductor at medium and extreme bias.

In the photovoltaic detector, the noise takes another form because the shot noise across the photodiode, as a result of the photo-induced currents, predominates for background-limited photodiodes. Thus, the photodiode noise takes the form,

$$I_n^2/\Delta f = 4kT/R_Z + 2qI_{sc} + K_1 I^2/f + K_2 I^2/f, \tag{12-23}$$

where R_Z is the real part of the parallel impedance for the detector–preamplifier combination taken at its operating point, and K_1 and K_2 are constants of proportionality for the forward and the reverse current in the photodiode. The *G–R* term from the photoconductor does not appear, though actually the generation of carriers is the source of I_{sc}. Recombination usually occurs in the majority layer, where P_i for Eq. (12-18) is nearly unity; so the recombinations do not enter as a significant noise source.

Photoconductive Detector

As its name suggests, the photoconductive detector depends on the variation of electrical conductivity with exposure to radiant energy. This variation can occur in several ways.

In intrinsic semiconductor material, a photon, to be sensed, must elevate an electron from the valence band to the conduction band. This creates an electron–hole pair, each component of which contributes to the conductivity of the material. Because of the relationship between photon energy and wavelength [see Eq. (1-9)], the width of the forbidden energy band sets the longest wavelength to which the intrinsic photoconductive detector will be sensitive.

In p-type extrinsic semiconductor material, the sensed photon must elevate an electron from the valence band to the acceptor level. While the electron is held at the acceptor level, its counterpart, the hole, in the valence band increases the conductivity of the material. In n-type material, the photon elevates an electron from the donor level to the conduction band to increase the conductivity. The locations of the impurity energy levels within the forbidden bandgap now determine the wavelength sensitivity.

Shortly after each photon–electron interaction has taken place, the electron recombines with a hole to restore the material to its initial condition. The recombination time contributes significantly to the time-constant characteristic of the detector. Because both thermal and photon excitation can create electron–hole pairs and affect conductivity as already described, cooling is often required to keep thermally induced changes from masking the photon-induced changes.

The photoconductive element is connected in series with a biasing voltage supply and a resistor, across which voltage variations (due to conductivity changes) are sensed by an appropriate electronic amplifier (Fig. 12.8).

As indicated above, both intrinsic (band-to-band) and extrinsic (band-to-impurity) photoconductors are stimulated by photons, and an *excess* number of carriers is generated. (The number of carriers over the equilibrium value set by thermal generation is the excess.) The conductivity σ of the semiconductor filament is

$$\sigma = \mu q N. \tag{12-24}$$

For detectors having the ultimate sensitivity, the detector resistance $R_c = 1/\sigma$ is set by radiant energy received from the background. A change ΔR_c caused by a change of incident radiant power $\Delta\Phi$ from an object against a "constant" background causes a change in the current through the detector, which is often referred to as the *signal* current. With only background

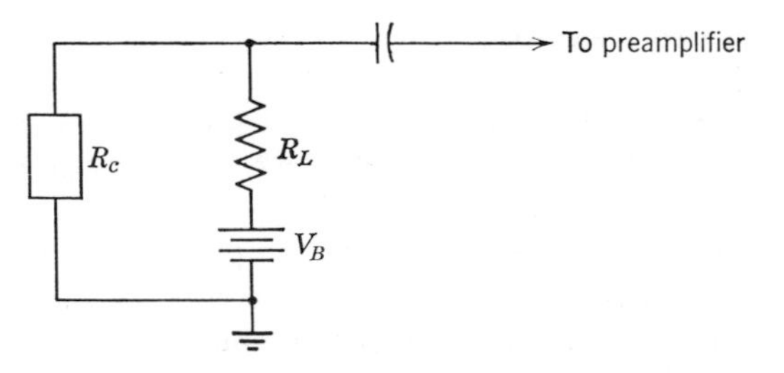

Figure 12.8 Biasing circuit for a photoconductive detector.

photons, the current passing through the detector with a biasing circuit as shown in Fig. 12.8 is

$$I_{dc} = V_B/(R_c + R_L). \tag{12-25}$$

If ΔR_c is small, a good approximation for the signal current is

$$I_s = \Delta I = I_{dc}\Delta R_c/(R_c + R_L). \tag{12-26}$$

By a straightforward derivation (not given here) one can arrive also at the following expression for the signal current (again assuming that the detector resistance is set by the background):

$$I_s = Ew\mu q \mathcal{N} t_c \,\Delta J_s/\{1 + (R_L/R_c)[1 + (\omega t_c)^2]^{1/2}\}, \tag{12-27}$$

where ΔJ_s is the signal photon flux within the detector bandgap and the other symbols are defined as they were for Eq. (12-19).

A signal-to-noise expression can be formed by setting up the ratio of I_s, as given by Eq. (12-27), and the noise current whose square is given by Eq. (12-19). To simplify the final form of this ratio, we assume that $R_L \ll R_c$. Even for applications where this assumption cannot be made, the simplified ratio is still useful to indicate how various parameters influence the signal-to-noise properties. The simplified ratio is

$$\begin{aligned} I_s/I_n &= Ew\mu q \mathcal{N} t_c \,\Delta J_s/(E^2 w \mu^2 q^2 \mathcal{N} t_c^{\,2} 4J_r \,\Delta f/l)^{1/2} \\ &= \Delta J_s (A\mathcal{N})^{1/2}/(4J_r \,\Delta f)^{1/2}. \end{aligned} \tag{12-28}$$

The area A of the detector element has been substituted for the product of l and w. It is implied in the derivation that the operating conditions make the generation–recombination noise predominant.

The photoconductor has a built-in gain mechanism for both the signal and the noise. The carrier velocity v is

$$v = \mu E, \tag{12-29}$$

and the carrier transit time is $l/v = \mathrm{T}$. If these substitutions are made in the noise expression of Eq. (12-19),

$$I^2_{G-R} = 4Aq^2 J_r \mathcal{N} \,\Delta f(t_c/\mathrm{T})^2. \tag{12-30}$$

Transit time decreases with increasing electric field, so the ratio t_c/T can be thought of as the photoconductive gain. Certain limitations apply that do not appear in Eq. (12-30). Full gain cannot be realized if any recombination takes place at the detector contacts. The upper practical bound is reached where the carrier approaches thermal velocities at which secondary ionization usually begins, and the noise becomes excessive. The signal expression equivalent to Eq. (12-30) is

$$I_s = q\mathcal{N}A\,\Delta f_s t_c/\mathrm{T}. \tag{12-31}$$

Thus, it is evident that, although the responsivity of the detector can be improved by increasing the photoconductive gain, the signal-to-noise ratio will not be affected and will remain at the value given in Eq. (12-28).

The most common photoconductive detectors are fabricated from the following intrinsic materials: CdS, CdSe, TlS, PbS, PbSe, PbTe, Si, Ge, GaAs, InAs, InSb, HgSe, and $Hg_xCd_{1-x}Te$. Probably more PbS detectors are used today than all the other photoconductors combined. Many other detectors approach both the wavelength response and the sensitivity of lead sulfide when they are cooled, but no other detector is as sensitive out to 3 μm while operating at room temperature. Still sought is a fast, more stable, uncooled detector for long-wavelength operations. In the small-energy-gap (longer-wavelength response) detectors, thermal carriers are responsible for the cooling requirements. The optically generated carriers must predominate for the photoconductors to be background limited.

The common extrinsic photoconductors are Ge:Au ("gold-doped germanium"), Ge:Hg, Ge:Cd, Ge:Cu, Ge:Zn, Ge:Ga, and Ge:Zn,Sb. As indicated, all are impurity levels in the germanium lattice. Photons with enough energy to excite the impurity are absorbed (with a much lower absorption coefficient than in the intrinsic semiconductors) to generate either a hole or an electron in the valence or the conduction bands. Hole-electron pairs are not generated. The smaller absorption coefficients require that these impurity-doped photoconductors be thicker and require more cooling than intrinsic photoconductors with response to the same wavelength.

Mercury-doped germanium has a near-theoretical sensitivity, which decays at the upper edge of the important 8- to 14-μm atmospheric transmission window. It requires cooling below 50 K for a 300 K background and a complete 2π detector solid angle. More cooling is required as the backgrounds are reduced in size by cold shields. This detector has been used in astronomy for thermally mapping the shaded portion of the moon and of Jupiter, and it has been used in military and industrial infrared mapping of the earth—an operation commonly called remote sensing.

Photovoltaic Detector

A photovoltaic detector element consists of a single intrinsic semiconductor crystal in which a diode junction has been formed between p and n regions. The junction is usually made by starting with a crystal of one type and then diffusing an impurity into the detecting surface to form a shallow layer of the opposite type. The junction between the two regions must be close enough to the detecting surface that photons can reach the junction through the thin layer.

At a *p–n* junction, the carriers are predominantly holes on one side (Fig. 12.6b) and electrons on the other (Fig. 12.6c). Under these conditions, the carriers tend to diffuse across the junction, holes into the *n* region and electrons into the *p* region. Without diffusion, each region would be neutral electrically; with diffusion, the displaced charges set up a field across the junction opposing the normal diffusion. Dynamic equilibrium is reached when the flow rate due to the field is equal to the diffusion rate. Under these conditions, the voltage from end to end of the detector element, perpendicular to the junction, is zero. Now if photons are allowed to create electron–hole pairs near the junction, the field set up by the diffusion process will sweep the minority charge of each across the junction, electrons from the *p* region to the *n* region and holes from the *n* region to the *p* region. This action on both sides of the junction tends to make the *p* region positive and the *n* region negative, so, if an external electrical circuit is closed from the *n* region end to the *p* region end, electrons will flow. This current, appropriately amplified, measures the photon rate or radiant power.

Because the photon–electron coupling mechanism is essentially the same in photovoltaic and photoconductive detectors, the two classes have similar cooling requirements. A notable exception is extrinsic material used in photoconductive but not in photovoltaic detectors. Because the photon is sensed in the extrinsic detector by exciting an electron to or from an impurity level, rather than across the whole forbidden band gap, thermal energy has to be suppressed by cooling to the vicinity of liquid helium temperatures (8–26 K). Intrinsic material is used typically at room temperature (about 300 K), at the sublimation point of solid carbon dioxide (about 200 K), or at the boiling point of liquid nitrogen (about 77 K), depending on the nature of the material and the application.

The best designed photovoltaic detectors have quantum efficiencies greater than 80%, and their performance is limited only by their quantum efficiencies. The action of a photovoltaic detector, often referred to as a photodiode, can be conveniently studied by first considering its diode behavior and then superimposing the photoinduced effect. When the junction region is shielded from photons, the current I_d' resulting from an external voltage V applied to the diode is

$$I_d' = I_0\{\exp[qV/(2kT)] - 1\}, \qquad (12\text{-}32)$$

where I_0 is the saturation current. A plot of I_d' is the familar diode I–V characteristic where the current increases rapidly with increasing positive voltage (forward direction) and approaches a saturation value I_0 with greater and greater negative values of voltage (reverse direction). The coefficient 2 of kT in the denominator of the exponent is determined by the nature of the junction. Unity and other values are sometimes given for this coefficient depending on the assumptions made in deriving the diode equation.

The reciprocal of the slope of the diode I–V characteristic is the dynamic resistance (a-c resistance) r_d of the diode:

$$r_d = 1/(\partial I_d'/\partial V) = [2kT/(qI_0)] \exp [-qV/(2kT)]. \tag{12-33}$$

When photons are allowed to create electron-hole pairs as already described, the minority carriers are swept across the junction to produce a current (or a voltage if the external circuit has appreciable resistance) in the "reverse direction" as defined above. The magnitude of this current I_{sc}, if the external circuit has no resistance, is

$$I_{sc} = q\mathcal{N}AJ_r, \tag{12-34}$$

where $\mathcal{N}$ is the quantum efficiency, A is the effective intercepting area in the vicinity of the junction, and J_r is the photon flux density for background photons within the bandgap for the photodiode. The total photodiode current is then seen to be I_d:

$$I_d = I_d' - I_{sc}. \tag{12-35}$$

The value of r_d, as given by Eq. (12-33), is unaffected by the photons because I_{sc} is assumed independent of V.

To find the open-circuit voltage V_{oc} caused by the photons, I_d is set at zero in Eq. (12-35), the expression given by Eq. (12-32) is substituted for I_d', and $V = V_{oc}$ is solved for in the resulting equation:

$$V_{oc} = 2(kT/q) \ln [(I_{sc}/I_o) + 1]. \tag{12-36}$$

Superimposed upon the photon action so far described is the effect of signal photon flux ΔJ_s. As one might expect, the short-circuit signal current expression is similar to the expression in Eq. (12-34) for the corresponding current due to background photons. In the following equation for the short-circuit signal current,

$$I_s = q\mathcal{N}A\,\Delta J_s, \tag{12-37}$$

where ΔJ_s is the signal photon flux within the detector bandgap.

To compute the electrical output of a photovoltaic detector, either of the two circuits shown in Fig. 12.9 may be used. The responsivity of the detector is obviously sensitive to the value of r_d, which is a function of T, I_0, and V in accordance with Eq. (12-33). Because of the relationship between T and I_0, which will not be developed here, r_d increases with decreasing temperature

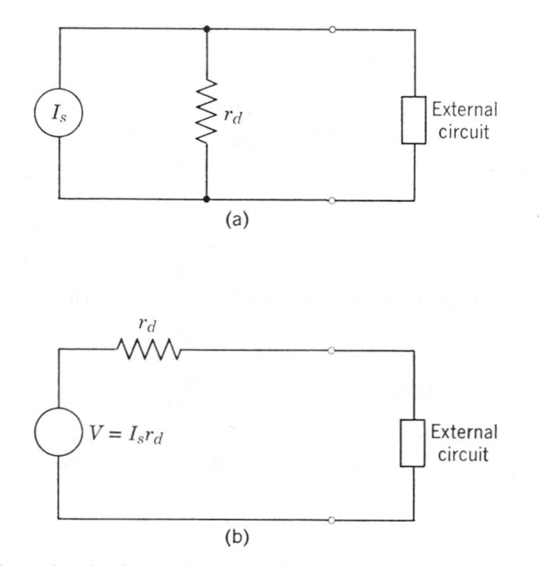

Figure 12.9 Equivalent circuits for a photovoltaic detector: (a) constant-current source in parallel with r_d, (b) constant-voltage source in series with r_d.

T. The equivalent resistance r_d is affected through V in two ways. First, by derivations like the one leading to Eq. (12-36), it is evident that V is a function of I_{sc} and, therefore, through Eq. (12-34), a function also of J_r, the background flux. So, in summary, r_d is affected by the background. The second way of influencing r_d through V is by including a bias voltage in the external circuit.

Unlike the photoconductive detector, the photovoltaic detector does not require a bias voltage to operate; but biasing is often practiced to minimize noise. A slight "reverse" bias, that is, in the direction to increase the absolute value of I_{sc}, is commonly used.

As noted earlier, shot noise predominates in photovoltaic detectors that are background-limited [see the second term on the right-hand side of Eq. (12-23)]. When the indicated substitutions are made, the signal-to-noise ratio comes out as follows:

$$I_s/I_n = (\mathcal{N}A)^{1/2}\, \Delta J_s/(2J_r\, \Delta f)^{1/2}. \qquad (12\text{-}38)$$

The most common infrared photovoltaic detectors are fabricated from germanium, indium arsenide, and indium antimonide. The silicon and germanium cells can operate at normal ambients with high sensitivity; however, the indium arsenide and the indium antimonide must be cooled to reduce the saturation currents.

Semiconductor Detector Performance

Table X gives the performance of various semiconductor detectors. Photoconductive (PC), photovoltaic (PV), and one phototransistor (PT) detectors are represented. The phototransistor detector is a photovoltaic detector modified by the addition of an emitter to give it internal gain.

TABLE X RESPONSIVITY, RESPONSE TIME, AND OPERATING TEMPERATURES FOR POPULAR INFRARED PHOTON DETECTORS*

Detector Type	Operating Temperatures, K	Response Time	Responsivity, volts watt^{-1} (500 K Source)
PbS (PC)	300	0.3 msec–30 msec	3×10^3–4×10^3
PbS (PC)	200	3–6 msec	9×10^3–8×10^4
PbSe (PC)	200	13–30 μsec	3×10^3–1×10^5
PbSe (PC)	77	40–140 μsec	3×10^5–2×10^6
InAs (PV)	200	<1 μsec	4×10^2–1×10^3
InAs (PT)	200	10–100 μsec	3×10^3–5×10^4
InSb (PC)	77	2–10 μsec	5×10^3–2×10^4
InSb (PV)	77	<1 μsec	5×10^3–2×10^4
Ge:Hg (PC)	26	0.3 μsec	2×10^5–1×10^6
HgCdTe (PC)	77	<0.3 μsec	2×10^2–1×10^4
Ge:Cu (PC)	8	0.1 μsec	2×10^5–1×10^6

* Average values taken from NOL-Corona reports [Ref. 19].

The value D^* for a number of the common detectors is plotted against wavelength in Fig. 12.10. As indicated on the ordinate scale, the chopping frequency for the tests was 1000 Hz, and the bandwidth was 1 Hz. The theoretical limit of performance indicated by the broken line is established by the fluctuations of a 300°K, hemispherical background.

Many considerations enter into the decision one must make when choosing a detector for a given application. Among these are where transmission windows occur in the atmosphere, the strength of signal that one anticipates from the target, and how much one is willing to pay to approach optimum performance. Two particularly important values that contribute to the decision are the temperature of the target (source of signal energy) and the temperature of the background. As the study of blackbodies in Chapter 3 indicates, the peak of the radiant energy curve moves toward shorter wavelengths and increases in value as the radiating source increases in temperature. This suggests that as the source gets hotter, one should choose a shorter cut-off wavelength to get maximum signal-to-noise ratio. As the curves of

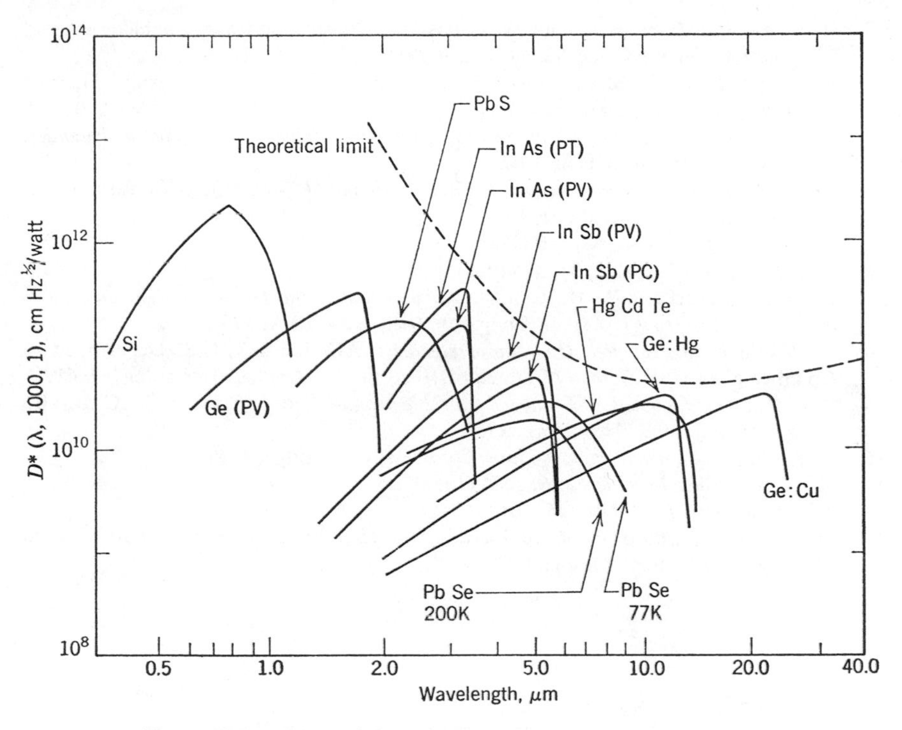

Figure 12.10 Spectral detectivities of available photon sensors.

Fig. 12.10 indicate, choosing a relatively long wavelength detector for a short wavelength (hot) target not only sacrifices D^* but also unnecessarily accepts background (noise) power at the long wavelengths where there is relatively little target power.

REFERENCES

1. R. A. Smith, F. E. Jones, and R. P. Chasmar, *The Detection and Measurement of Infra-Red Radiation*, Second Edition. Clarendon Press, Oxford, 1968.
2. R. D. Hudson Jr., *Infrared System Engineering*. Wiley–Interscience, John Wiley & Sons, New York, 1969.
3. W. L. Wolfe, *Handbook of Military Infrared Technology*. Office of Naval Research, Department of the Navy, Washington, D.C., 1965. (Obtainable from the Superintendent of Documents, U.S. Government Printing Office.)
4. J. A. Jamison et al., *Infrared Physics and Engineering*. McGraw-Hill Book Company, Inc., New York, 1963.
5. T. S. Moss, "Modern Infra-Red Detectors," *Advances in Spectroscopy*, Vol. 1, pp. 175–213. Edited by H. W. Thompson. Interscience Publishers, New York, 1959.
6. J. N. Shive, "Heating and Cooling of Bolometer Elements," *J. Appl. Phys.*, **18**, 398 (1947).

7. B. H. Billings, E. E. Barr, and W. L. Hyde, "Construction and Characteristics of Evaporated Nickel Bolometers," *Rev. Sci. Instr.*, **18,** 429 (1947).
8. R. De Waard and E. M. Wormser, "Description and Properties of Various Thermal Detectors," *Proc. IRE*, **47,** 1508 (1959).
9. R. A. Smith, "Thermoelectric Power," *Physical Principles of Thermodynamics*, Chapter 9. Chapman & Hall, 1952.
10. RCA Phototubes and Photocells, *Technical Manual PT-60*, Radio Corporation of America, Lancaster, Pa. 17604, 1963.
11. H. Nyquist, *Phys. Rev.*, **33,** 110 (1928).
12. J. B. Johnson, *Phys. Rev.*, **32,** 97 (1928).
13. W. B. Davenport and W. R. Root, *An Introduction to the Theory of Random Signals and Noise*. McGraw-Hill Book Company, Inc., New York, 1959.
14. W. Wolfe et al., *Infrared Quantum Detectors*. AD 326 487, U.S. Department of Commerce/National Bureau of Standards/Institute for Applied Technology, 1961.
15. L. P. Hunter, *Handbook of Semiconductor Electronics*, Second Edition. McGraw-Hill Book Company, Inc., New York, 1962.
16. R. A. Smith, *Semiconductors*. University Press, Cambridge, 1959.
17. K. M. van Vliet, *Proc. IRE*, **46,** 1004 (1958).
18. A. van der Ziel, *Proc. IRE*, **46,** 1019 (1958).
19. "Properties of Photoconductive Devices," a continuing series of reports, Naval Ordnance Laboratory, Corona, Calif.

13

Image Recording

Two methods of making a permanent record of the image formed by an optical system are discussed in this chapter: (1) ordinary photography, and (2) recording by a scanning process.

In the first method, a photographic film, or plate, is placed in the image plane of the optical system; and, after exposure and processing, the image appears as a photograph on the film. In the second method, both object field and image field are scanned, point by point—in the object field, to measure the radiance (or luminance) of the object and concurrently, in the image field, to expose a film to an illuminance proportional to the radiance in the object.

In preparation for quantitative analyses of both methods, our discussion begins with a simplified description of the photographic process [Refs. 1–5]. Included is a review of *densitometry*, which resembles, in many respects, the scanning of the object field in image recording. The function of densitometry however, is to measure the characteristics of the film, including film noise in terms of *granularity*, which is also discussed.

A mixture of units occurs in this chapter. Power radiated by the terrain and received by an aerial mapper is given in radiometric units; on the other hand, quantities related to exposure of film or to the strength of the light source making the exposure are in photometric units.

The Photographic Film

The basic material of photography is the film or plate made up of a thin photosensitive layer (usually less than 0.001 in. thick) coated on a mechanical support: glass, flexible transparent plastic film, or paper. The thin layer, called the emulsion, is a suspension of finely dispersed, minute particles of silver compounds—usually silver halide salts (chlorides, bromides, and iodides)—in gelatin. Each particle, or grain, is believed to be a silver halide crystal having a crystal lattice in which silver and the halide exist as ions. Only the crystals are light sensitive, the other materials being essentially transparent and unaffected by light. The more sensitive emulsions, the relatively *fast* films requiring the least luminous energy for exposure, contain silver bromide mixed with silver iodide; the less sensitive emulsions are usually silver chloride. The sensitivity (*speed*) and other photographic properties (like granularity and modulation transfer function) depend upon the size of the crystals and the size distribution. The crystal size—ranging from less than 0.1 μm in slow emulsions to several micrometers in fast emulsions—and size distribution are determined by the way the emulsion is precipitated.

The preparation of a photographic emulsion is still largely an art in spite of extensive scientific study of photosensitive materials over many years. The preparation starts with dissolving silver in nitric acid to produce silver nitrate. Then the light-sensitive emulsion is formed by mixing the solution of silver nitrate with a solution of an alkali halide, such as potassium bromide, in the presence of gelatin. The following double decomposition reaction is typical:

$$AgNO_3 + KBr \rightarrow AgBr\downarrow + KNO_3.$$

The silver bromide, which is only slightly soluble, is precipitated; myriads of minute crystals immediately coagulate, or clot together, as a pale yellow curd. The gelatin, which was dissolved in the alkali halide solution, prevents the precipitate from settling out by keeping the grains in suspension, each physically and electrically insulated from its neighbors. The mixture can be handled in most of the film-preparation processes as a homogeneous liquid.

The halide crystals are thin platelets with triangular, hexagonal, and a variety of other shapes. In addition to their normal crystal structure, the crystals contain imperfection centers—either molecules of some foreign material (probably silver sulfide) or fissures, dislocations, or grain boundaries—called *sensitivity specks*, which are formed on the crystal surfaces during emulsion preparation.

Exposure: Formation of Latent Image

When light is allowed to fall on a photographic film, the film is said to be *exposed;* the luminous energy produces chemical changes in those crystals that absorb photons. These changes can be made visible at once, under special conditions, as a darkening; but in normal practice, they are invisible. In the latter instance, a *latent image* is said to be stored, and the effects of exposure are not apparent until the emulsion is developed by a chemical reducing agent [Refs. 6–9]. In fact, light is not required to do all the work of producing the ultimate impression (image) of silver; subsequent chemical processes actually *develop* the image. Generally, light only triggers the work, producing changes so slight that no measureable chemical or physical difference can be detected between the exposed and unexposed silver halide, other than their respective responses to a photographic developer.

Latent image formation involves the movement of both electrons and silver ions; electron mobility is relatively high, the silver ion mobility is low. Silver halide crystals are believed to respond as photoconductors. Light causes halide ions to become halide atoms by ejecting electrons. The electrons move freely until trapped in the vicinity of sensitivity specks, causing negative charge accumulations. The resultant electric field produces an electrolysis of the silver halide. The current in this process, which is much slower than the prior motion of electrons, is carried by the interstitial silver ions. The trapped electrons and the silver ions combine to form silver atoms, which are deposited at the sensitivity centers. The chemical changes can be represented by the following equations:

$$Br^- + h\nu \rightarrow Br + e^-,$$

$$Br + Br \rightarrow Br_2,$$

$$Ag^+ + e^- \rightarrow Ag.$$

As a result of this process, the latent image consists of specks located at isolated points on the surfaces of individual crystals, the speck separation being much greater than the typical speck size. Since the crystals themselves are dispersed in the emulsion, the specks have a random distribution throughout the emulsion. The light-produced silver atoms at the sensitivity centers initiate development, during the development process, provided the silver specks have grown to sufficient size during exposure.

After exposure, the film must, of course, be protected from all other illumination until the latent image is made visible and permanent by developing and fixing processes.

Development

To develop the latent image of the exposed film, the emulsion is treated with a solution that reacts with the silver halide crystals, freeing more silver and "precipitating" it as tiny irregular grains. The development process begins at the sensitivity centers, and the silver specks "grow" during development until all of the host silver halide crystal is converted. Except when special processing is used, the size and shape of the silver grains produced during development are not at all like the original crystals of silver halide.

The silver is formed in a finely divided state and displays none of the bright metallic luster of a polished silver surface. The shape and size of the silver particles formed depend upon the kind of developer and other factors, but generally the silver has a filamentary, somewhat spongy, structure that can be described as a "tangle of filaments." The silver is, in fact, usually jet black and accounts for the darkening of the film. During development, a given emulsion blackens to a degree that depends on the extent of both exposure and development.

In chemical parlance, materials that form a metal from its salt or its oxide are called *reducing agents;* and the metallic salt is said to be *reduced* to the metal. A photographic developer is a compound, or a mixture of compounds, that not only can reduce the silver halide to silver but can also distinguish between the exposed and unexposed crystals. Relatively few unexposed crystals are affected by normal development, since the developer reacts far more slowly with unexposed—as compared with exposed—crystals. However, *fogging* of the film can occur if the unexposed areas are developed for too long a time.

Although practical developers have many different formulas, almost all of them are based on four primary kinds of constituents: developing agents, preservatives, accelerators, and restrainers [Ref. 10]. The most commonly used *developing agents* are organic compounds that function as reducing agents. *Preservatives* hinder oxidation of the developer (for example, by oxygen of the air) and thus prevent formation of colored oxidation products in the developer solution. *Accelerators* keep the pH of the solution high for rapid development. Finally, *restrainers* reduce the retarding effect of alkali halides accumulated during development.

Among the factors determining the properties of the developed film, the choice of emulsion–developer combination is the most important. Important, but of lesser significance to film characteristics, are the amount of liquid stirring, the solution temperatures, and the development period. Developers are formulated for various purposes and vary widely in their properties. The following is a typical developer formula for black-and-white film (Kodak Developer D-72) (reprinted with permission from Kodak, see Ref. 4):

Water, at about 125°F (50°C)	500.0 ml
Kodak Elon* (*p*-methylaminophenol sulfate)	3.0 g
Sodium sulfite, desiccated	45.0 g
Hydroquinone	12.0 g
Sodium carbonate, monohydrated	80.0 g
Potassium bromide	2.0 g
Water to make	1.0 *l*

The chemicals should be dissolved in the order listed.

* Trademark of Eastman Kodak Company.

When an exposed crystal starts to develop, it does so at the sensitivity specks; as suggested earlier, the speck itself accelerates the reduction process. Thus development begins at a number of isolated points, having a chance distribution on the crystal surface. If a crystal develops at all, and if the process continues long enough, the crystal ultimately develops completely.

The manufacturer recommends the temperature of the developer solution and time interval for development for his product when it is used to develop a given type of film. Such temperatures and times usually result in compromises between various characteristics of the completely processed film. Since the temperature of the developer affects the reaction rate of the developer chemicals, departures from the recommended temperature can be compensated for, over a limited range, by adjusting development time. As development time is increased, more and more grains are developed to completion. Since a higher speed for a given type of film requires greater darkening of the completely processed film for a given exposure, a prolonged development time has the effect of increasing speed. Increased graininess is also associated with higher speeds because high-speed emulsions have relatively larger crystals, and, to get the high speed, a larger proportion of them is developed to completion. Contrast also increases with development time until it reaches a maximum, after which fogging reduces the contrast. The recommended temperature and development time, therefore, do not allow maximum possible speed and contrast, but are rather a compromise combination chosen to keep fog and graininess reasonably low.

The degree of darkening in the film is determined quantitatively in terms of density D, which is defined in Chapter 10 by Eq. (10-4). However, D is used here for *density* where A occurs, in the earlier chapter, for *absorbance:*

$$D = \log_{10} (1/\tau) = \log_{10} (\Phi_0/\Phi). \tag{13-1}$$

When an exposed film is developed, an initial interval, known as the induction period, usually passes during which no visible effect appears. After this period the density increases, rapidly at first and then more slowly, until further time in the developer produces no more darkening [Refs. 6 and 11]. If one assumes, for simplicity, an emulsion–developer combination in which no fogging occurs on prolonged development (which requires, in practice, a low concentration of free halide on the surface of the crystal), the relationship between density and development time is:

$$D = D_0[1 - \exp(-kt)], \tag{13-2}$$

where D_0 and k are development constants.

Fixing

After development and subsequent rinsing, the residual undeveloped silver halide, which is still light sensitive, must be removed without disturbing the metallic silver image. Otherwise the film would ultimately darken all over by further exposure, thus obscuring the image. In a less common procedure, the silver halide is converted to colorless, stable compounds that are insensitive to light. These treatments stabilize the film against further chemical or physical change and are referred to as *fixing*, and the film is said to be *fixed*.

Since the solubility of silver halides is extremely low, these compounds must be converted to soluble compounds before they can be dissolved and washed away. To accomplish this, a solution of either ammonium or sodium thiosulfate, or a mixture of both (incorrectly called hyposulfite and referred to as *hypo* in traditional photographic usage) is applied to the emulsion. The detailed chemical reaction is complicated and probably not completely understood. Generally, a sequence of complex argentothiosulfates, which contain varying ratios between silver and thiosulfate, are produced. The simplified essential ionic reaction is

$$Ag^{+} + 2(S_2O_3)^{2-} \rightarrow [Ag(S_2O_3)_2]^{3-}.$$

These new ions (or complexes) are dissociated and water soluble.

Three processes must take place in the development tank during fixing (1) diffusion of the thiosulfate into the gelatin, (2) chemical formation of the soluble complexes, and (3) diffusion of the complexes and the alkali halides out of the gelatin into solution [see Ref. 10]. The reactions have been observed to be comparatively rapid, so fixing time is established by the diffusion rates. If a photographic material is left undisturbed in the fixer, a stagnant layer of partially exhausted solution accumulates near the surface and acts as a barrier to rapid diffusion. To hasten fixing, this layer can be removed by agitation, that is, by rocking the container, stirring the solution, or brushing the surface.

The following is a typical formula for a fixing solution (Kodak Fixing Bath F-5) (reprinted with permission from Kodak, see Ref. 4).

Water at about 125°F (50°C)	600.0 ml
Sodium thiosulfate (hypo)	240.0 g
Sodium sulfite, desiccated	15.0 g
Acetic acid, 28%	48.0 ml
Boric acid crystals	7.5 g
Potassium alum	15.0 g
Water to make	1.0 *l*

If the fixer remains in the emulsion, it slowly decomposes and attacks the image, causing discoloration and fading. Hence, the fixer must be removed by washing the film thoroughly in water, after which the film is dried.

Characteristic Curves for Films

After processing, the film is a plastic transparency containing a varying distribution of black, metallic silver which absorbs luminous energy. The amount of film darkening, which corresponds to the density, is measured by determining the transmittance (see Chapter 10). The density in a particular element of area depends upon the type of emulsion, the previous exposure of the element, the developer, and the development time. The bright parts of the subject (object in the optical system) produce more exposure than the dark parts, so the processed film is a negative: bright parts of the subject appear dark on the film.

One can explore the relationship between exposure and density by exposing several separate emulsion patches (each very large compared with an individual grain) in succession so that the exposure becomes progressively greater from one patch to the next. Exposure ξ is defined as the illuminance E (in lumens per centimeter^2) on the emulsion multiplied by the exposure time t:

$$\xi = Et. \tag{13-3}$$

The transmittance of each patch of the processed film is measured with a *densitometer*. Application of Eq. (13-1) gives the density. In each patch, the density is tentatively regarded only as a function of the exposure. By repeating the patch experiment with variations in developing procedures, a law of constant density ratios can be demonstrated: When successive areas of a film are subjected to a program of different exposures, the ratio of densities in the processed film for corresponding pairs of exposures is constant and independent of development time. Except for gross departures from accepted procedures to achieve certain special photographic effects (some of which are discussed in a later section), this law is generally quite accurate.

The science of measuring the sensitivity of photographic film is known as *sensitometry*, and instruments for subjecting a photographic material to a series of precisely known graduated exposures are called *sensitometers* (see Ref. 1, Chapter 9). Sensitometry includes all quantitative measurements relating the image produced to the processes to which the photographic material is subjected, including exposure and development.

Sensitivity is commonly expressed graphically as a plot of density versus

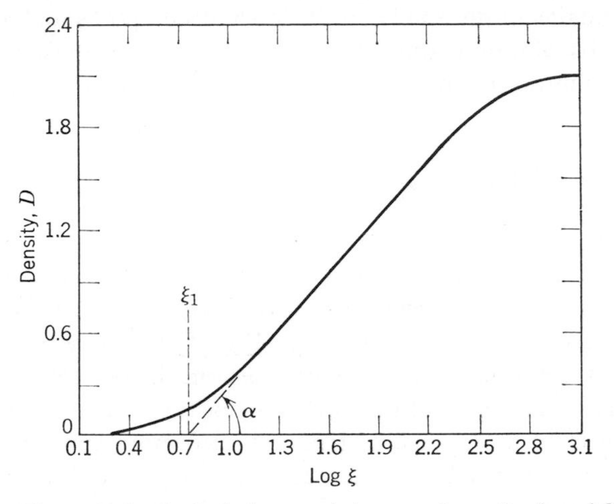

Figure 13.1 Typical characteristic curve for a developed film.

$\log_{10} \xi$; the plot is the characteristic curve for the emulsion. Figure 13.1 is a typical characteristic curve, this kind of curve is also known as the Hurter–Drifield (H–D) curve.

A characteristic curve is not dependent on the emulsion alone; it is characteristic of a combination of the emulsion plus the following:

(1) Conditions during exposure, that is, the spectral character of the light source, the average illumination level, and whether the exposure was intermittent or continuous.
(2) The developer, that is, the formula for the mixture and particularly the reducing agent.
(3) Temperature of the mixture during development, and development time.
(4) Conditions of agitation during development.
(5) The optical arrangement and character of the densitometer that measures the film.

Therefore, to provide full information about the characteristics of the emulsion, all details about exposure and development must be given. Also, to be meaningful, the condition of exposure and development used while preparing the film for measurements should resemble those normally used with that emulsion. The effects of the densitometer measurements on the resulting characteristic curve can be indicated briefly as follows [Refs. 12–14]: If a beam of parallel rays is incident on the film and transmitted through it, only a portion of the unabsorbed light is transmitted undeviated, and a measurement of only the undeviated portion gives *specular density*. *Diffuse density* is obtained by measuring the unabsorbed light including the light that

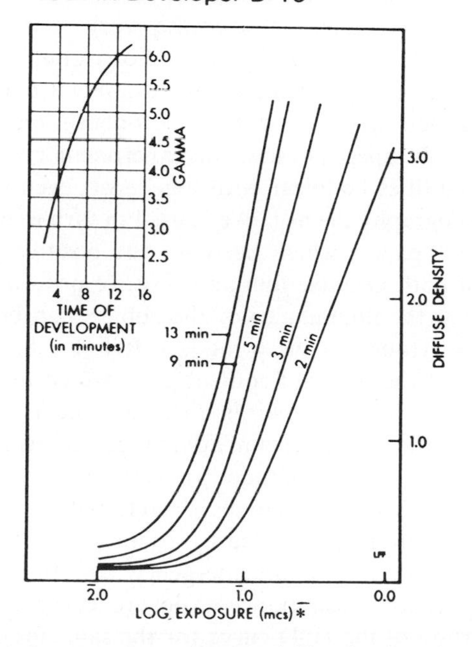

Figure 13.2 Characteristic curves for a developed film showing changes in the general characteristic and in gamma with changes in development time. (The film is Kodak Spectroscopic Plate and Film, Type IV-F, exposed by a tungsten filament source for 1 sec and developed in Kodak Developer D-19) [Ref. 22]. *1 meter candle second (mcs) = 1 lm m^{-2} sec.

is scattered out of the beam by the film. The grain size affects the amount of light scattered [Ref. 15]. In a following discussion on scanning optical systems, certain aspects of a densitometer that affect its response [Ref. 16] are described.

The equation for the linear portion of the characteristic curve is

$$D = \gamma(\log_{10} \xi - \log_{10} \xi_1). \tag{13-4}$$

In this equation, γ (equal to tan α of Fig. 13.1) is the slope of the linear portion and is called *gamma* of the emulsion; $\log_{10} \xi_1$ is called the *inertia point* and is the linear intercept on the $\log_{10} \xi$ axis. Equation (13-4) can also be written as

$$D = \log_{10} (\xi/\xi_1)^{\gamma}. \tag{13-5}$$

Characteristic curves for a given film illustrating the variation of γ with different development times are shown in Fig. 13.2.

A study of Fig. 13.1 suggests that the average exposure should be adjusted so that all exposure values on a processed film occur within the "straight" portion of the characteristic. In scientific work such as photometry and spectrophotometry, this is generally true. For example, if a film is placed in the exit slit plane of a spectrometer (see Chapter 10) and the exposed and processed film is then measured with a microdensitometer to determine the relative heights of the spectral lines, the experimenter would prefer that the densitometric quantities be linear with $\log_{10} \xi$ or, even better, with ξ itself. However, in photography the negative is used in turn to control the exposure pattern on a print paper, which becomes the positive picture. Thus, the characteristics of both negative film and positive print have to be combined in such a way that the illuminance of the subject can be related directly to density differences (tone) on the print [see Ref. 1, Chapter 22, and Ref. 4, p. 182]. Density of the print, of course, is measured in terms of reflectance rather than transmittance (unless it is a transparency).

In holography, there are certain advantages to using another kind of characteristic curve: the amplitude transmittance versus exposure curve illustrated in Fig. 13.3. This curve is preferred in holographic analysis because wavefronts are reconstructed by alterations of the amplitude, rather than changes in flux density, upon emergence from the film. The linear portion of an amplitude transmittance–exposure curve does not correspond to the linear portion of the H–D curve for the same processed film.

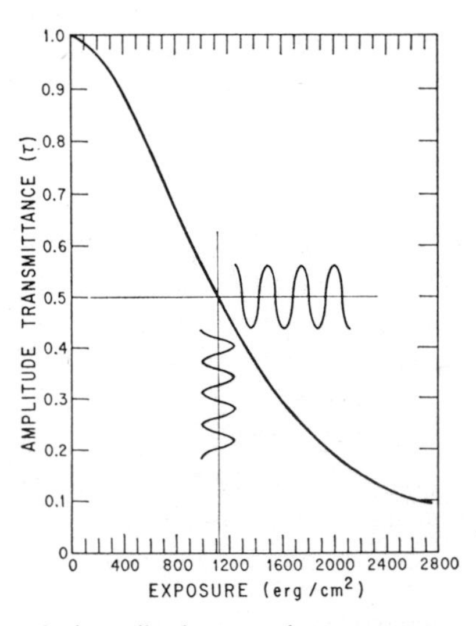

Figure 13.3 Typical amplitude transmittance versus exposure curve.

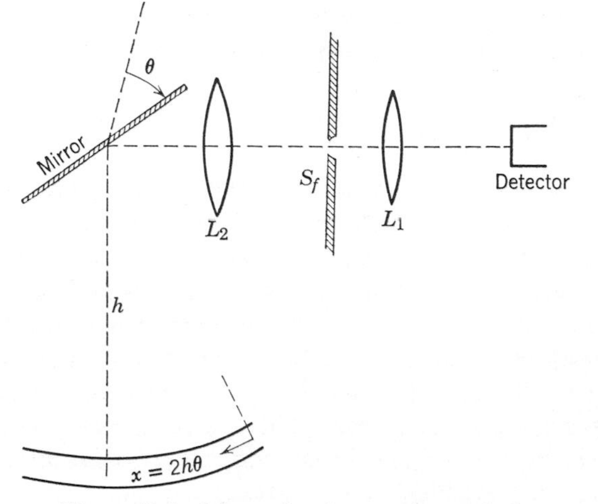

Figure 13.4 Schematic of a scanning optical system.

The Scanning Optical System

It is expedient here to interrupt the film discussion to describe the optical measuring system that scans the film and to explain certain optical problems in making measurements. Moreover, the following discussion has been deliberately extended beyond densitometers to include scanning optical systems in general as reflected in the section title. The discussion of films is continued in a later section.

In a densitometer, a small illuminated area is moved (scanned) across the film, and the transmitted (or reflected) light is detected by a photomultiplier tube. The resulting electrical signal is usually made proportional, by deliberate circuit design, to the optical density of the film.

In an airborne infrared mapper, the radiant energy received is radiated by the terrain being scanned. The radiant-energy–electrical-signal transducer is an infrared detector whose output signal (current or voltage) is proportional to the total radiant power received from the instantaneous field of view (scanning spot).

The schematic of one type of scanning optical system is shown in Fig. 13.4. A field stop S_f centered on the optic axis limits the instantaneous field of view to a small angular field (see Chapter 7), typically a minute of arc, of the object field being scanned. All radiant energy that passes through the stop is incident on the detector that is placed just "behind" the stop. The detector is shielded, that is, its environment is cooled and the incident energy filtered, if necessary, so that the only significant radiant energy that reaches the detector originates in the field of view and passes through the field stop.

First, only the scanning system, shown schematically in Fig. 13.4, is discussed. The recording system, which also scans, is deferred to a later section.

The Optical Response Function

A mirror that oscillates about an axis perpendicular to the optic axis is interposed between the optical system and the field. The motion of the mirror scans the image of the object field across the field stop. At each instant, the distribution of radiant energy in the image plane is found, as in previous chapters, by a convolution of: (1) a function representing the flux density (radiance) distribution in the field with (2) the flux density point-source response function defined in Chapter 9. For convenience, a square field stop and a square aperture stop, each with a pair of sides parallel to the direction of a scan line, are assumed so that, by choosing an appropriate set of coordinates, the variables can be easily separated.

The nature of the computation is determined by whether the radiant energy passing through the stops is from a coherent or an incoherent beam. If the system is an airborne mapper (in which the forward motion of the aircraft provides field scanning in a direction parallel to the axis of mirror oscillation shown in Fig. 13.4), detected energy is radiated from the terrain and the beam can be considered incoherent. In the measurement of film, on the other hand, because the microdensitometer instantaneously observes a very small area that is illuminated (rather than radiating its own energy), the beam is relatively coherent.

For coherent radiant energy, the instantaneous system amplitude distribution function in the image plane is given by an equation similar to Eq. (9-29):

$$U_i(x',y') = U_r(x',y') * U_f(x',y'), \tag{13-6}$$

in which $U_r(x',y')$, equivalent to $U_x(P')$ of Eq. (9-14), is the optical system point-source amplitude response function and $U_f(x',y')$ is the amplitude distribution function in the object field (translated to the image coordinates, as in Chapter 9). As in previous chapters, (x,y) and (x',y') are coordinates in object space and image space, respectively, with the origin on the optic axis. The x-axis is parallel to the direction of scan. For coherent energy, the flux density distribution, corresponding to Eq. (9-32), in the image is

$$W_{ic}(x',y') = U_i(x',y') \cdot U_i^*(x',y'). \tag{13-7}$$

For incoherent energy, the flux density distribution, corresponding to Eq. (9-31), in the image is

$$W_{ii}(x',y') = W_r(x',y') * W_f(x',y'), \tag{13-8}$$

where $W_r(x',y')$ is the optical system flux density point-source response function and $W_f(x',y')$ is the flux density distribution in the field translated to image space. For a circular aperture, $W_r(x',y')$ is given by Eq. (9-30). For a square aperture, $W_r(x',y')$ is obtained from Eq. (9-14) and the first right-hand expression of Equations (9-30):

$$W_r(x',y') = |U_0|^2 \operatorname{sinc}^2 2\pi k_x a \operatorname{sinc}^2 2\pi k_y a, \tag{13-9}$$

in which $2a$ is a side of the square exit pupil, $k_x = x'/(\lambda s')$, $k_y = y'/(\lambda s')$, and s' is image distance.

The field stop is represented by a transmittance function defined as

$$\begin{aligned} \tau_f(x',y') &= 1 \qquad \text{for } 0 \leq x' \leq |b| \text{ and } 0 \leq y' \leq |b|, \\ \tau_f(x',y') &= 0 \qquad \text{for other values of } x' \text{ and } y', \end{aligned} \tag{13-10}$$

in which $2b$ is one side of the square field stop. Then, at each instant, the radiant power passing through the field stop is the integral of the product of the field stop function and the flux density image distribution:

$$\Phi_d = \iint_{-\infty}^{+\infty} [\tau_f'(x',y')][W_i(x',y')]\, dx'\, dy'. \tag{13-11}$$

The image distribution function W_i is given by Eq. (13-7) or Eq. (13-8), whichever is appropriate according to the coherence.

The Radiant Power Signal

The basic equation, if incoherence is assumed, for evaluating radiant power received by the detector is Eq. (7-10):

$$\Phi = \pi L\, \delta a \sin^2 \varphi. \tag{13-12}$$

The geometry defining the symbols of Eq. (13-12) is shown in Fig. 7.8. The area δa is the area, in object space, of the Gaussian image of the field stop (when the optical system is considered in reverse). A comparison of $\pi \sin^2 \varphi$ with Eq. (2-6) shows that $\pi \sin^2 \varphi$, which is approximately equal to $4\pi \sin^2 (\varphi/2)$ for small angles, has the units of solid angle, so the product $\pi L \sin^2 \varphi$ turns out to be radiant flux density. The field function $W_f(x',y')$ of Eq. (13-8) is, in general, equal to $(\pi \sin^2 \varphi)\, L_f(x',y')$, by translating the coordinates to image space.

If the radiance in a diffusely radiating field has a constant value L, the field distribution is $W_f = \pi L \sin^2 \varphi$. The power is constant with time and can be calculated by the equation

$$\Phi = W_f \iint_{-\infty}^{+\infty} \tau_f W_r \, dx' \, dy'. \tag{13-13}$$

The intersection of the optic axis with the object field scans across the terrain, as shown in Fig. 13.4, with a velocity $v = 2\,h\omega$, where $\omega = d\theta/dt$ is the angular velocity of the oscillating mirror. The position, in a fixed field coordinate system, of the optic axis intersection point is $x_f = 2h\theta = 2h \int \omega \, dt = \int v \, dt$. To simplify the development and to show explicitly the functional dependence on time, v is considered constant so that $x_f = vt$. If the instantaneous image flux density distribution is described by $W_i(x,y)$, the optical system sees a change in this factor with time as follows: $W_i = W_i(x - vt, y)$. Consequently, Eq. (13-13) becomes:

$$\Phi(v't) = \iint_{-\infty}^{+\infty} \tau_f W_i(x' - v't, y') \, dx' \, dy'. \tag{13-14}$$

The right-hand side of this equation is recognized as a convolution integral of the field-stop function with the image-distribution function.

Application of Eq. (13-14) to three particular sets of operating conditions suggests how this relationship can be employed in practical problems. First, let us assume a "hot" object whose image is very small compared with the size of the field stop and which is in an otherwise uniform field. Also, we assume that the object passes through the center of the field of view along a path parallel to the x axis. Since the source approximates a delta function, the image distribution W_i given by Eq. (13-8) is the same as the point-source response function W_r, which is given by Eq. (13-9). Equation (13-14), with the variables separated for the example being considered, can be written

$$\Phi(v't) = \iint_{-\infty}^{+\infty} \tau_f(x')[W_r(x' - v't)][\tau_f(y')W_r(y')] \, dx' \, dy'. \tag{13-15}$$

The integration with respect to y' yields a factor that is constant with time. In the convolution integration with respect to x', the result depends on the relative magnitudes of $2a$ (the side of the square exit pupil) and $2b$ (the side of the square field stop). The nature of $\Phi(v't)$ can be explored by studying specific examples. In selecting values for such an example, Eq. (13-9), which gives the flux density point-source response function, is convenient; the $\text{sinc}^2\, 2\pi k_x a$ factor describes how the flux density due to a point source varies as a function of the coordinate x' in image space. The most prominent

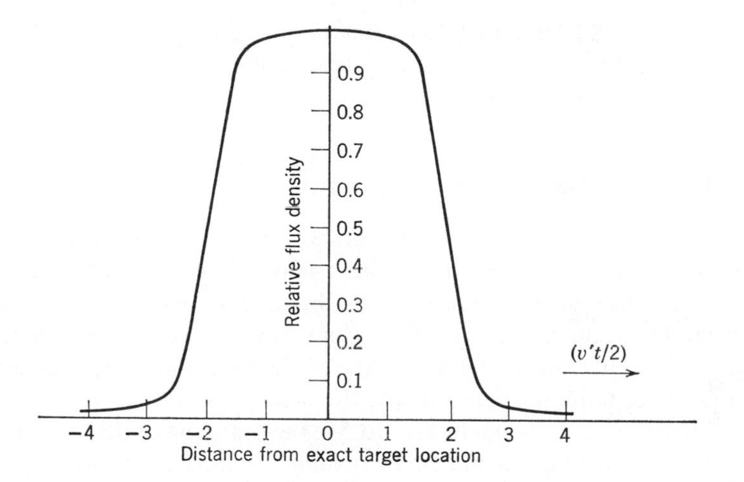

Figure 13.5 Result of convolution of a rectangular function with the sinc squared function when the rectangular function is twice the width of the sinc squared function. (The sinc squared function width is from $-\pi$ to $+\pi$ when distance is measured in units of $x'u_a = 2\pi a\, \Delta x'/(\lambda s')$.)

feature of this function is its major lobe, which extends from $2\pi k_x a = -\pi$ to $2\pi k_x a = \pi$ or for a total $\Delta x' = \lambda s'/a$ since k_x is defined as $x'/\lambda s'$. For an example, the side of the field stop is assumed twice the $\Delta x'$ just described. The function $\Phi(v't)$ for this particular example has been determined by computer and is plotted in Fig. 13.5. The abscissa values have been normalized by dividing the distance from the exact target image location by $(v't/2) = (\Delta x'/2) = (\lambda s'/2a)$.

In a second example, a similar outcome results when one scans a line source that is perpendicular to the x-axis, which is the scanning direction. Integration with respect to y', as in the preceding example, gives a constant. The line source can be represented by a delta function of x'. The result is again the same as for the preceding example except for a change of constant.

In a third example, the scanned region is described by a radiance distribution that contains a step function, which represents a sharp boundary between cool and warm regions; the boundary is parallel to the y-axis and perpendicular to the scanning direction. Integration with respect to y' again produces a constant. Equation (13-8) in this example is the convolution of the step function with the sinc^2 function. The result would then be convolved with $\tau_f(x')$, according to Eq. (13-11), to obtain an edge trace.

The function $\Phi(v't)$ in Eq. (13-15) represents only the radiant power that passes through the field stop; the complete system response depends also on the time characteristics of the detector–amplifier combination. From the time function $\Phi(v't)$, a frequency function $\Phi'(f)$ can be derived by a Fourier transform. The overall system response is $\Phi'(f)$ multiplied by the detector–amplifier frequency response function.

Spatial Resolution of a Scanning System

The effect of the convolution integrals, applied in the preceding developments, is that the region in the field from which radiant energy is being received instantaneously has a blurred boundary. Some radiant energy from outside the Gaussian image of the field stop is received, and some radiant energy from just inside the boundary of the Gaussian image passes through the entrance pupil but falls outside the field stop.

A curve like that shown in Fig. 13.5 is the scanning optical system response function. The width of the curve (usually defined as the width at the half-power points) is often given as the spatial resolution of the scanning system.

Graininess and Granularity

When a sufficiently magnified image is observed, inhomogeneities in processed film become apparent; with still greater magnification, individual minute masses of metallic silver become visible. The negative or print is said to be *grainy*. The *graininess* first becomes visible at a relatively low magnification. With a comparatively great magnification of about 2500× (with an electron microscope) a filamentary structure becomes apparent. Therefore, when a film region having an apparent uniform blackness is scanned with a microdensitometer, an electrical signal with an erratic amplitude results. Amplitude deviations about the mean, in general, constitute noise; when the noise is caused by the film inhomogeneities—the granular structure—it is referred to as *granularity*.

In densitometry, the distribution function $W_f(x',y')$ [see Eq. (13-8)], representing the transmitted exitance in the field, depends upon the nature of the silver grains and their distribution in the film. The granularity recorded for a particular film depends also on the instrument system response function which, in turn, depends on the size of the Gaussian image of the field stop relative to grain size and on the point-source response function, and on other densitometer characteristics like noise, response time, and responsivity of the detector.

Besides the qualitative noise significance of granularity, quantitative relationships describing this property can also be developed [Refs. 12–16]. If we assume that A is the area of an illuminated, instantaneous field of view, N is the number of grains within the area A, a is the average projected area

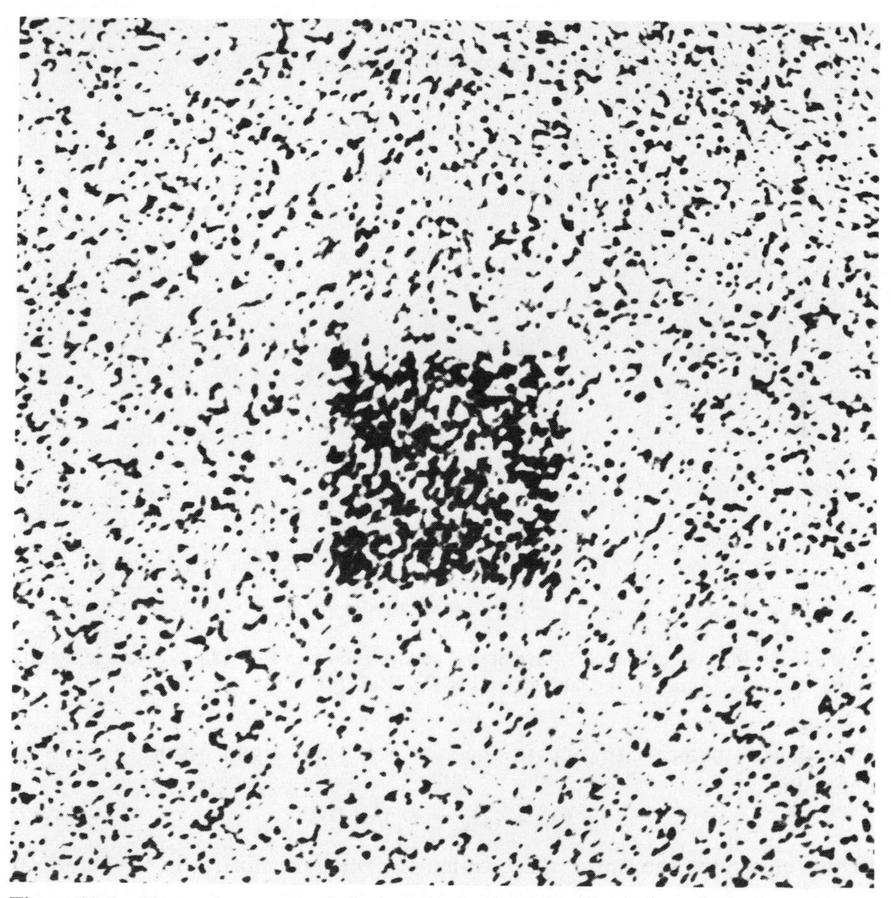

Figure 13.6 Photomicrograph of the image of a small illuminated square on a developed emulsion showing the granular structure [Ref. 14].

of a grain, and $K = \log_{10} \xi$, the density is approximately

$$D = KNa/A. \tag{13-16}$$

A magnified photograph of a developed film containing the image of an illuminated square on a uniform background is shown in Fig. 13.6. The difference $(\bar{D}_s - \bar{D}_b)$ between the average density of the background and the average density of the square is the signal. Granularity causes noise in the uniform background. The time required for an observer to recognize the square has an inverse relationship to the signal-to-noise ratio. The signal and noise are illustrated in Fig. 13.7, which shows the signal produced by a

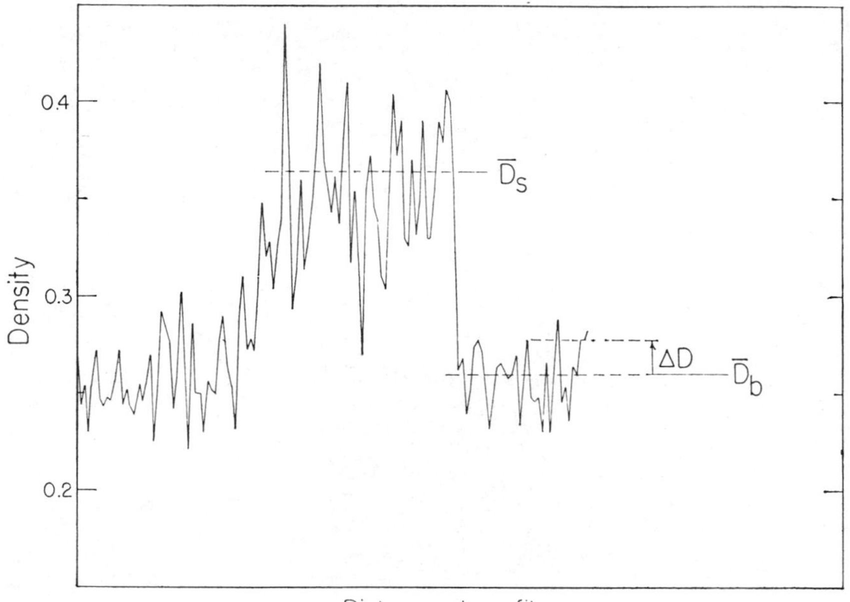

Figure 13.7 Microdensitometer trace across square shown in Fig. 13.6. (Average density in the square is $\bar{D}_s$ and in the background is $\bar{D}_b$) [Ref. 14].

microdensitometer scan across the square. The signal-to-noise ratio is defined as:

$$R_{sr} = (\bar{D}_s - \bar{D}_b)/\sigma(D). \tag{13-17}$$

In this equation, the noise $\sigma(D)$, called the *rms granularity*, is the standard deviation and is defined by the equation:

$$\sigma(D) = \left[(1/n)\sum_{i=1}^{n}(\Delta D_i)^2\right]^{1/2}, \tag{13-18}$$

in which ΔD_i is the deviation about the mean of each of a large number n of density readings. Granularity G is sometimes defined in terms of the area a of the Gaussian image of the field stop (the instantaneous field of view) by a relationship known as Selwyn's law:

$$G = \sigma(D)a^{1/2}. \tag{13-19}$$

Edge Trace and Transfer Function

To develop the concept of the edge trace and the transfer function of a film, we consider a film exposed in the following manner: an opaque material

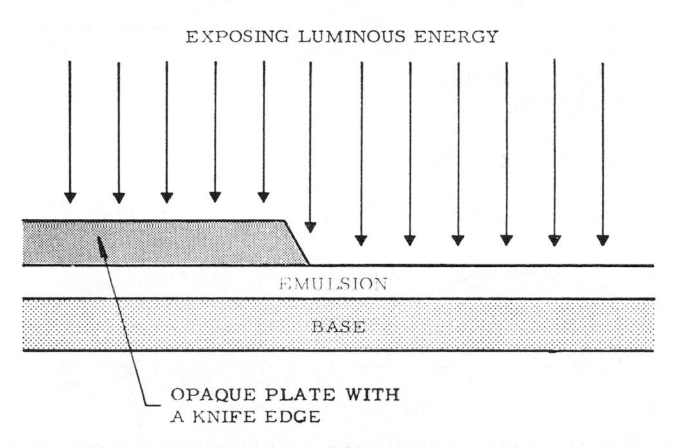

Figure 13.8 Exposure of film beneath a knife edge.

with a sharp edge is placed on top of the film as shown in Fig. 13.8 to form a step function of illuminance when the whole is subjected to uniform illumination. Since the emulsion is a turbid medium [see Ref. 1], light will diffuse by scattering in the emulsion, and the processed film will not show a step function of density but will change gradually in the vicinity of the knife edge. If the negative is measured with a densitometer, the density plot versus distance, to show this change, is called the *edge trace* for the film. This trace would be similar to the sigmoid shape shown in Fig. 9.3, but an active film plot would typically (but not necessarily) undershoot at A and overshoot at B.

An expression for the modulation transfer function of a film can be derived from the line-spread function, which could be derived by a "deconvolution" of the edge trace and the step function [Ref. 17]. It turns out, however, that a relatively simple mathematical procedure will yield the spread function: If $\tau_1(x)$ is the transmittance as a function of distance across the knife edge, an expression for the *line-spread function* $\tau_s(x)$ is

$$\tau_s(x) = d[\tau_1(x)]/dx. \tag{13-20}$$

Then a Fourier transform of the line-spread function gives the modulation transfer function. Figure 13.9 shows the modulation transfer functions for three typical films.

The role of the modulation transfer function (MTF) in the analysis of optical systems has already been indicated in Chapter 9. However, application of this function to photographic emulsions has certain difficulties [Refs. 17–21]. One basic problem is the nonlinearity of the characteristic curves (see Fig. 13.2). Calculations based on assumptions of linearity can be made only for very small amplitudes of modulation. When the amplitudes

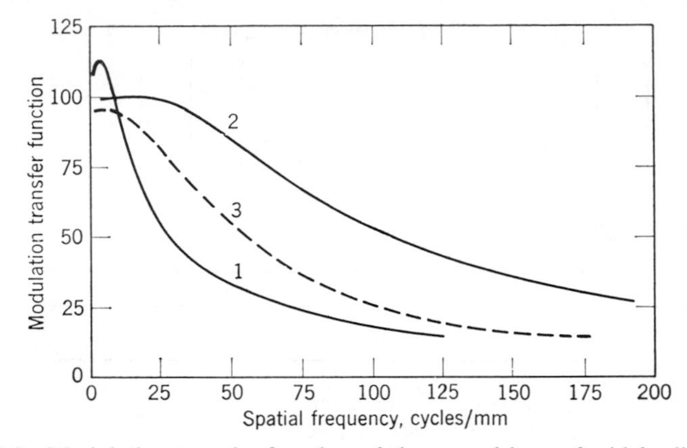

Figure 13.9 Modulation transfer function of three emulsions of widely different types: (1) a fast pictorial negative, (2) a fine-grained duplicating positive, and (3) a fine-grained pictorial negative [Ref. 14].

have to be relatively large, an approach to linearity can be made by working with change in transmittance as a function of change in exposure as shown in Fig. 13.3. The incremental exposure in this approach is usually referred to as "effective" exposure.

A conceptually satisfying, although somewhat impractical, way of measuring the MTF of an emulsion is to proceed as described in the discussion on spatial frequency in Chapter 9: An object consisting of a sinusoidal test pattern is recorded photographically on the emulsion to be measured. The processed emulsion and the original pattern are then measured to obtain the contrast on each [contrast was defined by Eq. (9-13)]. The frequency response is then found by Eq. (9-25). Part of the required information is the frequency response of the lens, which can be determined independently of the film [see Ref. 19]. By subtracting the response of the lens from the contrast ratio of the image to the object, one gets the response M of the emulsion [see Ref. 18]. The response at various spatial frequencies is required to define the complete MTF.

As suggested in Chapter 9, the utility of the MTF is that this characteristic of the entire system—object, optical system, optical image, and emulsion—can be calculated from the MTFs of the individual components by multiplying the several curves together ordinate by ordinate at corresponding abscissa values. However, this simple cascading of functions is valid only when all the elements have linear responses; the nonlinearity of photographic emulsions will introduce harmonics [Ref. 22]. The specific nonlinearity of these materials can be illustrated by equating Equations (13-1) and (13-5) to obtain for the transmittance:

$$\tau = (\xi_1/\xi)^\gamma. \qquad (13\text{-}21)$$

Though Eq. (13-21) shows that the film is inherently nonlinear, it has been shown [Refs. 13 and 14] that if the gamma of the negative is held to about 0.6 and the modulation in transmittance to about 60%, the waveform may be distorted but the magnitudes of the harmonics cancel one another. As a result, the amplitudes are not greatly affected by the nonlinearity.

Other photographic effects that influence the MTF of a film are adjacency, reciprocity failure, and halation as discussed in the paragraphs that follow.

Selected Adverse Photographic Effects

Many photographic effects tend to degrade the image in pictorial photography or interfere with quantitative measurements in scientific photography [see Ref. 3, Chapter 15; Refs. 6 and 23]. Three of the most significant have been selected for discussion here.

A type of edge trace often obtained by a densitometer scan across a knife edge image is shown in Fig. 13.10. This is quite different from the edge trace discussed earlier (see Fig. 9.3) and is an example of *adjacency effects*. When two contiguous areas of a photographic material receive greatly different exposures, their respective densities (after development) near the boundary may not truly represent the distribution of light flux within the emulsion at the time of exposure. Adjacency effects can be explained as follows: At the edge of the more highly exposed portion of the image, fresh developer is available not only from above the emulsion surface but also from the adjacent part of the emulsion, which, having received little exposure,

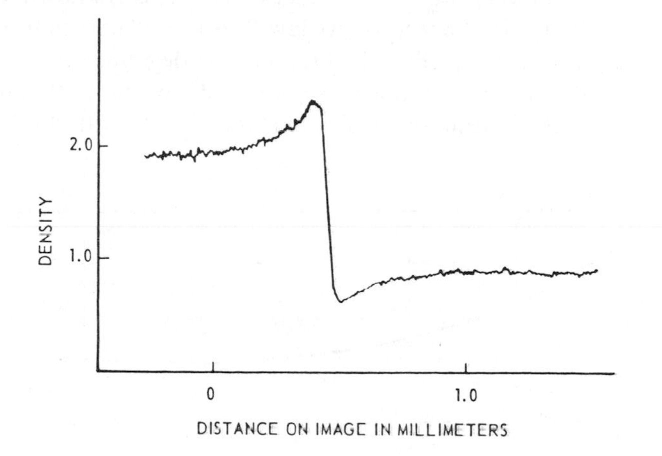

Figure 13.10 Microdensitometer trace across the image of a knife-edge showing adjacency effects [Ref. 22].

will not have "used up" the developer. Conversely, the little-exposed part of the emulsion at the boundary of the image is invaded by development reaction products from the developing image on the other side of the boundary, and these products tend to inhibit development. These effects, therefore, account for the overshoot and undershoot, when they occur, on edge traces like that of Fig. 9.3. These effects show up as an exaggeration of the frequency response function at very low frequencies as shown on Curve (1) of Fig. 13.9; the measured MTF is actually greater than 1! The bright portions of the sinusoidal object are overdeveloped and the dark portions underdeveloped because of the adjacency effects. These effects probably exist at high frequencies too, though they are not as evident as at low frequencies.

Exposure was defined earlier as simply the product of illuminance and the exposure time, but the actual exposure depends to some degree on the ratio, as well as the product, of these two variables. In sensitometry, a sequence of graduated exposures can be produced either by exposing all patches for a constant time and varying the illuminance from patch to patch or by keeping the illuminance constant and varying the exposure time from patch to patch. Although the density produced by a given combination of illuminance and exposure time is almost the same as that produced by half the illuminance acting for twice the time (or for twice the illuminance acting for half the time), large changes in illuminance though compensated by inverse changes in time—say, by ratios in the neighborhood of a hundred—may upset the invariance of the density.

Most photographic materials suffer some loss in sensitivity when exposures are made at very low or very high illuminances. This loss is known as reciprocity, short for "failure of the reciprocity law." A typical curve illustrating reciprocity effects is shown in Fig. 13.11. The 45-deg lines are lines of constant time, and the vertical lines are lines of constant illuminance. The abscissa scale at the bottom of the figure is for $\log_{10} E$, with illuminance

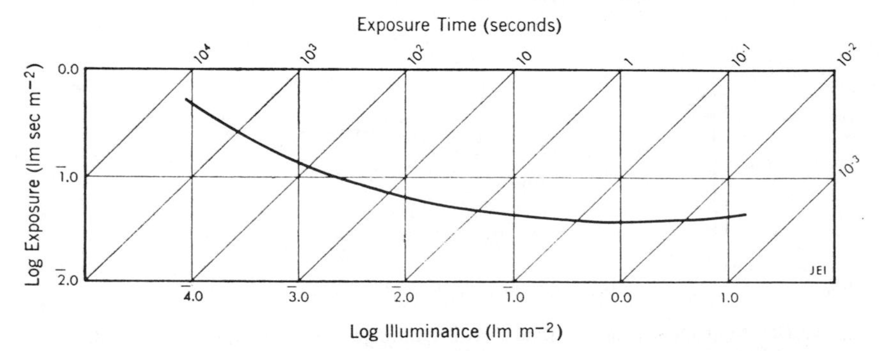

Figure 13.11 Typical reciprocity effect curve [Ref. 22].

E in lumen per square meter increasing from left to right. The values on the scale along the top, labeling the 45-deg lines, are in seconds, increasing from right to left. When the logarithm of the exposure ($\log_{10} Et$) required to produce a given reference density is plotted as a function of $\log E$, a plot similar to Fig. 13.11 results. If the reciprocity law held, this plot would be a straight horizontal line. The upward curvature at each end indicates that a greater exposure is required to produce the reference density at high and low luminance. Scientific photographic work is often done at either very low or very high illuminances, so for this kind of work departures from the reciprocity law become significant.

Some light passes through the emulsion and the supporting base (the plastic film or glass plate) and then is reflected at the back surface to return for a second encounter with the emulsion. The consequent spreading of the light, by reflection as well as by scattering in the emulsion, tends to produce halos about images of small sources. This effect is called *halation* and may be reduced by coating the back side of the base with a layer of gelatin containing dye that minimizes reflection. Curve (3) in Fig. 13.9 shows the MTF to be less than unity at frequencies near zero, which is a result of halation.

Recording the Scanned-Object Image

The infrared aerial mapper is an example of a scanning imaging system. The object-field scanning system, shown schematically in Fig. 13.4, scans the terrain line by line as the aircraft moves along its path. The distance d between consecutive scan lines is:

$$d = v_y \, \Delta t, \tag{13-22}$$

where v_y is the ground speed and Δt is the time interval between successive scans across the ground track (which is traced by a point vertically downward from the aircraft). If the "oscillating" mirror in the figure is actually rotating at the angular velocity ω, the velocity of the optic axis in the direction of a scan line (as described earlier), as it crosses the ground track, is $v_x = 2h\omega$, where h is the altitude. While the line is scanned on the terrain, the mirror rotates through the angle θ_s. Rotation through a further angle θ_r, during a "dead time" in which the instrument is not recording, brings the mirror back into position to begin the next line scan. The time interval between scans is, therefore:

$$\Delta t = \theta/\omega, \tag{13-23}$$

where $\theta = \theta_s + \theta_r$. If Δt is eliminated between Equations (13-22) and (13-23),

$$d = v_y\theta/\omega. \tag{13-24}$$

The optics of the aerial mapper system can be represented by the optical diagram in Fig. 6.5a. The object distance h is assumed so large compared with the focal length F that the image of the field is located very near the focal point. The boundary of the infrared detector's sensitive area generally serves as the field stop for the system. For convenience, $\sqrt{A}$ is taken as the dimension of the square detector so that A becomes its area. The Gaussian image of $\sqrt{A}$ is the dimension d' on the terrain. To make the scan "lines" just wide enough to be contiguous, the breadth of the optical system response function, the spatial resolution as defined earlier, must be d units wide. The d' dimension is related to d but is generally slightly larger. Thus the four parameters of the aerial optical system are related as follows:

$$(d/h) \approx 2a/h = \sqrt{A}/F. \tag{13-25}$$

This equation combined with Eq. (13-24) gives

$$v_y/h \approx \omega\sqrt{A}/F\theta. \tag{13-26}$$

The ratio of velocity over height, $v_y h$, is a significant performance parameter for aerial mappers. As indicated, highest performance is required on high-speed, low-level missions. Unfortunately, design convenience tends to make the numerator factors small and the denominator factors large: High rotational rates for mirrors bring with them bearing and other problems, so a low ω is desirable. Reduction in detector size reduces noise and increases resolution, so $\sqrt{A}$ should be small. On the other hand, optical design is considerably eased if a short focal length F can be avoided. A broad lateral dimension in the map requires long scan lines and, therefore, a large θ.

To record the image of a scanned object, a light beam illuminates a spot on the film, and the spot is scanned across the film in synchronism with the object scanning system. In an aerial mapper, roll film is moved at a rate proportional to the plane's ground speed, and the light spot is scanned across the film line by line as the terrain is scanned. The power in the light beam is modulated according to the instantaneous power $\Phi'(t)$ received from the field of view.

Only a few light sources are suitable for this application. The zirconium arc is sometimes used, even though it responds poorly to power modulations at high frequencies. A moving image of the arc is projected on the film by a beam-scanning mirror system that is synchronized by mechanical coupling to the object scanning mirror.

A laser beam may eventually be scanned by the described mechanical system or by acoustooptical deflection cells, and the beam might be power-modulated by electrooptical cells [Refs. 24–27]. However, the cost and weight of an appropriate laser system is still impractically high.

A third film-illuminating system employs a cathode ray tube in which the electron beam is controlled to provide both the power modulation and the scanning of the light source. The entire tube face is imaged on the film [Refs. 28 and 29]. While a linear sweep of the electron beam moves the light source across the tube face, its image exposes a line on the film.

A Fourier transform of the line spread function [like Eq. (13-20) for film] for the scanning optical system described earlier gives a spatial frequency response function (like the MTF for film, Fig. 13.9). For any specified minimum response, this function will give the maximum possible spatial frequency κ_x. The frequency κ_x is called the spatial high-frequency cut-off for the system; it is so related to the spread function that when the spread function is in terms of distance on the terrain, κ_x is approximately the reciprocal of twice the spatial resolution.

The significance of κ_x as a system parameter can be appreciated by returning to Eq. (13-14) and developing an equivalent time–frequency concept. If $\Phi(v't)$ is represented now by $\Phi''(t)$ and the Fourier transform is applied as follows:

$$\Phi(f) = \int_{-\infty}^{+\infty} \Phi''(t) \exp(-2\pi i f t)\, dt, \tag{13-27}$$

the function $\Phi(f)$ of frequency results. Without pursuing a rigorous development, certain relationships between $\Phi(f)$ and $\Phi''(t)$ can be inferred. A period (at a spatial frequency of κ_x cycles per millimeter) in the spatial domain of $\Phi''(t)$ is

$$l_x = 1/\kappa_x. \tag{13-28}$$

The time required to scan this distance on the terrain is

$$t_d = l_x/v_x = 1/v_x\kappa_x. \tag{13-29}$$

This time is also the time period of the corresponding electrical frequency f_x in the frequency domain of $\Phi(f)$, so

$$f_x = 1/t_d = v_x\kappa_x. \tag{13-30}$$

The bandpass for the electrical system—including detector, amplifiers, cathode ray tube, and associated circuits (or other light source and associated circuits)—must, therefore, extend at least to the electronic frequency f_x of Eq. (13-30). Otherwise, the resolution of the overall system will not achieve the spatial resolution of the scanning optical system.

Finally, let us examine how the light spot exposes the film. The size of the instantaneous spot on the cathode ray tube face (which is the light source), the luminous exitance of the spot, and the variation of exitance within the spot are determined by the cathode ray tube and its control circuits. The size of the spot illuminating the film in turn depends on the magnification of the intervening optical system. For example, if the cathode ray tube is 5 in. in the direction of a scan line and the image is recorded on 70-mm roll film, the optical reduction must be approximately 1/2. The Gaussian image would therefore have half the diameter of the cathode ray tube spot. As in earlier similar developments, the actual distribution of luminous flux in the spot image must be found by a convolution of the Gaussian image of the cathode ray tube distribution function and the flux density response function of the associated optical system.

As an example, we arbitrarily assume the distribution of illuminance in the illuminating spot to be:

$$E(r) = (2\Phi_w/\pi\sigma^2) \exp[-2r^2/\sigma^2], \tag{13-31}$$

where Φ_w is the total luminous power in the beam proportional to $\Phi''(t)$, r is radial distance from the "center"—the point of maximum illuminance—and σ is a constant that determines the size of the distribution. Equation (13-31) is recognized as representing a normal distribution; it also represents generally the distribution in a laser beam for the simplest single mode, with σ being the width of the beam at its waist (see Chapter 6 and Refs. 27 and 28). If $E(r)$ is multiplied by a time δt—an effective exposure time during which an element of area on the film, in the path of the scanning spot, is illuminated—the product is exposure. By substituting $r^2 = x^2 + y^2$ (where x and y are coordinates parallel and perpendicular respectively to a scan line on the film), by indicating the dependence of Φ_w on t, and finally by letting x be equal to $v_0 t$ (where v_0 is velocity of the spot on the film), Eq. (13-31) can be written in terms of exposure as

$$\xi(x,y) = [2\Phi_w(t)\,\delta t/\pi\sigma^2] \exp\{[-2y^2 - 2(v_0 t)^2]/\sigma^2\}. \tag{13-32}$$

The exposure, as a function of y in one scan line, can be found by integrating Eq. (13-32) with respect to t:

$$\xi(y) = (2\,\delta t/\pi\sigma^2) \exp(-2y^2/\sigma^2) \int_{-\infty}^{+\infty} \Phi_w(t) \exp[-2(v_0 t)^2/\sigma^2]\,dt. \tag{13-33}$$

This equation cannot be integrated without explicitly specifying $\Phi_w(t)$. An expression for the exposure as a function of y can be obtained, however, by assuming Φ_w constant and performing the integration with respect to time:

$$\xi'(y) = (\Phi_w\,\delta t/\sigma v_0)\sqrt{(2/\pi)} \exp(-2y^2/\sigma^2). \tag{13-34}$$

Since this is a normal distribution, edges of scan lines on the film are relatively underexposed. The power in the beam is usually adjusted to give the desired exposure in the center of the line where the illuminance is relatively high; then, to compensate at the edges, some overlap is allowed between adjacent lines.

REFERENCES

1. C. E. K. Mees and T. H. James, *The Theory of the Photographic Process*, Third Edition. Macmillan, New York, 1966.
2. C. B. Neblette, *Photography: Its Materials and Processes*, Sixth Edition. D. Van Nostrand Company, Inc., New York, 1962.
3. H. Baines, *The Science of Photography*, Second Edition. John Wiley & Sons, Inc., New York, 1967.
4. "Some Chemical Reactions in Photography," *Kodak Consumer Service Pamphlet*, No. AJ-15, Consumer Markets Division, Eastman Kodak Company, Rochester, N.Y. 14650, 1969.
5. F. H. Perrin, "The Photographic Emulsion," in *Applied Optics and Optical Engineering*, Vol. II. Edited by R. Kingslake. Academic Press, New York, 1965.
6. F. W. H. Mueller, "Review of Progress on Photographic Material (Silver Halide) in 1968," *Phot. Sci. Eng.*, **14**, 157 (1970).
7. J. Malinowski, "Latent Image Formation in Silver Halides," *Phot. Sci. Eng.*, **14**, 112 (1970).
8. J. F. Hamilton, "Investigation of a Latent Image Model by an Analytical Approximation," *Phot. Sci. Eng.*, **14**, 102 (1970).
9. J. H. Webb, "A Mathematical Model for Photographic Exposure and Its Experimental Verification," *Phot. Sci. Eng.*, **14**, 217 (1970).
10. J. Q. Umberger, "Rate-Limiting Steps in Photographic Development: Adsorption, Surfactant, and Electron Transfer Effects," *Phot. Sci. Eng.*, **14**, 131 (1970).
11. T. H. James, "The Initiation and Continuation of Development—Some Views and Correlations," *Phot. Sci. Eng.*, **14**, 371 (1970).
12. K. S. Weaver, "Measurement of Photographic Transmission Density," *J. Opt. Soc. Am.*, **40**, 524 (1950).
13. J. H. Altman, "The Measurement of rms Granularity," *Appl. Optics*, **3**, 35 (1964).
14. G. C. Higgins, "Methods for Engineering Photographic Systems," *Appl. Optics*, **3**, 1 (1964).
15. "USA Standard Diffuse Transmission Density," (USAS PH2.19-1959) United States of America Standards Institute, 10 E. 40th St., New York, 10016.
16. D. Galburt, R. A. Jones, and J. W. Bossung, "Critical Design Factors Affecting the Performance of a Microdensitometer," *Phot. Sci. Eng.*, **13**, 205 (1969).
17. R. A. Jones and E. C. Yeadon, "Determination of the Spread Function from Noisy Edge Scans," *Phot. Sci. Eng.*, **13**, 200 (1969).
18. R. Williams, "A Re-examination of Resolution Prediction from Lens MTF's and Emulsion Thresholds," *Phot. Sci. Eng.*, **13**, 252 (1969).
19. E. W. H. Selwyn, "Combination of Lens and Film," in *Applied Optics and Optical Engineering*, Vol. II. Edited by R. Kingslake. Academic Press, New York, 1965.
20. E. S. Blackman, "Effects of Noise on Determination of Photographic System Modulation Transfer Function," *Phot. Sci. Eng.*, **12**, 244 (1968).
21. I. J. Launvesch et al., "Threshold Modulation Curves for Photographic Films," *Appl. Optics*, **9**, 875 (1970).
22. G. C. Higgins, "Methods for Analyzing the Photographic System, Including the Effects of Nonlinearity and Spatial Frequency Response," *Phot. Sci. Eng.*, **15**, 106 (1971).
23. "Kodak Plates and Films for Science and Industry," Kodak Publication No. P-9, Department 454, Eastman Kodak Company, Rochester, New York 14650.

24. E. I. Gordon, "A Review of Acousto-optical Deflection and Modulation Devices," *Proc. IEEE*, **54,** 1391 (1966).
25. A. Korpel, A. Adler, P. Desmares, and W. Watson, "A Television Display Using Acoustic Deflection and Modulation of Coherent Light," *Proc. IEEE*, **54,** 1429 (1966).
26. O. H. Schade Sr., "Modern Image Evaluation and Television," *Appl. Optics*, **3,** 17 (1964).
27. G. Goubau and F. Schwering, "On the Guided Propagation of Electromagnetic Wave Beams," *Trans. IRE*, **AP-9,** No. 3, 248 (1961).
28. A. G. Fox and Li Tingye, "Resonant Modes in a Maser Interferometer," *Bell Syst. Tech. J.*, **40,** 453 (1961).
29. M. J. Cowan et al., "Device Photolithography: The Primary Pattern Generator, Part I—Optical Design," *Bell Syst. Tech. J.*, **49,** 2033 (1970).

14

Vision

Introduction

Because engineers and scientists deal generally with physical quantities, the objective and concrete, a confrontation with the physiology and psychology of vision may be found, for them, unusually subjective. The study of the relationships between physical quantities and the subjective, mental responses to these quantities as stimuli is *psychophysics*. To help clarify a later discussion of the nature of *psychophysical methods* and *psychophysical functions*, we resort to a bit of fantasy. A little man inside of us is often alluded to in discussions concerning the science of vision. He receives and interprets the evidence supplied by our eyes and thus does our seeing for us. He does the best he can, with incomplete evidence, to tell us what the physical world is like.

Each of us does live in two worlds at the same time: the actual physical world outside of us, which generally is wholly independent of our existence, and the inner world of consciousness, which each of us largely makes for himself and which is vastly different from reality. We casually speak of "blue light" and "red light" with little thought that the sensation of hue—for example, the "blueness" or the "redness"—is something that happens *within* us. In fact, all direct information that we get concerning the physical world comes to us through our sense organs, which, besides being limited in number, are susceptible only to certain special kinds of stimulation and are often rather mediocre detectors.

Because we address ourselves to engineers and physical scientists, we should probably emphasize the psychological aspect of seeing with some implication that things are not always what they seem. Accordingly, we dwell a bit on this theme of two worlds, which was elegantly discussed by Southall many years ago [Ref. 1].

One of the allegories of John Bunyan relates the eventful history of the famous town of Mansoul, which was advantageously situated in the spacious continent of the Universe and was protected from assaults from the outside by the well-nigh impregnable wall completely surrounding it. Access to this stronghold was possible only through the five gates of the city, which were known far and wide as ear-gate, eye-gate, mouth-gate, nose-gate, and feel-gate. Therefore, whenever the curious inhabitants of Mansoul desired to know what was going on in the great world around them, they would repair to these gateways and converse with travellers from afar who chanced to pass that way. They congregated most at eye-gate and ear-gate because these served the two principal highways, and here gossip and rumor abounded. Much of the information the townspeople gained in this way was useful and reliable; yet sometimes, the reports were misleading and false—as Mansoul learned to its cost on more than one occasion.

In spite of physiological and psychological complication, the eye is an optical system and can be analyzed as such. Furthermore, a pair of eyes and the associated human system often function as the extension of a more conventional optical system. A microscope, for example, forms an image only to be retransmitted by an eye; hence design considerations like the optimum positions of the image and the exit pupil depend upon the characteristics of the eye. A display system, in general, forms (though not necessarily by optical means) a "picture" of information that is to be viewed, comprehended, and acted upon by its human observer. Sometimes, as in the theater, the transfer process is so effective that the observer feels directly involved in the "display." Thus the engineer or scientist concerned with effective display design must appreciate the capabilities of the eye (and the man behind it) before he can make competent design decisions.

Not only do the characteristics of the eye and general human responses vary considerably from individual to individual, but these characteristics for a given individual vary with changes in environment and with the individual's physical and emotional state. Normal aging processes also cause significant changes in eye characteristics. Because of these non-uniformities, statistically averaged characteristics are the rule for any quantitative discussions involving the eye. Fortunately, a number of excellent references are available [Refs. 2–7] that treat the subject of vision at greater length than is possible here.

Structure of the Eye

Figure 14.1 shows a horizontal cross section of the eye. The eyeball is nearly spherical except for a bulge at the anterior surface. The eye is enclosed by three membranes. The outermost one consists of the transparent cornea,

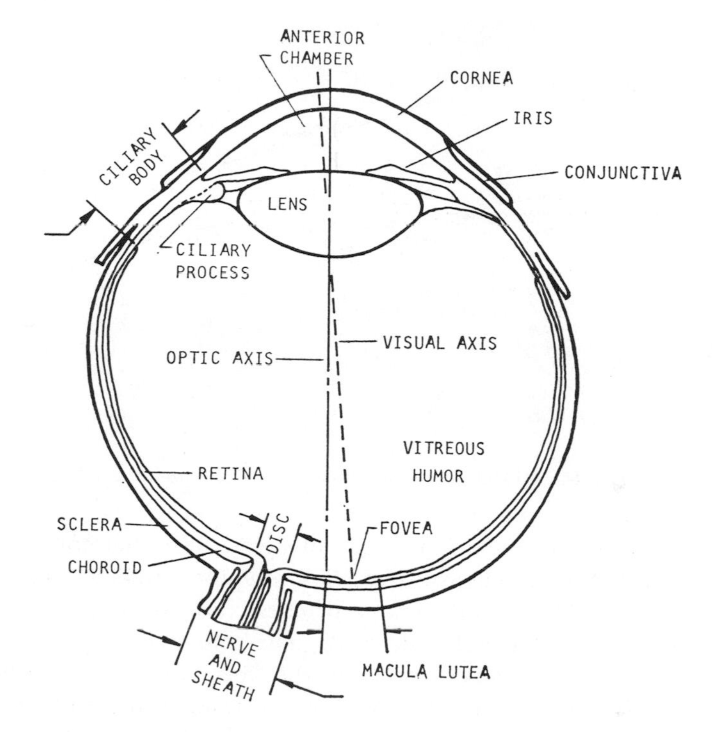

Figure 14.1 Horizontal section of the right human eye [Ref. 8].

forming the bulge, and the opaque sclera, enclosing the remainder of the eyeball. The choroid coat, containing many nerves and blood vessels, is immediately under the sclera. The innermost membrane of the eye is the retina, which lines all of the posterior wall.

The opening near the center of the eye, seen as a large black dot, is the pupil through which light enters. Surrounding the pupil is a colored ring called the iris. The front of the eyeball is covered by a transparent membrane called the conjunctiva, which begins at the edge of the upper eyelid, covers the inside of the upper eyelid, folds and continues across the front of the eyeball, and then folds again to cover the inside of the lower eyelid.

Just back of the iris is the crystalline lens, which is suspended by zonal fibers attached to the ciliary body. A portion of the eye is shown in greater detail in Fig. 14.2. Muscles in the iris change the size of the pupil, and other muscles change the shape of the crystalline lens. Two chambers, one anterior to the lens and the other posterior, are filled with transparent material, the first with the aqueous humor and the second with the vitreous humor. The aqueous humor has a watery consistency similar to blood plasma; the vitreous humor is jellylike.

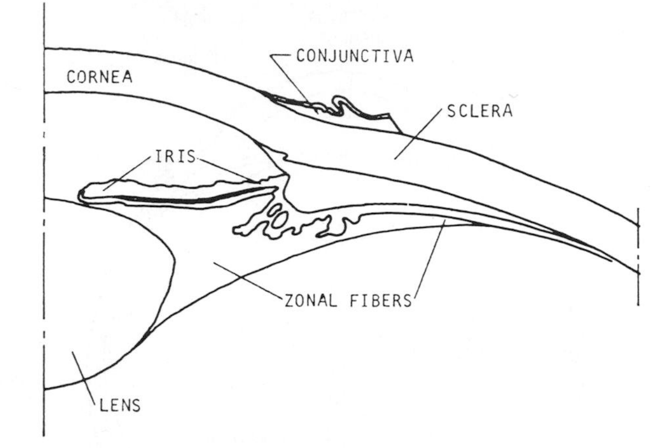

Figure 14.2 Detail of anterior segment of the human eye [Ref. 9].

In the order encountered by a ray of light, the retina consists of blood vessels, nerve fibers, light-sensitive photoreceptors (rods and cones), and a pigment layer. Although the retina includes blood vessels, the major blood supply of the eye is the choroid coat. The rods and cones are connected through a complicated network of nerve fibers to the optic nerve. The nerve fibers leave the eyeball at a common point to form the optic nerve. No receptors are located near the region, called the optic disk, where the optic nerve emerges, which accounts for the well-known blind spot.

The Eye as an Optical System

To a first approximation the eye is a system of centered spherical surfaces with an internal aperture stop [Ref. 6]. Radii of curvature, surface positions, and refractive indices of the media for a typical adult eye are given in Table XI [Ref. 10]. The refractive index of the lens medium varies continuously from the lowest value near the anterior surface to the highest value in the middle and then back to the lowest value near the posterior surface. Because of this variation, the lens power is considerably greater than it would be if the index were uniform at even the highest (1.406) actual value. A uniform index of 1.42 would be necessary for the lens to equal its actual refracting power with nonuniform index. The lens changes its shape, and hence its power, under the influence of muscular forces that are controlled by the central nervous system.

The optic axis is defined as the line joining the centers of curvature of the cornea and the lens. Such a line can usually be established quite accurately

TABLE XI OPTICAL PARAMETERS OF A TYPICAL NORMAL HUMAN EYE*

Surface	Radius of Curvature, mm	Refractive Index: Anterior	Refractive Index: Posterior	Distance from Anterior Surface of Cornea, mm	Refractive Power, D
Anterior cornea	7.8	1.000 (air)	1.376	0.0	+48.2
Posterior cornea	6.8	1.376	1.336 (aqueous)	0.5	−5.9
Anterior lens	10.0†	1.336	1.386‡	3.6†	+5.0‡
Posterior lens	−6.0	1.386‡	1.336 (vitreous)	7.2	+8.3‡
Retina				24.0	

* From Gerald Westheimer, "The Eye," in *Medical Physiology*, V. B. Mountcastle, Ed., 12th Ed., The C. V. Mosby Company, 1968.

† During maximum accommodation the anterior surface of the lens has a radius of curvature of 5 mm and its anterior surface is moved forward to be nearly 3 mm behind the anterior surface of the cornea. Partial accommodation will produce values between these values and those given in the table.

‡ The index of refraction of the lens varies from 1.386 near each surface to 1.406 in the center. The indicated refractive power is for the lens surface only. The gradient of refractive index within the lens produces additional refractive power.

by aligning a small light source with its images reflected from the various spherical eye surfaces.

When the eye is reduced to a simple optical system and the principal planes located, their separation is only 0.31 mm, and the first principal plane is 1.54 mm behind the anterior corneal surface. Figure 14.3 is a greatly simplified model called the *reduced eye*, which consists of a single surface of radius 5.51 mm separating air from a medium having the refractive index of water, 1.333. The vertex of the single surface is located at the second principal plane. The reduced eye is a model designed to form an image in the same position as the typical real eye described in Table XI.

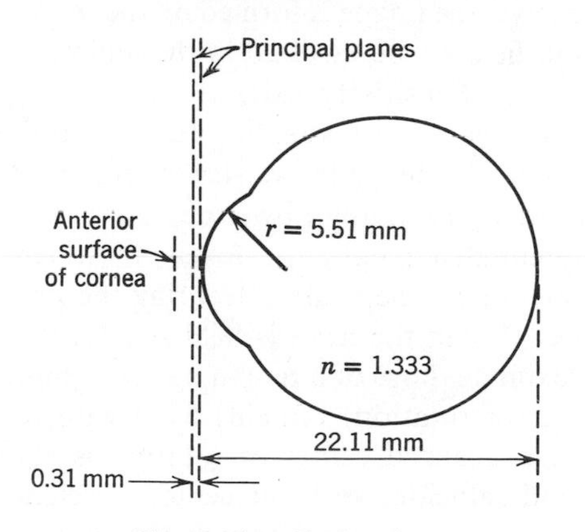

Figure 14.3 Reduced eye.

The aperture stop of the eye is the pupil, and the entrance pupil is about 3 mm behind the anterior surface of the cornea. The pupil is usually somewhat off center with respect to the optic axis, its center typically being 0.5 mm toward the nose from the optic axis.

Accommodation

It is well known that the eye can adjust its effective focal distance for viewing either near objects (as close as 4–8 in.) or distant objects. This adjustment is called *accommodation*. The refracting power at the anterior surface of the cornea is greater than that at any other surface of the eye, largely because that is where the largest refractive index change occurs, from the index of air to the index of the medium of the cornea. The refractive index for any one of the optic media is not much greater than that of water, 1.33. Although the contribution by the lens to the total focusing power of the eye is relatively small, the lens provides the primary focusing control mechanism by its ability to change shape. Both its anterior surface curvature and its thickness are variable. When the muscles connected to the lens are relaxed, the lens is its "flattest," and the eye is focused for distant viewing. The lens is thickest when the eye is focused for nearby objects. At this focus, the muscles are contracted, and curvatures of the surfaces, primarily the anterior surface, are greater [Ref. 11].

Eye Movements and Field of View

In a properly focused eye, the image is formed on the retina. Its curved image screen allows a large field to be covered with minimum optical problems from curvature of field. Sensitivity extends over a total angular field of 180 deg in a horizontal meridian (extending further on the outside than on the nose side) and about 130 deg in the vertical meridian. Not all parts of the field are equally sensitive; resolving power degrades rapidly away from the central, highly differentiated patch, the fovea, in which acuity is highest (see Fig. 14.16). The eye can be rotated to bring the image of a peripheral target onto the fovea. When the head is held steady, the kinematics of the eye movement approximate those of a ball in a socket joint, that is, one point in the eye (the center of rotation) remains fixed with respect to the bony orbit. This somewhat idealized center of rotation is about 15 mm behind the corneal vertex and coincides with the center of scleral curvature, but it usually lies somewhat toward the nose from the optic axis. Another important asymmetry in the eye is a displacement of the fovea about 1 mm outward from the intersection of the optic axis with the retina. As a consequence, the designation of reference points and lines in a typical human eye is rather complicated.

If an observer is asked to look, with his head fixed, at a "point" source, he will rotate his eye so that the light is imaged in the center of the fovea (this can be done with great precision). The target is then said to be *fixated*, and it becomes the *fixation point*. The line joining the target with the center of the entrance pupil is the eye's visual axis, usually called the primary line of sight. If one tries to align two targets at different optical distances, the centers of the blur patches of the two targets are superimposed, and the line joining the two targets will be the eye's chief ray or visual axis. This serves as a good operational definition of the primary line of sight; it can also serve as the basis for the determination of the actual kinematics of the moving eye because the line of sight thus defined remains fixed with respect to the eye, regardless of eye position.

The Neuro-Retinal Response

The light-sensitive part of the retina is an array of two kinds of "photodetectors" called rods and cones. The human eye has some 100 million of the thin rod-shaped cells and some 5 million of the slightly cone-shaped ones. Each cell is connected through a synapse, or junction, to a nerve fiber that forms part of the optic nerve to the brain.

In three remarkable photochemical processes—vision, photosynthesis, and photoperiodism—light participates by being absorbed in small, colored molecules called *chromophores*, each of which is associated with a large protein molecule. In vision, the light-sensitive molecules account for the pink and purplish color of the retina. Electron micrographs show that the outer end of each rod and cone is packed with thin membranous sacs and that the light-absorbing chromophores are associated with these sacs. Excitation of a chromophore by light causes some kind of change in the membrane, and this change produces a signal in the nerve fiber. Most workers in the field of physiological optics believe that, to explain color vision, the cones must contain three kinds of color-pigments, or chromophores.

The signals in the nerve fibers are short, repetitive electrical discharges. Because receptors outnumber the nerve fibers in the optic nerve one hundred to one, much of the integrative action of the visual system must occur within the retina rather than upon the arrival of impulses at the brain. Electrical engineers would say that a lot of information processing takes place within the retina. The message passing along an optic nerve fiber is related not only to the amount of luminous energy absorbed by the receptors feeding it, but also to the much more complex space–time variations of the luminous energy that constitute the total retinal stimulus.

In a region called the *chiasma*, near the brain, nerve fibers from the two eyes cross so that fibers from the left side of both eyes pass to the left side of the brain, and fibers from the right side of both eyes pass to the right side of the brain. Hence, the nerve fibers from the side of each eye near the nose are actually the ones that cross in the chiasma.

Adaptation

Most of us have suffered the embarrassment of stumbling into a dark theater seat on a sunshiny day and, later, of wondering whether we would ever be able to see again, without tightly squinting, after returning to the bright street. Actually, after one or two of these experiences, we learned that a few minutes would cure either deficiency. The automatic vision adjustment to the surrounding light level is called *adaptation*.

The ratio between the highest and lowest luminance levels that bound the adaptation range for comfortable seeing is about ten billion or 10^{10}. Only a small part of this adjustment is accomplished by the usual optical practice of varying the aperture stop diameter. Since the pupil diameter varies from approximately 2 mm to 8 mm, the ratio of minimum to maximum pupil area is only about 1 to 16. The actual aperture stop of the eye is inaccessible to the outside observer; what one sees is the entrance pupil, whose position

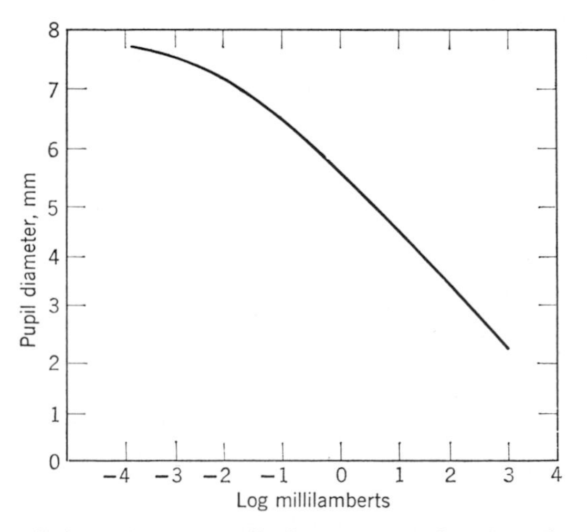

Figure 14.4 Typical steady-state pupil diameter as a function of retinal illumination [Ref. 6].

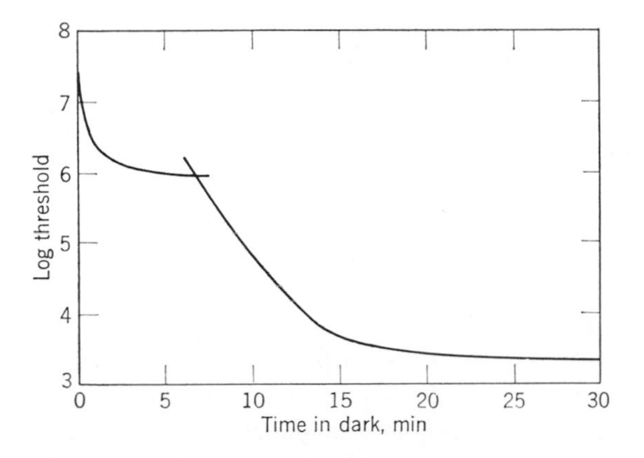

Figure 14.5 Dark adaptation following light adaptation to a luminance of 1550 mL. (The threshold was determined for a retinal field 2 deg in diameter and situated 7 deg nasally. The threshold-determining light is from the violet part of the spectrum, below 0.46 μm. The first section of the curve shows cone function, whereas the second section shows rod function) [Ref. 2, p. 75].

and size are close to those of the real stop. Calculations involving the eye invariably relate the entering bundle of rays to the entrance pupil.

Although the pupil does not adjust enough, even as an approximation, to keep the ingoing flux constant, the prevailing luminance level does have an effect on the average pupil size. Figure 14.4 shows the relationship between the two for steady-state conditions, but the relationship is only approximate because many factors (visual, physiological, and psychological) enter into the determination of pupil size in any given instance. The size of the pupils of a person's eyes, in fact, depends as much on his emotional state as upon luminance level (the pupil is larger during excitment and pleasure).

Brief changes of the light levels lead to more pronounced responses than the steady-state curve of Fig. 14.4 indicates. A delay of about 0.2 sec occurs between stimulus and response. Although the quantity of light entering the eye is little affected by the pupil compared with the range of light levels over which the visual system functions, the quality of the retinal image, particularly the resolving power, is significantly affected by the pupil size.

At least three mechanisms participate in adaptation to a lower prevailing luminance: (1) the action of the iris to enlarge the pupil, (2) an increase in the concentration of the photochemical substances in the receptors, and (3) a change in the neural processes associated with a reduction in retinal stimulation. Adjustment of the eye to darkness (dark adaptation) requires more time than the reverse (light adaptation). Approximately 30 min are required for the eye to become completely dark adapted (Fig. 14.5).

Vision in Bright and Dark Surroundings

The two kinds of photoreceptors in the retina, rods and cones, differ in a number of respects: threshold of sensitivity, dependence of the threshold on the state of adaptation, and distribution over the retina. Because seeing in near darkness is almost entirely with the rods and seeing in bright light is almost entirely with cones, the nature of sight under these extremes can be expected to be as different as the characteristics of the two receptors.

The fovea contains cones but no rods and occupies about 2.5 deg of the field of view. A few rods occur just outside the fovea. Then with increasing distance from the fovea, the density of cones decreases and the density of rods increases until, at the periphery of the retina, there are many rods but no cones. Consequently, vision at the center of the field differs markedly from peripheral vision.

When the eye is adapted to bright light, the rods become relatively insensitive, so what we see is with essentially cone vision. This means that we can see best near the fixation point. Cone vision is called *photopic*, which is characterized by a chromaticity, that is, we distinguish hue and saturation (see Chapter 15). (The relative spectral sensitivity is in close agreement with the curve of Fig. 2.5.) That we can see only poorly away from the region of the fixation point is demonstrated by two simple experiments: With your hand held at the side of your field of view, about 30 deg away from the axis, you will not be able to count the fingers. A small object that is bright red in the center of the field will become black when moved to the extreme edge of the field. During high light level adaptation, the eye is most sensitive to radiant energy at 0.555 μm.

When the eye is adapted to near darkness, the cones become relatively insensitive, and seeing is essentially rod vision. Then we see best, except for brief bright flashes, several degrees away from the fovea. Rod vision is called *scotopic*, which is color blind (only shades of gray). A different luminous efficiency function applies because the relative spectral sensitivity of the rods when fully dark-adapted is hundreds of thousands of times as great as when the eye is light-adapted. Only a few photons of light, such as from a "standard candle" 10 mi away, can be detected by the dark-adapted eye. Under these conditions, the eye is most sensitive at 0.510 μm.

The relative wavelength sensitivity of the eye for both photopic and scotopic vision is given in the curves of Fig. 14.6. The two curves, each of which is normalized to its own maximum, are plotted on the same graph to show only the respective wavelength ranges of sensitivity for the two conditions of eye adaptation. On an absolute scale, the scotopic peak would be many, many times the value of the photopic peak. The value of K_m

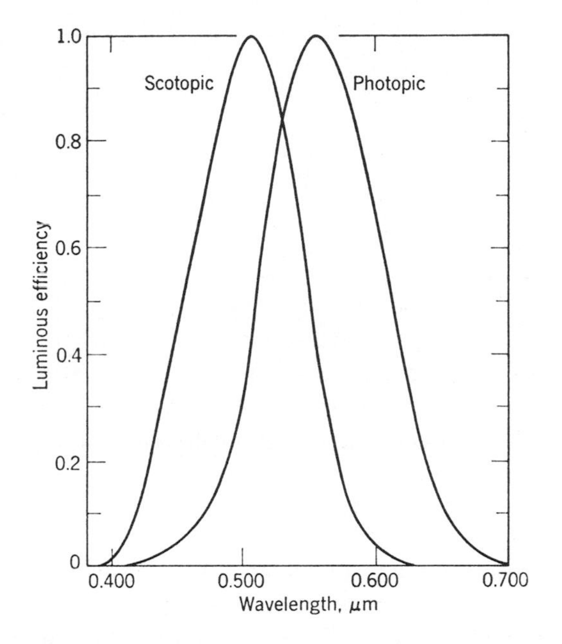

Figure 14.6 Comparison of photopic and scotopic spectral luminous efficiencies.

(680 lm/watt) as used in Equations (2-28), (14-2), and (15-1) apply only to the photopic function when the eye is adapted to a relatively bright luminance, greater than 0.1 millilambert.

The cones also adapt to darkness, but their relative sensitivity increases only by hundreds instead of by hundreds of thousands as with rods.

Spectral Transmission by the Ocular Media

The cornea absorbs light of all wavelengths shorter than about 0.32 μm. This means that the interior structures of the eye are safe from short-wave ultraviolet, but the cornea itself may be damaged by it. Since much of the ocular media is water and since water absorbs heavily at wavelengths longer than 1.40 μm, radiant energy in the infrared, except in the band near the visible, never gets far into the eye. However, radiant energy in the 0.68–1.40 μm region does enter the eye, although the observer may be unaware of its presence because it is poorly detected by the photoreceptors.

In the visible region, the transmission of the ocular media is not uniform over the spectrum (Fig. 14.7) [Ref. 12]. Blue light is attenuated heavily, particularly in older eyes, because of absorption and scattering in the crystalline lens.

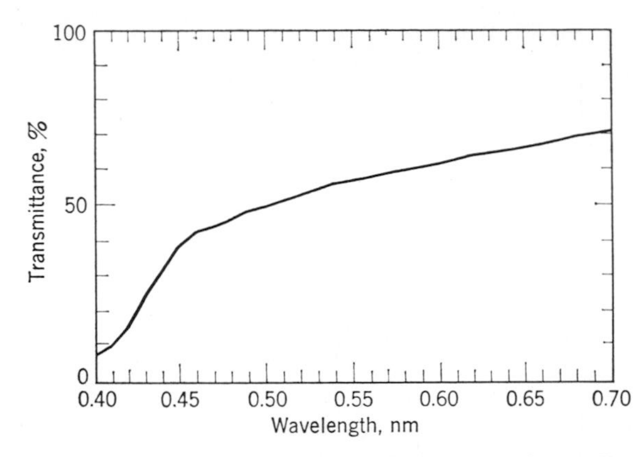

Figure 14.7 Transmission of visible light by the ocular media [Ref. 12].

Retinal Stimulation

As with other optical systems, the light flux entering the eye [given by Eq. (7-10)] varies with pupil diameter. However, several corrections must be made before this light can be regarded as the stimulus that reaches the receptors. Also, since psychological factors as well as luminance level affect pupil size, a precise determination of the actual diameter at the moment of measurement becomes a delicate procedure. Since the transmittance of the ocular media varies with wavelength, the spectral transmittance of the light has to be considered. Finally, the amount of light flux reaching the retina, unlike the assumption for Eq. (7-10), is not directly proportional to $\sin^2 \varphi$. This happens because light entering the marginal zones of the pupil is less effective for retinal stimulation than light passing through the center. (Light passing through both zones is assumed to reach the same point on the retina.) The variation of stimulation is known as the Stiles–Crawford effect. It applies mainly to cone vision (Fig. 14.8).

Both the physiologist and the psychologist consider light stimulus to be any change in radiant energy incident on the retina that is capable of exciting a response in the nerve tissue. Since the light arriving at the receptors is not usually a constant fraction of the light emerging from the field of view, even under very carefully controlled lighting conditions, an evaluation of the stimulus by determining the light flux reaching the retina is a difficult procedure. An often-used unit of retinal stimulation is the Troland, which is defined as the product of luminance at the corneal anterior surface in candelas per meter^2 and the area of the pupil in millimeters^2. This unit, as most often applied, is not corrected for absorption by the occular media or for the Stiles–Crawford effect.

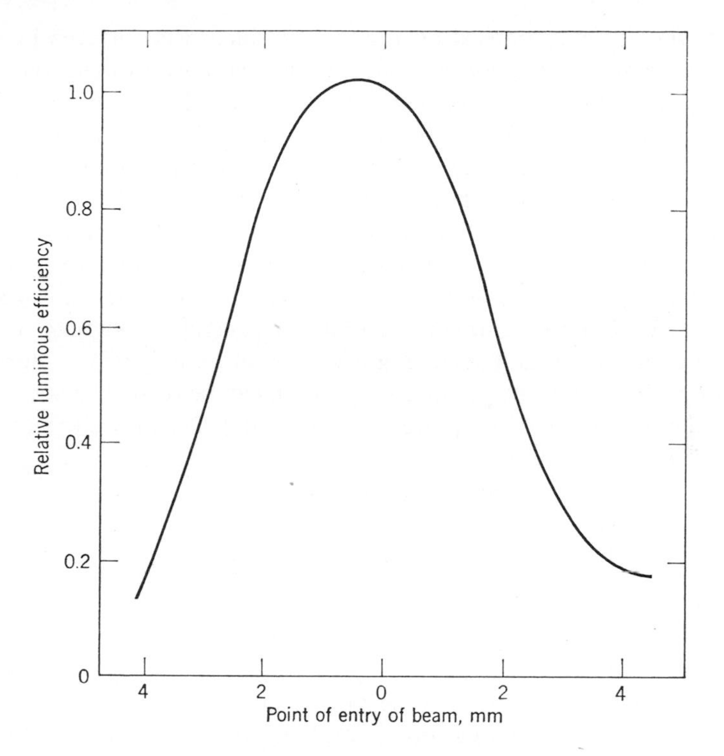

Figure 14.8 The Stiles–Crawford effect.

Methods of Testing

Psychology is strongly involved in seeing because seeing is a perceptual process affected by other sensations and associative mechanisms involving emotions, education, and experience. Seeing is just one of several possible contributors to experience, and the contributors rarely occur alone. Though seeing is generally associated only with luminous energy stimuli, it cannot, because of the other factors, be related to it with mathematical precision. Because experience particularly is many-faceted and subjective, the evaluation of visible stimuli, that is, of displays, is almost impossibly complex. Measurements generally can determine only statistical probabilities rather than absolute values.

Man is not in the habit of observing accurately with his senses except as they enable him to recognize specific external objects. On the contrary, he tends to disregard all those parts of a sensation that are not of conscious importance—we might say that he sees mainly through the spectacles of his memory. The remembered color of an object is always awakened if the memory of the object is aroused by any other of its aspects or even by only the name of the object; the remembered color is especially aroused when one sees the object itself or even when he thinks he sees it! The memory, in fact, partly determines the way one sees an object when he actually looks at it. Thus in color photography the colors need not be reproduced precisely. They need only be close enough to jog one's memory to "see" correctly.

A discussion of the words "luminance" and "brightness" brings out the differences between a quantity that measures stimulus (luminance) and the response made to the stimulus (brightness). The first can be measured in physically defined units while the second, in general, is subjective and cannot be quantified without averaging many sensation judgments. Spectral luminous flux $\Phi_{v\lambda}$ in lumens per unit wavelength can be related to a physical quantity, the spectral radiant power (radiant flux) $\Phi_{e\lambda}$ in watts per unit wavelength, by

$$\Phi_{v\lambda} = K_m V(\lambda)\Phi_{e\lambda}, \tag{14-1}$$

or the integrated flux Φ_v can be expressed as

$$\Phi_v = K_m \int_0^\infty V(\lambda)\Phi_{e\lambda}\, d\lambda. \tag{14-2}$$

With some change in notation, these are Equations (2-28) and (2-29) of Chapter 2. Thus the luminous flux stimulating the retina is expressed in physical quantities, but brightness, only one of the attributes of seeing, is related to luminance, source size, color, time duration of the stimulus, and to other factors. A person can say, for example, that one object is equally as bright as another provided the two are near each other in space and time—or that one is brighter than the other. He can say that one is blue and another green or even that one is greener than the other. However, uncertainty arises when he tries to compare quantitatively, say, by saying that one thing is twice as bright as another.

An approach to quantifying subjective responses like brightness can be illustrated by describing a typical experimental procedure for establishing a threshold. The procedure also is an example of a method for determining a psychophysical function, that is, the relationship between luminance level and a threshold of seeing. The subject, who has remained in the dark for about half an hour, is instructed to observe a fixed, lighted patch and to respond by saying "yes" if he sees another lighted patch in the field and "no" if he does not. The purpose is to determine the frequency of "yes" responses as a function of luminance of the second patch. The luminance of the second light is usually varied in not less than seven steps. The luminance levels are finally adjusted so that at the lowest luminance 100% negative

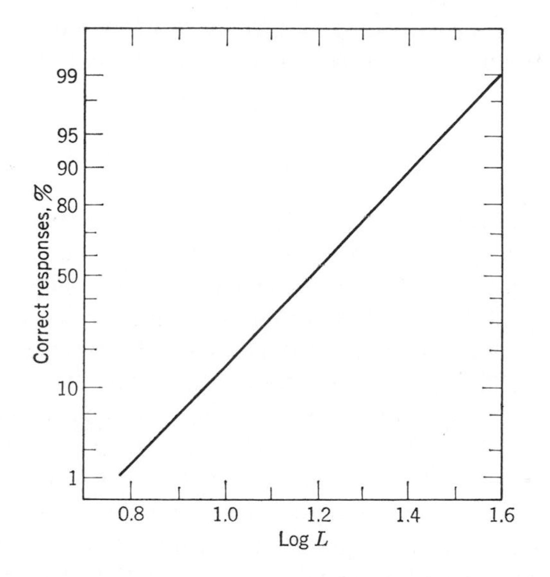

Figure 14.9 Data on "percent frequency of seeing" for various values of log L [Ref. 22].

responses occur and at the highest level 100% positive responses. Figure 14.9 is a plot of one set of data showing the per cent of "yes" responses against the log of luminance of the second patch. From such a set of data one could define the threshold of seeing as that luminance for which half the responses are positive. However, this is only the threshold for the one person being tested and for the given set of experimental conditions. After a number of subjects are tested, some kind of statistical average of all the thresholds obtained could be found and then offered as the standard of a subjective response.

To conduct experiments like the one described, the experimenter has a variety of instruments and techniques for controlling the stimulus and the measurement conditions. He typically has to build all or part of his test equipment. Some of the physical conditions that require control are:

(1) Luminous intensity—by varying the distance between subject and source and by using diaphragms (variable aperture stops), neutral density filters, and polarizers in the optical system.
(2) Spectral compositions—by using colorimeters, transmission filters, combinations of monochromatic sources, and selective reflectors (colored papers).
(3) Stimulus duration—by using mechanical shutters, flash sources, and flicker vanes (rotating sector disks).

In addition to these, the psychological factors—fatigue, emotional state, and motivation—would have to be controlled throughout a sequence of test sessions.

Resolving Power of the Eye

In addition to diffraction, the eye has defects that broaden the optical point-source response (spread) function. Spherical and chromatic aberration are appreciable in the normal eye [Refs. 13 and 14]. The amount of chromatic aberration is approximately what would be found for the reduced eye of Fig. 14.3; longitudinal chromatic aberration as a function of wavelength is shown in Fig. 14.10. The diopter, the unit given on the ordinate, is the focusing power [see Eq. (6-30)]. The power of a lens in diopters is the reciprocal of its focal length in meters. The total range of differences shown corresponds approximately to a range from 0 to 1-mm variation in focal distance. When scotopic vision predominates, no chromatic discrimination occurs, so chromatic aberration produces only a broadened spread function. Scattering—variable with wavelength and age of the observer, increasing with age—in the occular media also tends to broaden the spread function. Diffraction effects, of course, vary with size of the pupil.

Fairly consistent measurements of the eye as an optical system have been made by many observers. Generally, the fundal image (image on the retina)

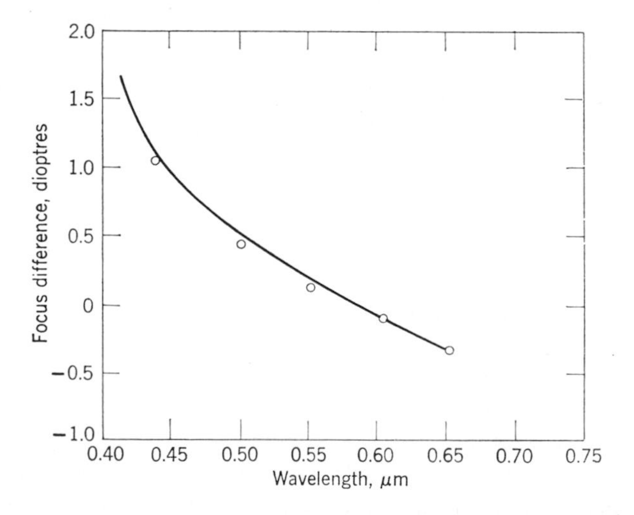

Figure 14.10 Chromatic aberration of the human eye and of an equivalent system composed of water. (Focus in diopters = 1/focal length in meters. Focus differences indicated measure the focusing power added to the eye to restore perfect focus at each wavelength (color)) [Ref. 14].

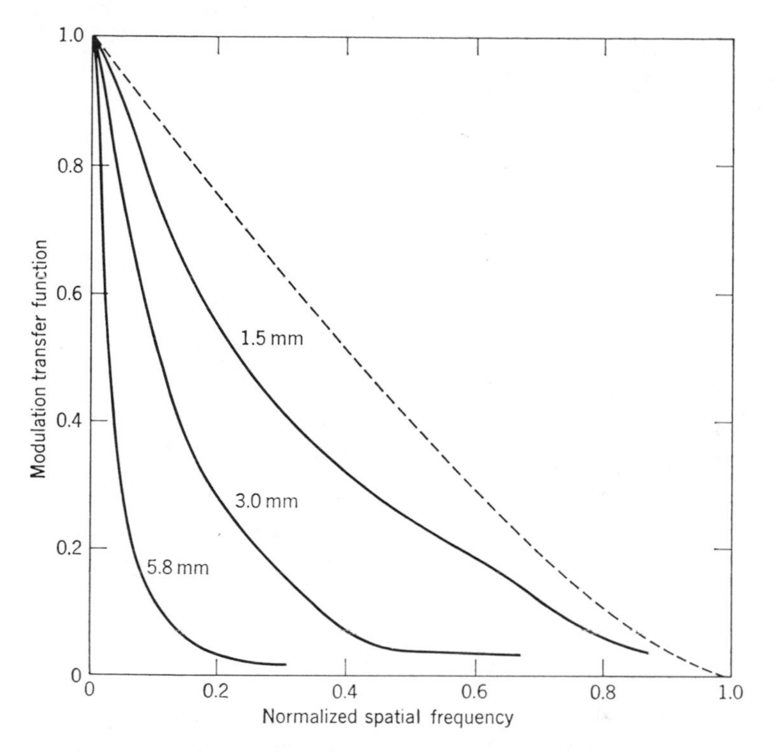

Figure 14.11 Normalized modulation transfer functions of the human eye averaged for three subjects. (The results are normalized to the highest spatial frequency transmitted by a diffraction limited optical system for light of wavelength 0.57 μm. The dotted curve is the diffraction limited MTF. Other curves are for pupil diameters of 1.5, 3.0, and 5.8 mm) [Ref. 15].

of a line source was found to be at least twice as broad as the line's diffraction image for each given pupil size. Recent results [Ref. 15] by a more accurate measurement procedure give the results shown in Fig. 14.11. A line source (slit illuminated by a high-pressure xenon arc) projects a line-spread function on the retina. The retina functions as a somewhat diffuse reflector. The external image of the retinal image formed by light reflected from the retina and transmitted back through the eye's optical system—is scanned with a narrow slit to obtain the (double passage) line-spread function. Application of the Fourier transform to this line-spread function produces a modulation transfer function. After correction for the double passage, the modulation transfer functions (single passage) of Fig. 14.11 result (shown for a number of different apertures). Spread functions for a single passage are obtained by taking the inverse transform of the modulation transfer functions; the results are shown in Fig. 14.12. The curves in both figures

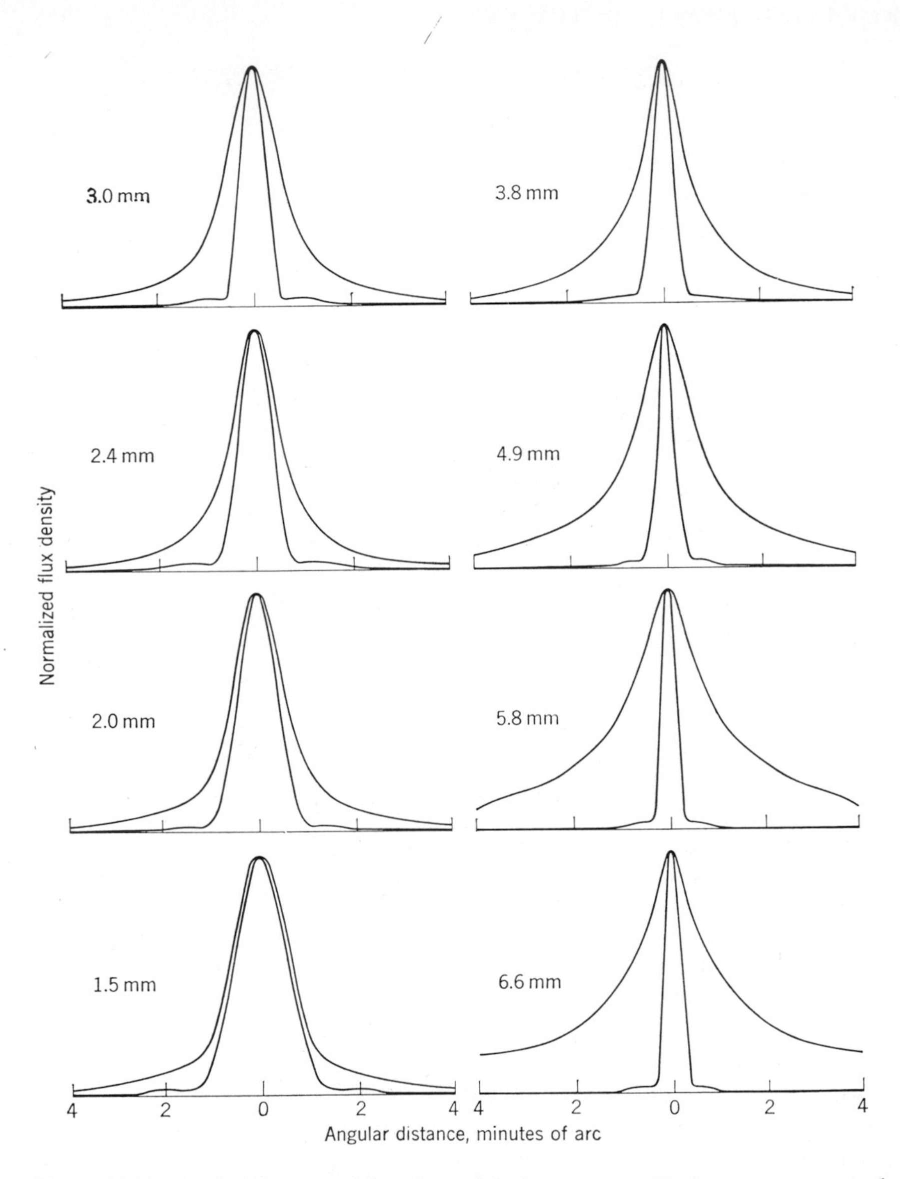

Figure 14.12 Optical line-spread functions of the human eye. (Each curve represents the normalized distribution of illuminance occurring on the fundus for a thin line source of light. Narrower curves are for calculated diffraction images of a line at the given pupil diameter) [Ref. 15].

are averages of data obtained with three subjects. Calculated line spreads for diffraction alone are shown for comparison in Fig. 14.12.

Although the actual line spreads depart increasingly from the diffraction limited spread as the pupil diameter increases, a minimum half width of the line spread occurs at the intermediate diameter of 2.4 mm. This is reasonably consistent with the rule of thumb that the eye is diffraction limited for pupil diameters of 2 mm or less.

If the sharp edge of a large bright field is imaged near the center of the eye, the 50% point in the edge trace will correspond to the Gaussian image of the edge, but the separation between the 90% and 10% points in the flux distribution function may be as great as one or two minutes of arc. In a precise subjective determination of an edge, the observer will probably not even come close to the 50% point because the relationship between retinal illuminance and subjective brightness is nonlinear; it is more nearly logarithmic. Therefore, it is too much to expect subjective alignment of, say, a bright edge with a fiducial mark to better than one minute of arc.

The microscope is extensively used for measuring small nonperiodic objects, measurements being made on the magnified image. As shown in Chapter 9, a gradient of flux density—an edge trace—occurs across the Gaussian image of the sharp boundary between a bright and dark region of the object. Observed through the eyepiece of a microscope, the distribution of the subjective brightness is shifted from the distribution represented by the edge trace because the visual response is nonlinear. If the observer attempts to set a fiducial line at the image of the boundary, he is not likely to set it at the true Gaussian image. For example, when he sets out to measure the diameter of either an opaque circular disk or a circular aperture in an opaque screen by lining, first, one edge and then moving the object (by translating the microscope stage) a measured distance until he can line up the edge at the opposite end of a diameter, the errors will add, resulting in a measurement that is too large in both instances [Ref. 16]. The error is of the order of the resolution limit, that is, the half-width of the line spread function; the actual error in each instance depends on the retinal illumination, the object size, the magnification, and the eye adaptation.

The curves of Fig. 14.11 take into account only the optical properties of the eye—as if it were a fabricated system. Neglected are the all-important physiological and psychological factors: The transformation of radiant flux density to nerve impulses is nonlinear; the physiological and neural-gain processes of the retina are inhomogeneous; appreciable information processing takes place in the retina and nerve fibers; and each observer reacts from a unique background of experiences and consequent response patterns. Therefore, accurate predictions of visual response cannot be made from only measurements of line-spread functions and their Fourier analyses.

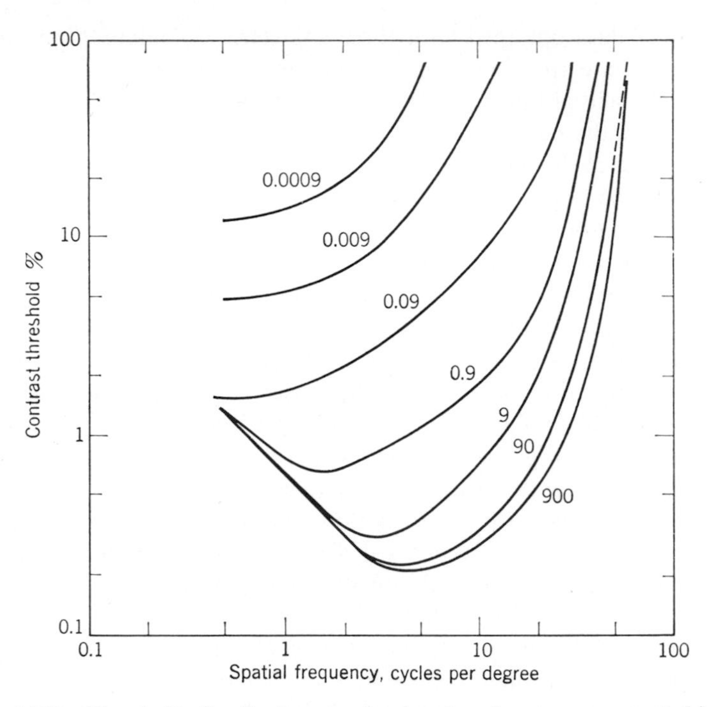

Figure 14.13 Thresholds for the human visual system for sine wave test objects with a 2-mm pupil and monochromatic light of wavelength 0.525 μm. (The number on each curve is the retinal illuminance level in trolands) [Ref. 17].

The omitted factors can be taken into account by a different modulation transfer function arrived at through the following test: The subject views a field consisting of a sinusoidal variation of flux density (set up by a sine-wave grating test object) and is instructed to respond when he sees the grating. If the contrast (as is defined in Chapter 9) and the spatial frequency (in cycles per degree) are varied, curves of the threshold as a function of frequency, similar to those of Fig. 14.13, can be plotted from the responses [Ref. 17].

If a field of illuminance varies slowly enough with position in the field, a common experience is that the variation cannot be detected by the visual system. This phenomenon is equivalent to a falling off of the modulation transfer function curves at low frequencies, the cause of which is unknown.

Visual Acuity

The straightforward evaluation techniques described in Chapter 9, applied to the optical part of the "seeing" mechanism, would grossly underestimate the actual capability of the visual system to distinguish fine detail. The

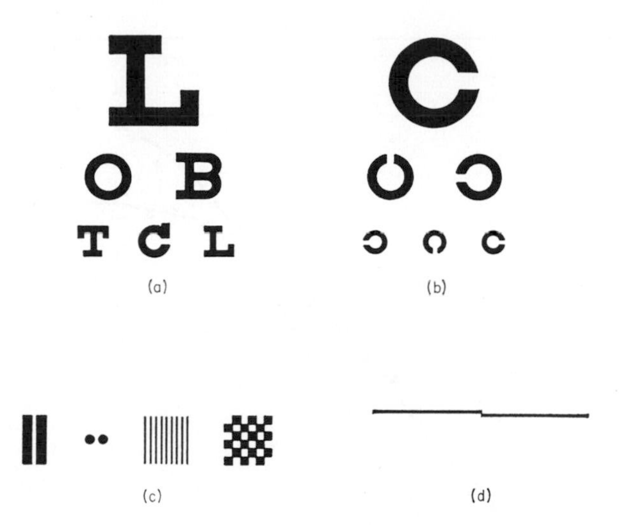

Figure 14.14 Several patterns used for testing visual acuity: (a) Snellen letters, (b) Landolt "C" rings, (c) bars, dots, gratings, and checkerboards, (d) displaced straight lines.

reason for this discrepancy is that the retina, the neural processes along the optic nerve, and the brain do considerable information processing.

In physiological optics, the term *visual acuity* is used instead of spatial resolution. Visual acuity is defined as the reciprocal of the angle in minutes subtended by the smallest discernible dimension in the field of view. Several test patterns with varying dimensions of patterns are used for testing visual acuity. Among these are the Snellen letters, Landolt "C" rings, gratings, checkerboards, and the vertically displaced horizontal line shown in Fig. 14.14. Acuity determined by detecting displacement in the displaced straight line is called *vernier acuity*. One minute of arc is usually considered to be an average visual acuity for the eye, but vernier acuity is just a few arc seconds. Acuity depends upon the light used during testing and upon other conditions (Figures 14.15 and 14.16).

Contrast

A highly developed sensitivity of the eye is its ability to detect a small difference in luminance. This difference is called contrast sensitivity, liminal contrast, or just contrast. Contrast for a given set of conditions is defined as

$$C = (L_B - L_O)/L_A. \tag{14-3}$$

where L_O is the luminance of an object (such as a circular disk), L_B is the luminance of the surrounding area (such as an annulus about the disk),

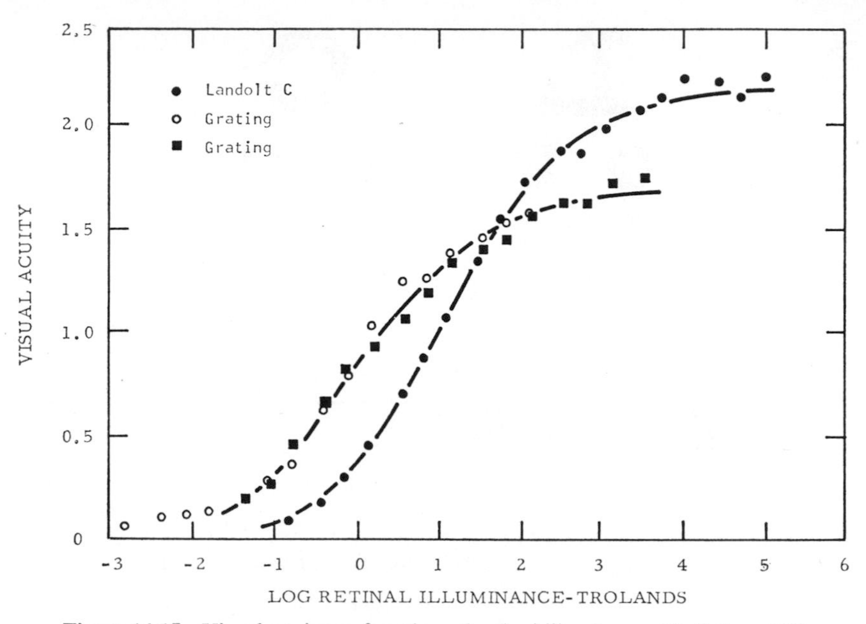

Figure 14.15 Visual acuity as function of retinal illuminance [Ref. 2, p. 337].

and L_A is the luminance to which the eye is adapted. When the object and its background essentially fill the field of view of the eye, L_A is determined by L_O and L_B. When object and the surrounding, illuminated area are approximately equal in size,

$$L_A \approx \tfrac{1}{2}(L_B + L_O). \tag{14-4}$$

If the object is small compared with the surroundings,

$$L_A \approx L_B. \tag{14-5}$$

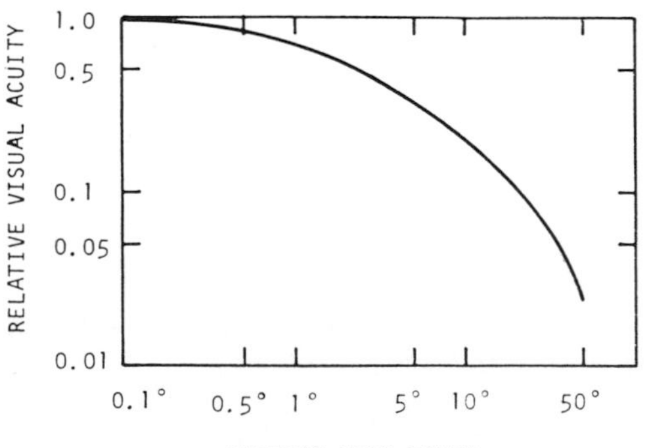

Figure 14.16 Variation of visual acuity with retinal position of image on retina. (Visual acuity is given relative to that at the fovea) [Ref. 18].

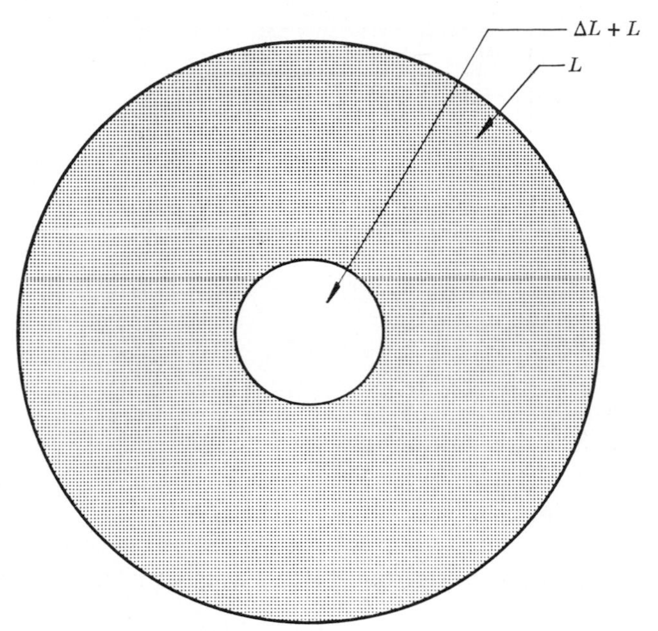

Figure 14.17 Field pattern for testing luminance difference threshold.

In many tests of contrast the difference between L_O and L_B is so small that L_A may be considered equal to either.

In an experiment to determine contrast, an observer is shown an area having uniform luminance L except for a portion to which an increment of luminance ΔL is added from time to time as a short pulse (Fig. 14.17). The luminance difference is first so small that the observer sees no change in brightness when the increment is added. The increment is increased until the threshold is reached where the observer can consistently sense the change of brightness. Figure 14.18 shows the variation of $\Delta L/L$ as a function of the luminance L to which the eye is adapted.

There are two luminance thresholds: The first, called the *absolute threshold*, is the minimum detectable stimulation. The second, the *differential threshold* called liminal contrast, is the smallest detectable luminance *difference* ΔL. Either threshold depends upon both past and present conditions in the visual field. The area of the stimulus–object, the stimulus time duration, the luminance level in the surrounding field, the relative motion between object and background, and the observer's adaptation all combine with luminance to bring about a particular response. Even "future" conditions in the field can influence the response; for example, a bright flash occurring a very short time later can mask the stimulus flash.

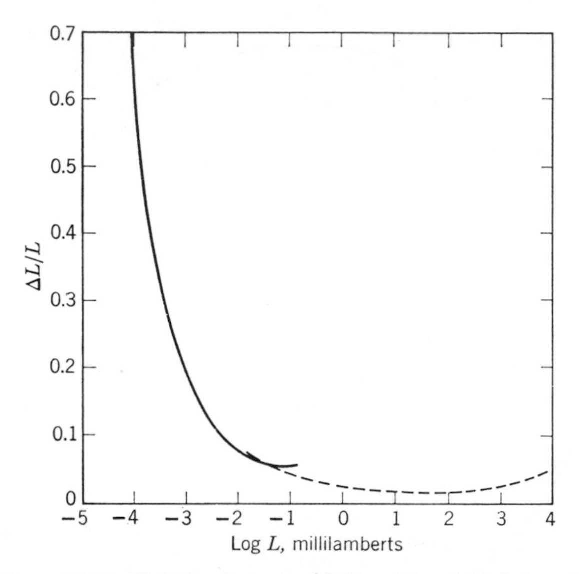

Figure 14.18 Relation between $\Delta L/L$ and log L [Ref. 2, p. 215].

Under favorable conditions, only a few photons of light reaching the retina at approximately the same place and time can produce a sensation of seeing in a completely dark adapted eye. This illustrates the absolute threshold. The size of an object and its luminance threshold are inversely related, so objects with small retinal images have high luminance thresholds and those with larger retinal images have lower thresholds.

The effectiveness of a light stimulus is altered by the introduction of a similar further stimulus either at the same time or nearly at the same time. This effect is called *meta contrast* or masking. The brightness of a constant-luminance test circle (actually a test zone between two concentric circles of different diameters), for example, can be decreased (depressed) or increased (enhanced) by surrounding the circle with a field of greater or less luminance respectively. If the surrounding field is eliminated (made dark), enhancement results from an increase in area of the test circle; for example, the threshold luminance necessary to see the circle will decrease if the area of the circle is increased.

Visual interaction is also illustrated by the change in brightness match between a test object and a comparison object when a third object, an *inducing* object, is brought adjacent to the test object. If A and B are the comparison and test objects, respectively, and are initially matched in brightness, the match is lost when an inducing object C is placed in the field. If the objects are arranged in the sequence ABC in the field and C is brighter than B, the luminance of B must be increased to restore the match between

A and B. If, on the other hand, B is brighter than C, the luminance of B must be reduced to restore the match between A and B.

Flicker

If a test patch in the field of view is alternately and abruptly made bright and dim ("square wave") with a gradually increasing repetition rate, the bright patch and dim patch are first seen as separately illuminated patches displacing each other. As the frequency increases, a peculiarly unpleasant coarse flicker is first observed. This changes to a fine tremulous appearance until, finally, complete fusion of the two conditions occurs, and a steady-state illuminated patch is seen. The brightness observed is somewhere between the brightnesses of the bright and dim luminances.

In a typical flicker experiment, a light beam illuminating a surface viewed by the observer is chopped by a rotating disk containing alternating transparent and opaque sectors. The repetition rate of light and dark phases at which the steady-state response (fusion) occurs—the *critical flicker frequency* (CFF)—is related to a number of variables, the most important of which is the illuminance produced by the interrupted light beam. The critical frequency is low at low luminances and rises as the luminance increases until essentially a constant critical frequency is reached over an appreciable range of luminances. Thereafter the CFF rises again until a second relatively constant value is reached. Analysis indicates that the low luminance branch of the curve is due to activity of the rods and the high luminance branch to cones.

Above the CFF it is of interest to know what steady luminance is equivalent to the fused bright and dark phases. When the dark phase is actually of negligible luminance, a simple equation relates the equivalent luminance L_m and the on-time of the light beam:

$$L_m = L_i[t_l/(t_l + t_d)], \qquad (14\text{-}6)$$

where L_i is the luminance of the flickering light, t_l is the on-time of the light, and t_d is the dark time.

The frame rate in television and motion pictures has been standardized at a frequency that is well above the CFF for the appropriate luminance conditions, the former at 30 and the latter at 24 frames/sec [Ref. 2, Chapter 10]. (The effect in television is helped further by interlacing 60 half frames/sec to produce the 30 full frames/sec.)

Space and Form Perception

The perception of space and the form and the distribution of objects in space involve many human factors [Refs. 19 and 20]. Our sensing of physical space depends upon psychological stimuli called cues, some of which are binocular. From these we perceive the following:

(1) Space—the positions of objects and the relative as well as approximate actual distances to and between objects.
(2) Form—the shapes and sizes of objects, relative as well as approximate actual sizes.
(3) Motion—directions and speeds of moving objects.

The cues for space perception are discussed at some length in Ref. 2, Chapters 18 and 19. The cues for space perception are:

(1) Monocular Cues
    (a) Relative size—discrimination of distance is dependent upon the size of the retinal image of an object and on past and present experience with objects of the same class.
    (b) Interposition—an overlapping object that cuts off the view of part of a second object is perceived to be nearer.
    (c) Linear perspective—a constant distance between pairs of points subtends a smaller and smaller angle at the eye as the points recede from the observer (like railway rails and telephone wires).
    (d) Aerial perspective—a loss of contrast (graying), a washing out of colors, and a loss of detail (degradation of resolution) go with increasing distance.
    (e) Accommodation—focusing the eyes on objects at different distances gives a feel for relative distances to the objects.
    (f) Monocular motion parallax—when we move sidewise, objects beyond the fixation point seem to move with us, while objects between us and the point do not move as fast; so they seem to move the other way.
    (g) Light and shadow—highlighted objects seem to be nearer than objects in a shadow. Green or blue objects against a light background seem to be nearer than orange and red objects. Red and yellow objects seem to be nearer when they are against a dark rather than a light background.
(2) Binocular Cues
    (a) Convergence—lines from the two eyes to a fixation point at a great distance are parallel; when the point is near, the lines converge sharply to the point. Convergence cues are effective for distances up to several yards.
    (b) Stereopsis (binocular disparity)—in stereoscopic photography, two pictures are taken simultaneously, from slightly different points of view. The differences between the two pictures are called *binocular*

disparities. When the right eye is made to see only the picture taken by the right-hand camera and the left eye the one taken by the left-hand camera, a vivid sense of depth is produced.

All cues present at a given moment act together to bring about an overall response to the "scene." It is a mistake, for example, to assume that visual space perception is rigidly determined by light ray directions and intersections. The effect of any given cue cannot be eliminated by either the absence or presence of others. If monocular cues, for example, are inconsistent with binocular cues, a conflict in interpretation is created, a conflict that can be resolved only by perceiving a distorted space and incorrect sizes and distances. Two very important cues are monocular movement parallax and stereopsis; both are present in a hologram, and neither is present in a single "flat picture" display. Although a vivid sense of depth is apparent with stereopsis, there is often an unnatural aspect of the picture in visual space. The scene sometimes seems to be presented on a set of flat cards, positioned at varying distances from the observer. With stereopsis and without parallax, we cannot move our head to see further around real objects; head movements give us the impression only that all objects are twisting and moving in a way to prevent our seeing around them. In viewing an actual scene, we find that other monocular cues also play a powerful role in space perception and contribute to a sense of space and solidity. Caution must be exercised in providing the various cues in entertainment displays. The improvement gained by stereopsis is obvious, but pitfalls occur because inconsistencies between cues and unsatisfying viewing conditions are sometimes created. Prolonged viewing of stereoscopic movies too close to the screen can produce "excruciating pain" [Ref. 21]. An attempt to compensate for the incorrect viewing distance by altering the separation between the two pictures causes distortions of relative sizes and shapes of objects.

Flat screen displays very often give us a much better feel for depth than we realize. For example, using only monocular cues an artist can create a realistic feeling of distance and form in a flat painting. In the absence of monocular motion parallax, a sidewise motion of the moving picture camera to show a scene from a continuously changing point of view produces the effect of parallax, enhancing the sense of depth. Moving the camera from a closeup during a "take" to a great distance creates a sense of distance to the scene.

Good Seeing

Measurements relative to seeing usually concern thresholds or minimums and are of ultimate interest to the illuminating engineers or display designer only insofar as thresholds tell something about "good seeing." Normally, people

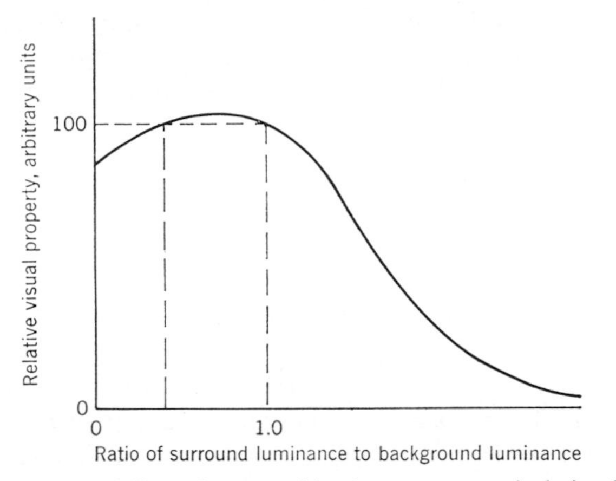

Figure 14.19 Effect of surround luminance on a typical visual property.

are seldom in testing situations observing thresholds. Without defining "good seeing" one can say that the conditions for seeing are good when the person is comfortable in his seeing environment and is able to perform his visual tasks accurately.

One of the factors in good seeing is ambient lighting. People have two distinct requirements for light. The first is to make seeing accurate, fast, and effortless. The second is to create and influence emotional response.

Illumination levels in thousands of lumens per foot^2 are required for adequately seeing dark, low-contrast tasks. Well lighted, colored tasks of high contrast can be done under considerably lower levels—in fact, in only tens of lumens per foot^2. Many observations suggest a minimum of 30 lm ft^{-2} for all areas where seeing is done regularly, even for the simplest seeing tasks.

The eye performs best when the luminance of the *surround* (the region outside the immediate background) is within a range of optimum values. Contrast sensitivity and visual acuity in particular are degraded when the surround luminance is either too high or too low. An optimum value for the ratio of luminance of surround L_S to luminance of background L_B is near unity or somewhat less. Figure 14.19 shows a generalized characteristic curve for the condition of any visual property as a function of the ratio L_S/L_B. The characteristic can be divided into three regions as follows:

(1) A region of low surround luminance, L_S/L_B less than approximately 0.1, where seeing is poor.
(2) A region of moderate surround luminance, L_S/L_B slightly less than or equal to 1, where optimum conditions exist.
(3) A region of high surround luminance, L_S/L_B considerably greater than 1, where seeing is poor.

REFERENCES

1. J. P. C. Southall, *Introduction to Physiological Optics.* Oxford University Press, London, 1937. (A paperback reprint is available from Dover Publications, Inc., 180 Varick Street, New York, City 10014.)
2. *Vision and Visual Perception*, Edited by C. H. Graham. John Wiley & Sons, Inc., New York, 1965.
3. Y. LeGrand, *Light, Colour, and Vision*, Chapman and Hall, Ltd., London, 1968. (Translated by R. W. G. Hunt and F. R. W. Hunt.)
4. Y. LeGrand, *Form and Space Vision*, Second Edition. Indiana University Press, Bloomington, 1967. (Translated by M. Millodot and G. G. Heath.)
5. M. L. Rubin and G. L. Walls, "*Fundamentals of Visual Science*," Charles C. Thomas, Springfield, Ill., 1969.
6. G. Westheimer, "Image Quality in the Human Eye," *Optica Acta*, **17,** 641 (1970).
7. R. M. Boynton, "Progress in Physiological Optics," *Appl. Optics*, **6,** 1283 (1967).
8. G. L. Walls, *The Vertebrate Eye.* Cranbrook Institute of Science, Bloomfield Hills, Mich., 1942.
9. F. W. Weymouth, "The Eye as an Optical Instrument," in *A Textbook of Physiology*, Seventeenth Edition, Chapter 23. Edited by J. F. Fulton. W. B. Saunders and Co., Philadelphia, 1955.
10. G. Westheimer, "The Eye," in *Medical Physiology*, Twelfth Edition, Chapter 66, p. 1532. Edited by V. B. Mountcastle. The C. V. Mosby Co., St. Louis, 1968.
11. F. M. Toates, "A Model of Accommodation," *Vision Research*, **10,** 1069 (1970).
12. D. G. Pitts, "Transmission of the Visible Spectrum Through the Ocular Media of the Bovine Eye," *Am. J. Optom.*, **36,** 289 (1959).
13. A. Ivanoff, "About the Spherical Aberration of the Eye," *J. Opt. Soc. Am.*, **46,** 901 (1956).
14. H. Hartridge, *Recent Advances in the Physiology of Vision.* J. & A. Churchill Ltd., London, 1950.
15. F. W. Campbell and R. W. Gubisch, "Optical Quality of the Human Eye," *J. Physiol. London*, **186,** 558 (1966).
16. W. N. Charman, "Visual Factors in Size Measurement by Microscopy," *Optica Acta*, **10,** 129 (1963).
17. F. L. Van Ness and M. A. Bouman, "Spatial Modulation Transfer in the Human Eye," *J. Opt. Soc. Am.*, **57,** 401 (1967).
18. W. J. Smith, *Modern Optical Engineering*, p. 105. McGraw-Hill Book Company, Inc., New York, 1966.
19. C. Blakemore, "The Range and Scope of Binocular Depth Discrimination in Man," *J. Physiol.* (London), **211,** 599 (1970).
20. G. A. Fry, "Visual Perception of Space," *Am. J. Optom.*, **27,** 531 (1950).
21. D. L. MacAdam, "Stereoscopic Perception of Size, Shape, Distance and Direction," *J. SMPTE*, **62,** 271 (1954).
22. W. J. Crozier, "On the Visibility of Radiation at the Human Fovea," *J. Gen. Physiol.*, **34,** 87 (1950).

15

Color

Not only do we learn the sizes and shapes of objects and their positions in the space about us by the visual process, but we also perceive their colors. Thus, color is commonly regarded as a property of an object. However, a particular color sensation is actually the observer's response to radiant energy of a particular spectral distribution received by the eye. Unlike the other properties, an object's color manifests itself only because the object selectively reflects—according to wavelength—the radiant energy that illuminates it.

Light is defined in Chapter 2 as "the aspect of radiant energy of which the human observer is aware through the visual sensations that arise from the stimulation of the retina of the eye" [Ref. 1, p. 220]. The particular sensation called color is not, however, just the physical or chemical effect on the retina but it includes the resultant effect in the mind of the observer [Ref. 1, Chapters 4 and 5]. Because color signifies other inhomogeneities of light than spatial and temporal, it is a psychophysical concept [Refs. 1–7].

Hue, Saturation, and Lightness

The subjective attributes of color are hue, saturation, and lightness. The *hue* of a color is the attribute that is described as red, yellow, green, blue, or similar designation. *Saturation* is described by terms like bluish white, moderate blue, strong blue, and vivid blue. Saturation, therefore, has something to do with the "grayishness" (or "whitishness") of a color; the "more grayish" and less vivid the color, the less saturated it is. *Lightness* describes to what degree a color is pale, light, or brilliant as contrasted with dark or deep, corresponding to shades of gray.

The Arrangement of Colors

One way to develop the accepted classification scheme for colors is to review the "desert island" experiment described by Judd [Refs. 2 and 8]. We set the stage by placing a person who has had no experience with color classification on a desert island surrounded by a large number of pebbles having a great variety of colors. With nothing more interesting to do, he undertakes to arrange the pebbles in some orderly way according to their colors. He might first separate them into groups according to the common hue names: red, blue, green, yellow, purple, etc. By this procedure, he ends up with a group left over consisting of grays—which are known technically as achromatic, that is, without hue—ranging in shades from black to white. Arranging the grays within their group provides a second logical order, which is according to *lightness* (other terms used for lightness are brightness, value, and luminance).

Besides the grays, the island experimenter finds a number of colored pebbles that do not belong any more in one group than in another—for example, a pebble that is as easily placed in the green as in the blue. He finds a place for this pebble by letting it occupy a new position between the green and blue groups called blue-green. Furthermore, by choosing certain pebbles from the green and blue groups, he forms a row of pebbles that changes gradually from blue to green. Similarly, he forms a row between each of the other pairs of hue groups. Then he finds that these rows, when placed in proper order end to end, form a complete circle from blue through green, then through a series of other hues, and finally through purple to blue again to cover the whole gamut of hue gradations.

The experimenter finds by trial that he can select pebbles that have the same hue but differ only by lightness, that is, their color variations correspond to shades of gray. He then arranges a row of pebbles at each hue perpendicular to the circle, each row ordered according to lightness.

The experimenter still finds that a number of different pebbles in each hue group cannot be arranged in any of the defined orders. Among these, for example, are several blues, which though distinctly different are neither more purplish or more greenish than the others; that is, they all have the same hue. Furthermore, they cannot be arranged in an orderly scale similar to the shades of gray because they all have the same lightness. However, he does find that he can arrange this sub-group of blue pebbles so that they go from vivid blue to a strong or brilliant blue, then to slate blue, and finally approach a gray or a white. This final arrangement is said to be according to *saturation* or *chroma*. In terms of mixing paints, saturation corresponds roughly to the proportion of a pigment that is added to white paint, or to a gray paint.

A Color Coordinate System

The described color sorting experiment indicates that three coordinates—for hue, saturation, and lightness—are needed for arranging all colors. A

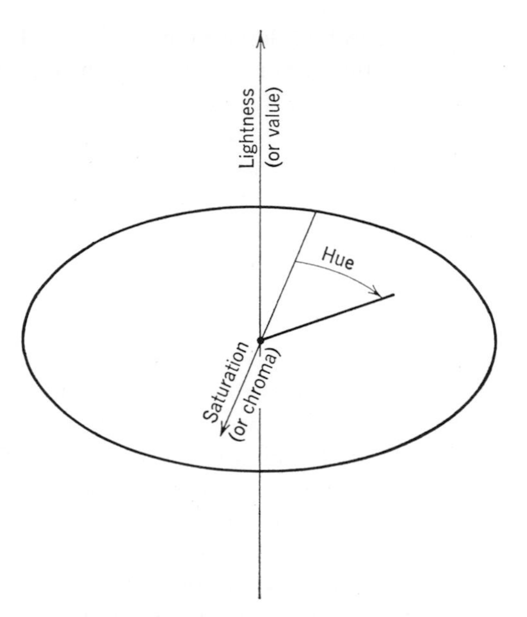

Figure 15.1 Color coordinate system.

convenient scheme is a cylindrical coordinate system (Fig. 15.1) with the radius representing saturation, the azimuth angle hue (with an arbitrarily chosen hue as the zero of azimuth), and the z-coordinate lightness (with black at $z = 0$). In a systematic arrangement of colors such as this, a color can be designated by its appropriate coordinate, that is, by (r,φ,z), but names, especially for the hues, are generally more readily comprehended. Consequently, points along the scale of hues are usually designated by hue names. Arbitrarily selected, simple and descriptive names—red, yellow, green, blue, and purple—are placed evenly around the scale of hues. Compound adjectival forms of these simple, principal hue names such as greenish yellow and purplish red are used to designate hues intermediate to the principal hues. In addition, at certain values of lightness and saturation, the names, orange, olive, violet, pink, and brown appear at appropriate positions around the hue scale. The recommended name of a color is its hue name preceded by a modifier that helps to place the color along the scales of lightness and saturation. A dictionary of recommended color names with an explanation of their arrangement in the color coordinate system has been published by the National Bureau of Standards [Ref. 9].

The arrangement of colors just described forms the basis for the Munsell color system [Ref. 9]. Figure 15.3 shows a color plane from the system of Figures 15.1 and 15.2. This plane intersects the plane of Fig. 15.2 in the

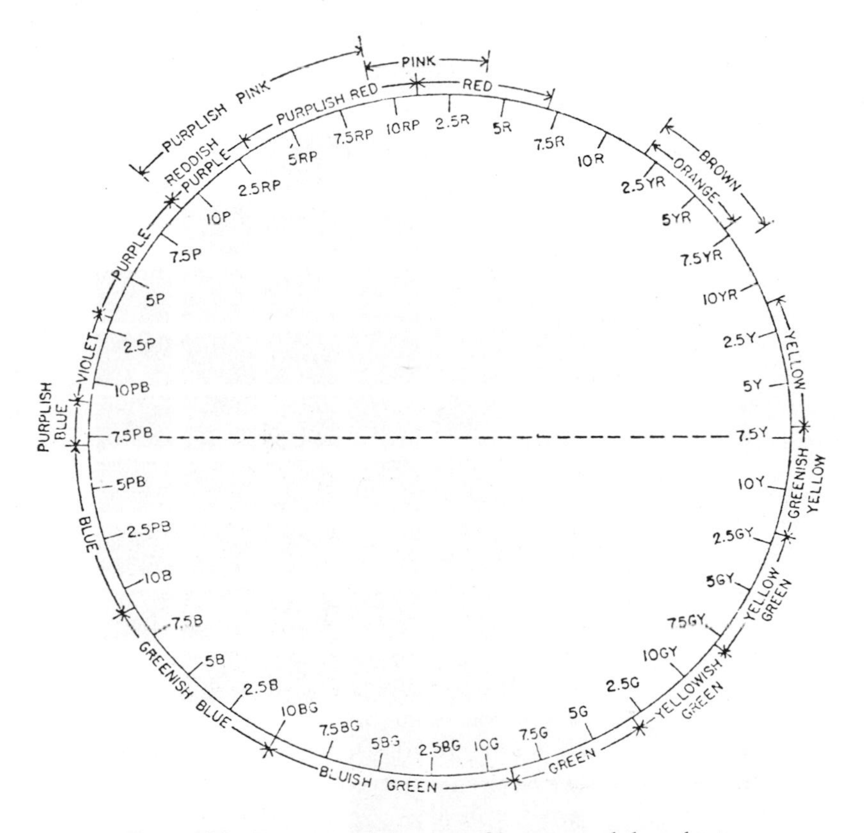

Figure 15.2 Recommended names of hues around the color tree.

indicated diametrical line. The chroma = 0 axis corresponds to the lightness ("value") axis of Fig. 15.1. This plane also appears as a page in the Munsell Book of Color. All similar planes (or pages) contain a column of gray shades at saturation (or chroma) = 0, which is often called the *trunk* of the *color tree*. In the plane of Fig. 15.3, the colors range from saturated purplish blue, 7.5 PB, on the right through the neutral grays at the trunk to saturated yellow 7.5 Y, at the left. The colors also range vertically from the darkest purplish blues, on a level with the dark grays at the bottom, to the lightest yellows on a level with the very light grays at the top. Some color coordinates are not filled. For example (Fig. 15.3), no vivid yellow is so dark as to be level with the very dark grays.

Recommended names for the colors in the color tree are given in a National Bureau of Standards circular [Refs. 9–11]. Figure 15.2 gives the recommended names for the hues and shows their respective positions in the

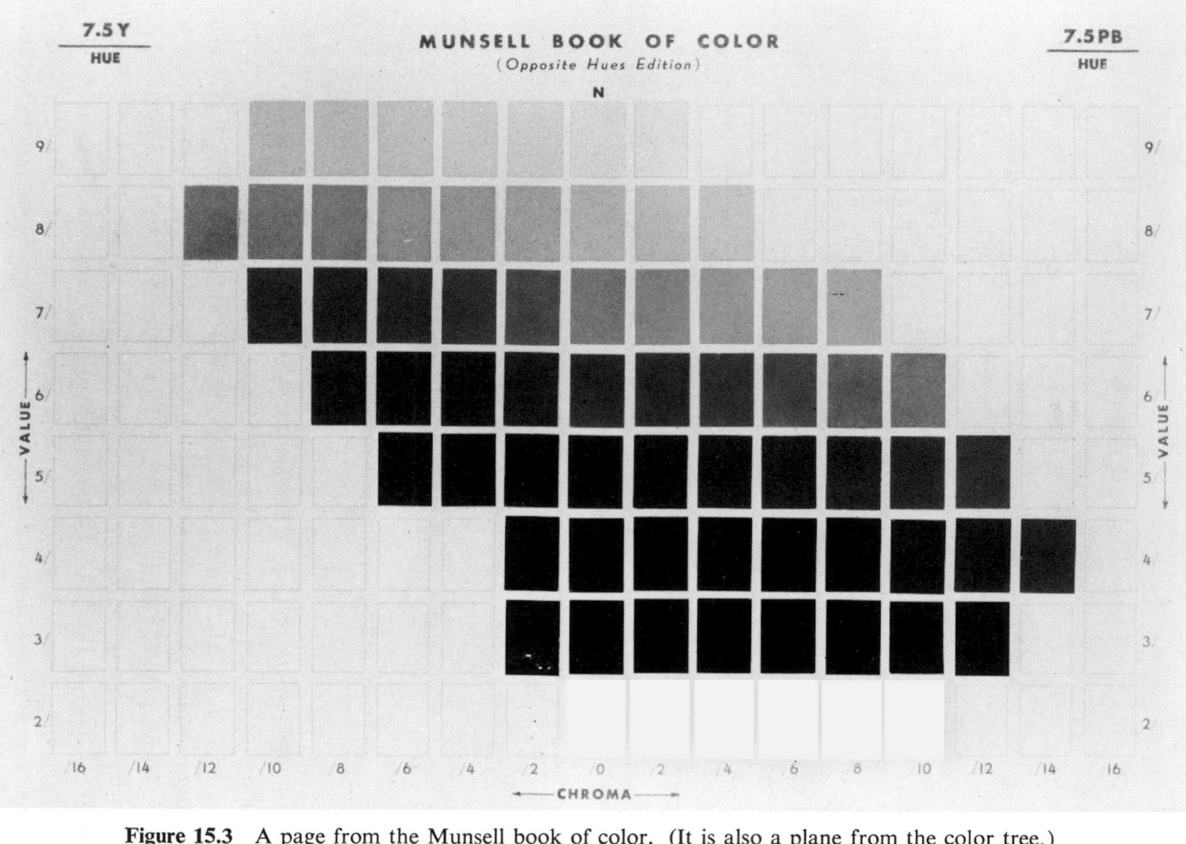

Figure 15.3 A page from the Munsell book of color. (It is also a plane from the color tree.)

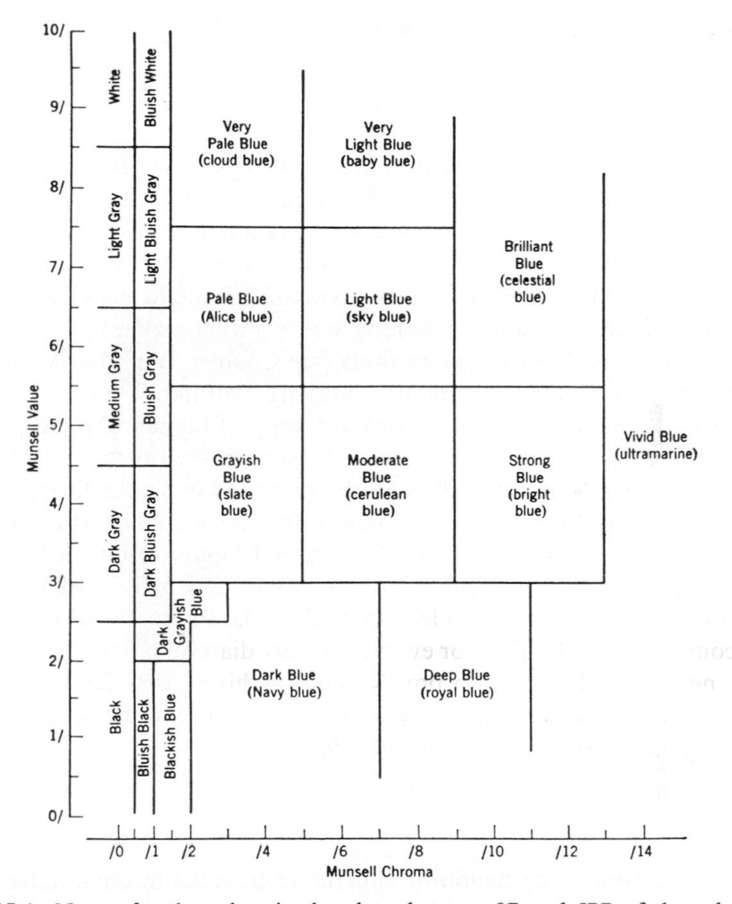

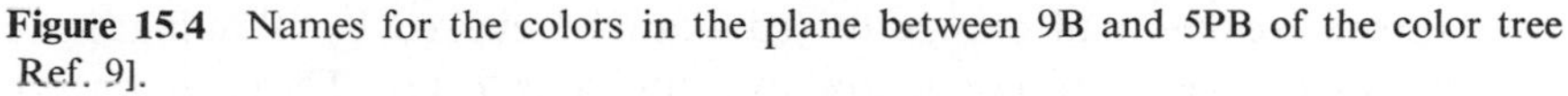

Figure 15.4 Names for the colors in the plane between 9B and 5PB of the color tree Ref. 9].

azimuth of hues [see Ref. 9]. Figure 15.4 is a page from the NBS circular just cited showing the arrangement of names assigned to various Munsell color values and chroma for the Munsell hues between 9B and 5 PB. The common names in parentheses were added by Billmeyer and Salzman [see Ref. 8].

Standard Observer

Most statements in this chapter tacitly assume an observer with *normal* vision. Even with this condition, the bridge required between psychology and mathematics presents difficulties. Some of these can be indicated by discussing Eq. (2-33) introduced in Chapter 2:

$$M_v = K_m \int_0^\infty V(\lambda) M_{e\lambda} \, d\lambda, \tag{15-1}$$

which is known in physiological optics as Abney's law. In Chapter 2 $V(\lambda)$ was defined, with specific values, as spectral luminous efficiency (Table IV and Fig. 2.6). When $M_{e\lambda}$ is the spectral radiant emitted exitance of a blackbody at the temperature of solidifying platinum and M_v is arbitrarily given the value of 60π lm cm^{-2} for this blackbody, K_m is mathematically determinable. The physics questions remaining are: (1) what are the correct values of the first and second radiation constants (see Chapter 3) in Planck's formula, and (2) what is the temperature of solidifying platinum?

When Abney's law is involved in the testing of human observers, experimenters find variations in the value of K_m and in the function $V(\lambda)$. These variations depend upon the different observers and upon the different testing conditions. For instance, an experiment with a 2-deg test field (thus involving only cones in foveal vision) would be expected to give results different from those with a 10-deg test field (involving cones and a few rods). Spectral luminous efficiency $V(\lambda)$ has been adopted by the International Commission on Illumination (CIE) as a convenient standard for evaluating radiant energy. However, an observer may never be found who precisely fits this standard function, so the mathematical constructions that follow are true only for a hypothetical person called a *standard observer*.

Spectrum Hues

Each of the hues of the rainbow—ranging from violet through indigo, blue, cyan, green, yellow, and orange to red—can be evoked by illuminating a screen with a sequence of monochromatic luminous fluxes. These are known as the spectrum hues. The common spectrum hues and the approximate wavelength range associated with each are given in Table XII. The hues in

TABLE XII SPECTRUM HUES

Hue	Wavelength Range, μm
Violet	Shorter than 0.450
Blue	0.450–0.490
Green	0.490–0.560
Yellow	0.560–0.590
Orange	0.590–0.630
Red	Longer than 0.630

the magentas and purples cannot be produced by luminous fluxes of single wavelengths (narrow wavelength bands). They are not spectrum colors and can be produced only by additive mixing of reds and blues.

Color of Objects

Although we commonly ascribe colors to objects, we know that nonluminous objects show color only by modifying, by selective reflection of wavelength, the luminous flux incident upon them. The pigments (of paints) and dyes have their particular hues because they absorb certain hues and transmit or reflect others. So, mixing paints or dyes is "subtractive" mixing; the spectral absorptance of one material is combined with that of the others and the observed color of the object is determined by the spectral characteristics of the light after portions have been absorbed. The color of an object is therefore dependent upon the spectral characteristics of the incident light as well as the spectral reflecting properties of the object surface. An object that appears colored under one light can look gray or black under another: A red automobile appears black under the blue–green light of mercury vapor street lamps.

Additive Mixing

What happens with additive mixing of colors can be described by following an experiment. A transparent diffusion screen is illuminated from the back by four light projectors (Fig. 15.5). The screen is uniformly transparent over all wavelengths. Three of the sources—usually a red, a green, and a blue—are *primaries* and illuminate the same patch on the screen. A fourth source, a sample to be matched, illuminates a second patch. The observer perceives in the sample a particular color that may be characterized by its lightness, hue, and saturation. He adjusts individually the red, blue, and green primaries, that is, their luminous flux densities, on the screen until their patch appears to be the same color as the sample. The following symbolism represents quantitatively the source conditions for matching:

$$M_s(S_s) \equiv M_1(S_1) + M_2(S_2) + M_3(S_3). \tag{15-2}$$

$M_i(S_i)$ stands for "M_i units of luminous transmitted exitance produced at the screen by the source S_i," and $\equiv$ means "will match."

To achieve a match, some samples require that one of the primaries be switched to illuminate the sample patch rather than the primary patch. If primary S_2, for example, had to be mixed with the sample to match the other

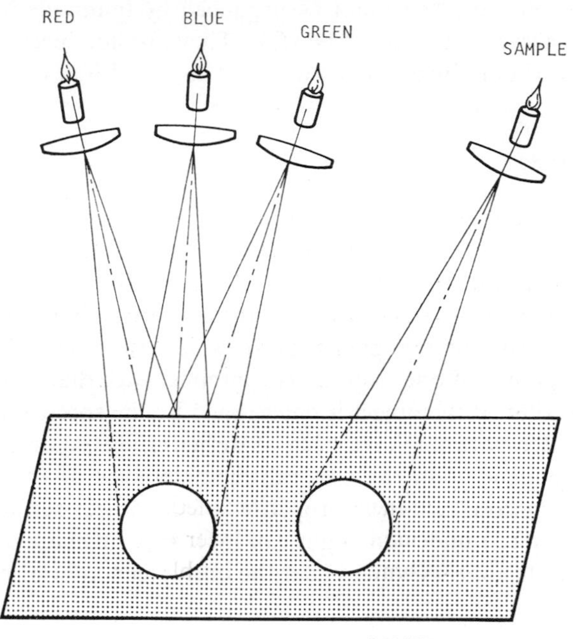

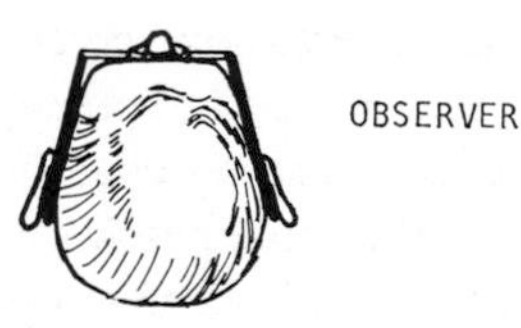

Figure 15.5 Matching of light source colors by additive mixing.

two primaries, the equality for matching would be written:

$$M_s(S_s) \equiv M_1(S_1) - M_2(S_2) + M_3(S_3). \tag{15-3}$$

With the option of negative quantities of this kind, any sample can be "matched" by, or specified in terms of, specific quantities of three primaries. This principle is known as Grassman's law.

Since a sample can be matched by "adding" specific amounts of three given primary sources, the amounts, as numbers, can designate the sample. Such numbers in general represented by the symbols r, g, and b are called *tristimulus* values.

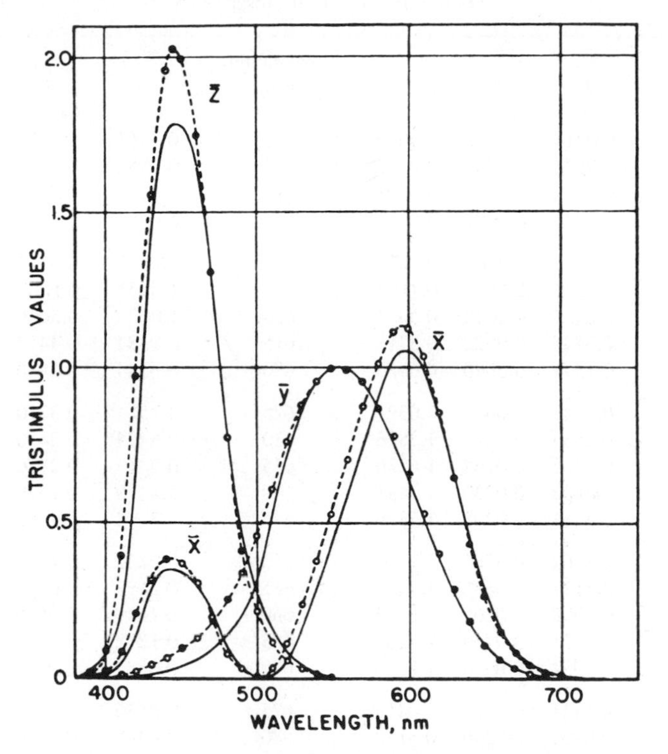

Figure 15.6 Comparison of spectral tristimulus values for 1931 CIE standard observer (2 deg field; solid curves) and 1964 supplementary observer (10 deg field; dashed curves with open dots). (Comparison is based on equal areas under the $\bar{x}$, $\bar{y}$, and $\bar{z}$ curves) [Ref. 20].

Matching Spectrum Colors

Each of the spectrum colors, like other colors, can be specified in terms of three primaries. The specifications—the tristimulus values designated $\bar{x}$, $\bar{y}$, and $\bar{z}$, based on a particular set of primaries adopted by CIE for all the spectrum colors—are shown in Fig. 15.6, (the solid curves) and are tabulated in Table XIII. The quantities $\bar{x}$, $\bar{y}$, and $\bar{z}$—the ordinates on the curves, and the values from the table—at a given wavelength λ, give the amounts (for example, the radiant exitances M_{x1}, M_{y1}, and M_{z1}) of the primaries needed to match, when mixed, the spectrum color evoked by a unit radiant exitance in a narrow wavelength range about λ_1.

The spectral distributions of the three CIE observer functions $\bar{x}$, $\bar{y}$, and $\bar{z}$, standardized in 1931 for a 2-deg field of view, are shown as solid curves in Fig. 15.6. In 1964 the CIE recommended a set of supplementary color-matching functions that apply to a 10-deg field of view; these are shown by

TABLE XIII TRISTIMULUS VALUES OF AN EQUAL-ENERGY SPECTRUM (1931 CIE Standard Observer)

Wavelength, nm	$\bar{x}$	$\bar{y}$	$\bar{z}$	Wavelength, nm	$\bar{x}$	$\bar{y}$	$\bar{z}$
380	0.0014	0.0000	0.0065	580	0.9163	0.8700	0.0017
385	0.0022	0.0001	0.0105	585	0.9786	0.8163	0.0014
390	0.0042	0.0001	0.0201	590	1.0263	0.7570	0.0011
395	0.0076	0.0002	0.0362	595	1.0567	0.6949	0.0010
400	0.0143	0.0004	0.0679	600	1.0622	0.6310	0.0008
405	0.0232	0.0006	0.1102	605	1.0456	0.5668	0.0006
410	0.0435	0.0012	0.2074	610	1.0026	0.5030	0.0003
415	0.0776	0.0022	0.3713	615	0.9384	0.4412	0.0002
420	0.1344	0.0040	0.6456	620	0.8544	0.3810	0.0002
425	0.2148	0.0073	1.0391	625	0.7514	0.3210	0.0001
430	0.2839	0.0116	1.3856	630	0.6424	0.2650	0.0000
435	0.3285	0.0168	1.6230	635	0.5419	0.2170	0.0000
440	0.3483	0.0230	1.7471	640	0.4479	0.1750	0.0000
445	0.3481	0.0298	1.7826	645	0.3608	0.1382	0.0000
450	0.3362	0.0380	1.7721	650	0.2835	0.1070	0.0000
455	0.3187	0.0480	1.7441	655	0.2187	0.0816	0.0000
460	0.2908	0.0600	1.6692	660	0.1649	0.0610	0.0000
465	0.2511	0.0739	1.5281	665	0.1212	0.0446	0.0000
470	0.1954	0.0910	1.2876	670	0.0874	0.0320	0.0000
475	0.1421	0.1126	1.0419	675	0.0636	0.0232	0.0000
480	0.0956	0.1390	0.8130	680	0.0468	0.0170	0.0000
485	0.0580	0.1693	0.6162	685	0.0329	0.0119	0.0000
490	0.0320	0.2080	0.4652	690	0.0227	0.0082	0.0000
495	0.0147	0.2586	0.3533	695	0.0158	0.0057	0.0000
500	0.0049	0.3230	0.2720	700	0.0114	0.0041	0.0000
505	0.0024	0.4073	0.2123	705	0.0081	0.0029	0.0000
510	0.0093	0.5030	0.1582	710	0.0058	0.0021	0.0000
515	0.0291	0.6082	0.1117	715	0.0041	0.0015	0.0000
520	0.0633	0.7100	0.0782	720	0.0029	0.0010	0.0000
525	0.1096	0.7932	0.0573	725	0.0020	0.0007	0.0000
530	0.1655	0.8620	0.0422	730	0.0014	0.0005	0.0000
535	0.2257	0.9149	0.0298	735	0.0010	0.0004	0.0000
540	0.2904	0.9540	0.0203	740	0.0007	0.0003	0.0000
545	0.3597	0.9803	0.0134	745	0.0005	0.0002	0.0000
550	0.4334	0.9950	0.0087	750	0.0003	0.0001	0.0000
555	0.5121	1.0002	0.0057	755	0.0002	0.0001	0.0000
560	0.5945	0.9950	0.0039	760	0.0002	0.0001	0.0000
565	0.6784	0.9786	0.0027	765	0.0001	0.0000	0.0000
570	0.7621	0.9520	0.0021	770	0.0001	0.0000	0.0000
575	0.8425	0.9154	0.0018	775	0.0000	0.0000	0.0000
580	0.9163	0.8700	0.0017	780	0.0000	0.0000	0.0000
				Totals	21.3713	21.3714	21.3715

dotted lines in the figure. The larger visual field is more representative of the viewing conditions encountered in commercial use. To avoid confusion over which set of functions is being used, one speaks of the 1931 CIE Standard Observer Functions or the 1964 CIE Supplementary Observer Functions. In symbolic notation, the subscript 10 is used to designate the larger field.

The CIE primaries are abstractions having certain convenient features. Since psychophysical experiments show that a set of actual primaries can match the spectrum colors, abstract primaries derived by a mathematical transformation (see Refs. 1–3) are assumed equally valid. The curves of Fig. 15.6 are on or above the zero axis, a convenience resulting from a judicous choice of the CIE primaries, which need be real mathematically but need not be physically realizable. In fact, the wavelength range of one or more of the abstract primaries can even extend beyond the wavelength range of visible light without compromising their matching properties.

To match a unit (radiance, radiant intensity, or radiant exitance) of 0.500-μm wavelength λ_s green light, the data of Fig. 15.6 or Table XIII indicate the following amounts (in the same units) of the three primaries:

$$\bar{x}_5 = 0.0049, \qquad \bar{y}_5 = 0.3230, \qquad \text{and } \bar{z}_5 = 0.2720.$$

The Chromaticity Diagram

Three quantities called *chromaticity coordinates* (also called *trichromatic coefficients*) are defined from the tristimulus values as follows:

$$x = \frac{\bar{x}}{\bar{x} + \bar{y} + \bar{z}}, \tag{15-4}$$

$$y = \frac{\bar{y}}{\bar{x} + \bar{y} + \bar{z}}, \tag{15-5}$$

$$z = \frac{\bar{z}}{\bar{x} + \bar{y} + \bar{z}}. \tag{15-6}$$

Since $z = 1 - x - y$, x and y suffice to specify a light source matched by the respective quantities $\bar{x}$, $\bar{y}$, and $\bar{z}$ of the primaries. Therefore, a two-dimensional coordinate system, with axes x and y, can be set up so that all light sources are represented by points with the general chromaticity coordinates (x,y). These points have been found for all the spectrum colors based on the CIE primaries. The locus of the coordinates for the spectrum colors is a curve that roughly resembles a horseshoe and begins at the point (0.1741, 0.0050), corresponding to about 0.380 μm, and ends at the point (0.7374, 0.2653), corresponding to about 0.780 μm. When a straight line

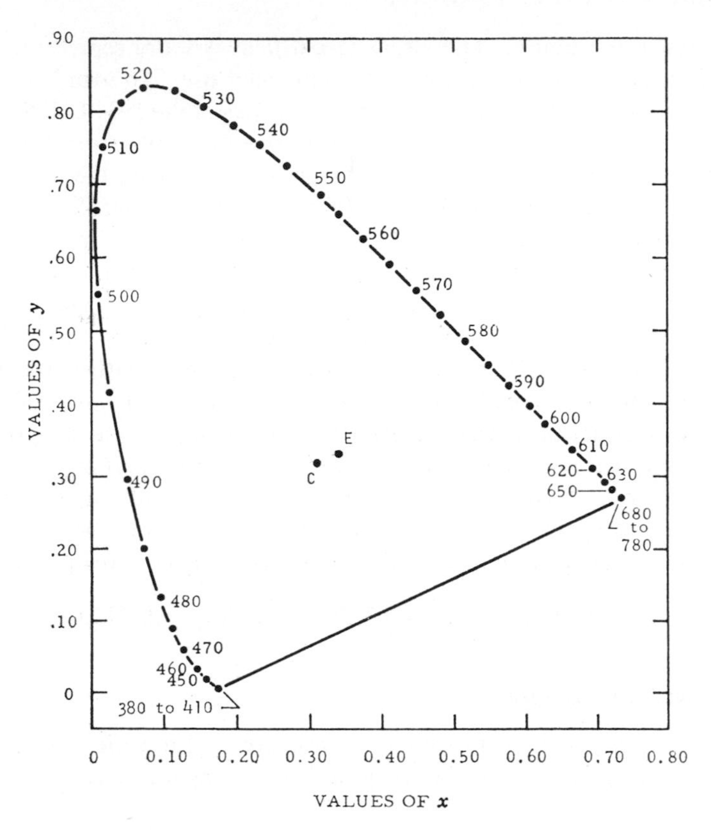

Figure 15.7 Chromaticity diagram based on the CIE standard observer and coordinate system for colorimetry.

is drawn between these two points, a closed curve called the *chromaticity diagram*, shown in Fig. 15.7, results.

The *chromaticness* of a light source combines the qualities of hue and saturation, but not lightness, so the chromaticity diagram is independent of lightness. However, one can resort to a three-dimensional space to include this property by plotting luminance (or lightness) perpendicular to the x–y plane.

Coordinates for a Spectrally Distributed Source

Chromaticity coordinates for a source of known spectral distribution Φ_λ are determined by a summation of spectrum color matches of Φ_λ over all wavelengths: The tristimulus values to match that portion of Φ_λ in a 5-nm wavelength interval about a wavelength λ_i are given in the table. That is, for matching at just the narrow interval about λ_i, the products,

$$\bar{x}_i\Phi_{\lambda i}, \bar{y}_i\Phi_{\lambda i}, \bar{z}_i\Phi_{\lambda i},$$

are required. The total quantities of the primaries (the tristimulus values), designated $\hat{x}$, $\hat{y}$, and $\hat{z}$, are obtained by summation:

$$\hat{x} = \sum \bar{x}_i\Phi_{\lambda i}, \tag{15-7}$$

with similar equations for $\hat{y}$ and $\hat{z}$. If continuous functions like those plotted in Fig. 15.6 are known, the tristimulus values are found by integrating with respect to wavelength:

$$\hat{x} = \int_0^\infty \bar{x}(\lambda)\Phi_\lambda \, d\lambda, \tag{15-8}$$

$$\hat{y} = \int_0^\infty \bar{y}(\lambda)\Phi_\lambda \, d\lambda, \tag{15-9}$$

$$\hat{z} = \int_0^\infty \bar{z}(\lambda)\Phi_\lambda \, d\lambda. \tag{15-10}$$

The chromaticity coordinates of this light source are:

$$x = \frac{\hat{x}}{\hat{x} + \hat{y} + \hat{z}}, \tag{15-11}$$

$$y = \frac{\hat{y}}{\hat{x} + \hat{y} + \hat{z}}, \tag{15-12}$$

$$z = \frac{\hat{z}}{\hat{x} + \hat{y} + \hat{z}}. \tag{15-13}$$

Standard Observer and Standard Source

The curves $\bar{x}$, $\bar{y}$, and $\bar{z}$ shown in Fig. 15.6 actually define a standard observer because these functions place a given light source at definite coordinates in the chromaticity diagram. The source so located is nearest the other light sources with which it is most closely associated by a "normal observer" in any ordered color system. Figure 15.8 shows the names of common colors and their positions in the chromaticity diagram [Ref. 11].

The function of $\bar{y}$ in the CIE system is arbitrarily made the function $V(\lambda)$ of Fig. 2.6, which was adopted as a standard for evaluating the effectiveness of radiant energy in producing "brightness," that is, for evaluating the *sensed quantity* of light. The functions $\bar{x}$ and $\bar{z}$ are analogous to the spectral

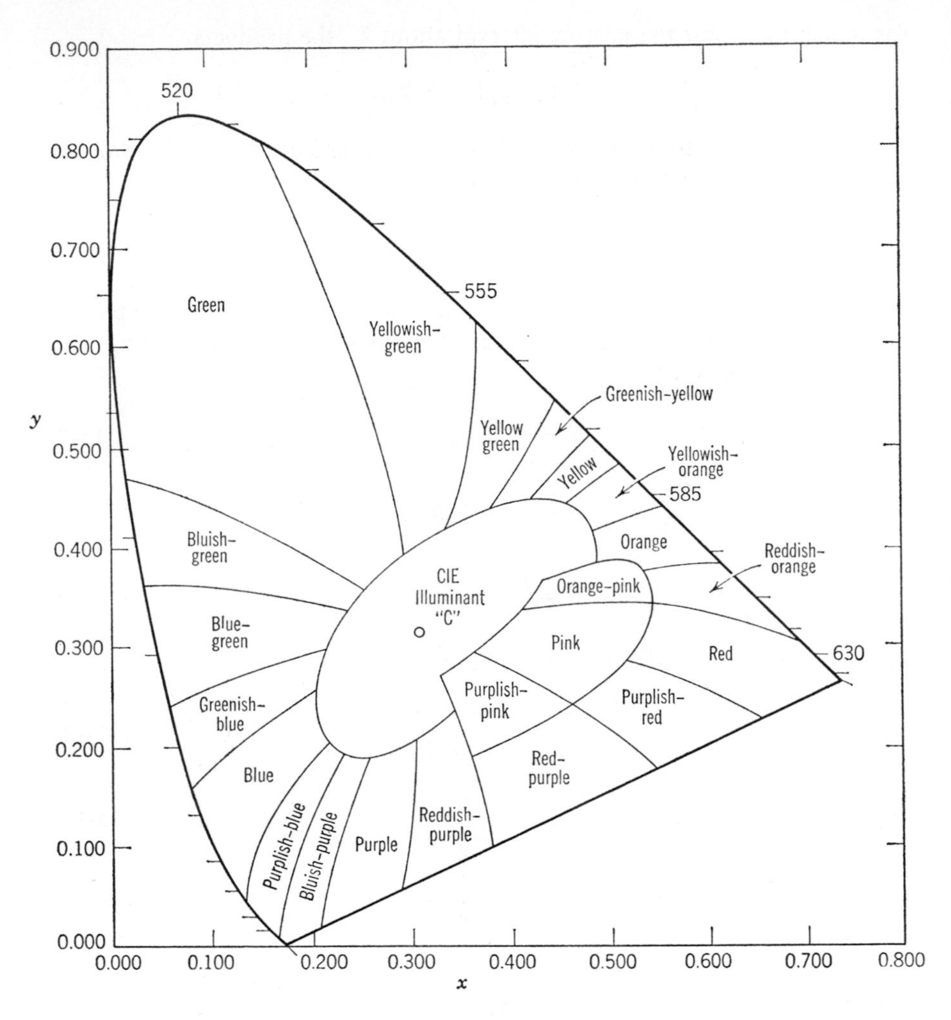

Figure 15.8 Suggested definitions of color designations for self luminous sources. (Spectrum chromaticities are shown, the wavelengths being indicated in nanometers) [Ref. 7].

luminous efficiency function $\bar{y}$ in that they are standards adopted for evaluating the effectiveness of radiant energy in producing the attributes of color other than brightness (or lightness). The functions $\bar{x}$, $\bar{y}$, and $\bar{z}$ still represent the three color mixture curves. The freedom to make one of them, $\bar{y}$, equal to $V(\lambda)$ reduces the number of functions by one and thus reduces the tedium in applying color theory to certain problems.

TABLE XIV CHROMATICITY COORDINATES OF VARIOUS SOURCES [REF. 1]

Source	x	y	Correlated Color Temperature
CIE (1931) Standard Source A	0.4475	0.4075	2,850
CIE (1931) Standard Source B	0.3485	0.3518	4,880
CIE (1931) Standard Source C	0.3101	0.3163	6,740
Equal energy per nanometer	0.3333	0.3333	5,500
Mean noon sunlight at Washington	0.3442	0.3534	5,035
Sunlight above atmosphere	0.3204	0.3301	6,085
Direct sunlight	0.3362	0.3502	5,335
Sun plus sky on horizontal plane	0.3213	0.3348	6,000
Overcast sky	0.3134	0.3275	6,500
North sky on 45-deg plane	0.2773	0.2934	10,000
Zenith sky	0.2631	0.2779	13,700
Carbon arc	0.3152	0.3325	6,400
Macbeth (6800 K) lamp	0.3081	0.3231	6,800
Macbeth (7500 K) lamp	0.2996	0.3123	7,500
Sunlight above atmosphere (moon's data for $m = 0$)	0.3179	0.3297	6,215
Direct sunlight at sea level (moon's data for $m = 2$)	0.3431	0.3567	5,080

A number of standard sources are listed in Table XIV with their chromaticity coordinates. Tabulated values of the products $\bar{x}(\lambda)\Phi_\lambda$, $\bar{y}(\lambda)\Phi_\lambda$, and $\bar{z}(\lambda)\Phi_\lambda$ are also available for these sources [see Ref. 1].

Achromatic Sources

Several sources—like sunlight, skylight, and the incandescent and fluorescent bulbs used in ordinary room lighting—are described by the general term "white light sources." One could question whether these sources are truly without hue, that is, with zero saturation. However, they are called achromatic sources and are thus considered to have zero saturation. They are all located on the chromaticity diagram in the "central" region near the point for the CIE illuminant "C" shown in Fig. 15.8.

A source having equal amounts of radiant power in each wavelength zone of its spectrum would have the chromaticity coordinates $x = 0.333$, $y = 0.333$. It is located at point "E" in the diagram of Fig. 15.7.

Additive Mixing of Sources

If two sources having the chromaticity coordinates (x_1,y_1) and (x_2,y_2) are mixed, the resulting mixture is located in the chromaticity diagram somewhere

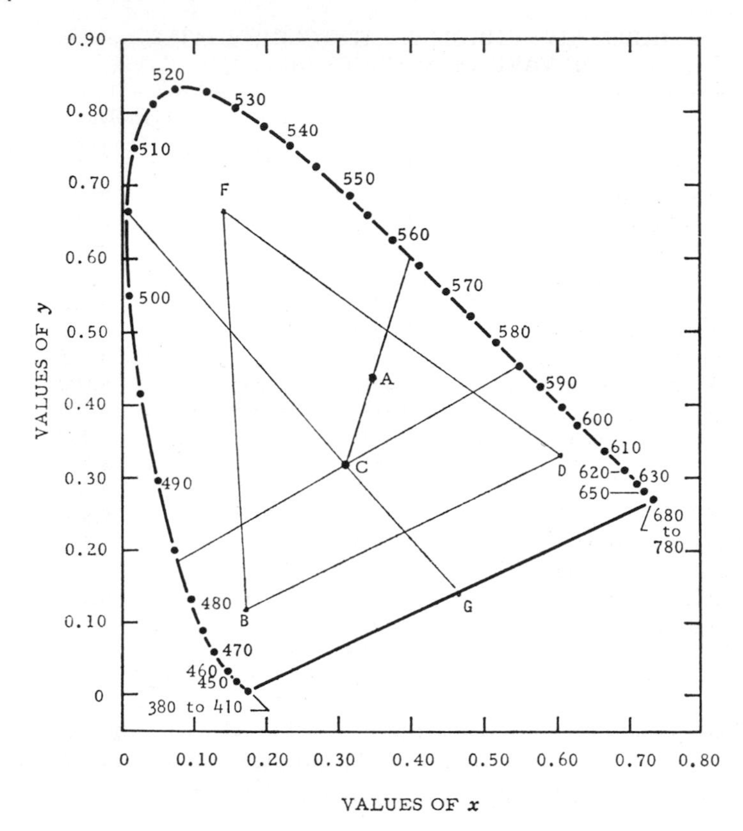

Figure 15.9 Chromaticity diagram showing by the triangle BDF the colors that can be matched by three primaries having the chromaticity coordinates of points B, D, and F, respectively.

on a straight line joining (x_1,y_1) and (x_2,y_2). For example, if the two points are B and D in Fig. 15.9, the mixture will be somewhere on the line joining B and D. It will be a fraction γ of the distance from B to D where γ is given by

$$\gamma = \frac{M_D}{M_B + M_D}. \tag{15-14}$$

If three sources are represented by the points B, D, and F, a mixture of the three sources will be somewhere within the triangle BDF. Colors within the diagram but outside the triangle cannot be matched by these primaries. This property of the triangle and the configuration of the chromaticity diagram indicate that three spectrum colors—a red, a green, and a purplish-blue—will match more samples by positive mixing than any other three primaries.

Dominant Wavelength and Purity

The subjective attributes of a light source color, which are hue, saturation, and lightness, can be associated with the physical or psychophysical quantities called dominant wavelength, purity, and luminance, respectively. Luminance as associated with lightness needs no further discussion. If a given light source is represented by the point labeled *A* in Fig. 15.9, for example, and a line is drawn from *C* through *A* until it intersects the curve of spectrum hues, the wavelength corresponding to the point of intersection is called the dominant wavelength of the source (with respect to the standard source *C*). The ratio of the length of the line from *C* to *A* to the total length of the line is called purity. If the intersection had occurred along the straight line joining the red to the violet, point *G*, there would be no dominant wavelength because the colors near *G* are not spectrum colors. Instead, the *complementary* wavelength would be given; it is found by extending the line back through the point *C* until it intersects the curve on the other side.

Complementary colors are pairs of colors that can be mixed in appropriate proportions to match an achromatic standard. If the *C* point on the chromaticity diagram is the standard, a pair of complementary hues are found at the ends of a line passing through the *C* point. The wavelength of a complementary hue, therefore, depends upon the choice of the source to represent the *C* point. The complements for the spectrum hues between 492 and 568 nm are not spectrum colors. The mixing of a spectrum hue with its complement, whether or not it is a spectrum color, can match the achromatic source *C*. After a prolonged observation of a high purity source, an after-image of the source in the hue of its complementary color is generally seen when the source is turned off.

According to the definition of purity just given, one would assume that all the spectrum colors have unit purity. Spectrum hues, however, seem to differ in saturation. For example, a spectrum yellow, when compared with a spectrum blue of the same luminance, will usually appear less saturated than the blue.

A surface with high reflectance in only a narrow wavelength band from about 560 to 570 nm, when illuminated with light from the source *C*, would appear a nearly saturated yellowish-green. A color match would fail, according to theory, if the yellowish-green surface color, observed when illuminated with white light, is compared with the color of the surface when illuminated by two spectrum-hue sources of high saturation (for example, one near 500 nm and one near 630 nm). The surface should appear black in the latter instance because light from both sources is nearly all absorbed; however, the observer's memory of the surface and its color helps him see it normally, and it still will usually appear yellowish-green when observed for an extended length of time under either kind of illumination.

The "Two-Color" System

Remarkably satisfying reproductions of certain color pictures can be made by using only two light sources [Ref. 13]. However, the two-sources system will not reproduce all scenes faithfully; in fact, if certain objects in a scene are unfamiliar to the observer, he may see them in incorrect colors [Ref. 4, pp. 469–472; Refs. 14 and 15].

In the two-source color system, two transparencies of a scene are made on photographic film, one through a green filter and the other through a red filter. The scene is reproduced by projecting both transparencies on the same screen. The "green" transparency is projected by an incandescent light with no filter, and the "red" transparency is projected by an incandescent light through a red filter.

Much psychology is involved in seeing color, and many experimental observations cannot be explained by either physics or physiology. When one is viewing a scene containing familiar objects, his memory helps him see them in true colors, just as a yellow object illuminated with predominantly green light may still be seen as yellow. One particular situation in the two-color system could reproduce an object as either gray, brown, or purple [Ref. 14]. However, because we know that an object is a flower, we can see it as purple rather than brown or gray. Knowing that an object is a tree trunk helps us to see it as gray rather than brown or purple, and knowing that an object is a stone turns it brown instead of gray or purple. Also, our color memory could specifically help to intensify, where needed, the blues in a familiar scene. These are the kinds of accommodations that have to occur to make a two-color system appear true.

REFERENCES

1. OSA Committee on Colorimetry, *Science of Color*, Optical Society of America, 1953.
2. D. B. Judd and G. Wyszecki, *Color in Business, Science, and Industry*, Second Edition. John Wiley & Sons, Inc., New York, 1963.
3. J. J. Shepard, *Human Color Perception, A Critical Study of the Experimental Foundations*. American Elsevier Publishing Company, New York, 1968.
4. *Vision and Visual Perception*, Edited by C. H. Graham. Chapters 12–16. John Wiley & Sons, Inc., New York, 1965.
5. Y. LeGrand, *Light, Colour, and Vision*. Chapman and Hall Ltd., London (Barnes and Noble, New York), 1968.
6. W. D. Wright, *The Rays Are Not Coloured. Essays on the Science of Vision and Colour*. Adam Hilger, 1967.

7. G. Wyszecki and W. S. Stiles, *Color Science*. John Wiley & Sons, Inc., New York, 1967.
8. W. Billmeyer Jr. and M. Saltzman, *Principles of Color Technology*. Interscience Publishers, New York, 1966.
9. K. L. Kelly and D. B. Judd, "ISCC-NBS Method of Designating Color and Dictionary of Color Names," NBS Circular 553, U.S. Department of Commerce ($2), 1956.
10. A. Chapanis, "Color Names for Color Space," *Am. Sci.*, **53,** 327 (1965).
11. K. L. Kelly, "Color Designations for Light," *J. Opt. Soc. Am.*, **33,** 627 (1943).
12. D. B. Judd, "The 1931 CIE Standard Observer and Coordinate System for Colorimetry," *J. Opt. Soc. Am.*, **23,** 359 (1933).
13. E. H. Land, "Color Vision and the Natural Image, Part I," *Proc. Nat. Acad. Sci. U.S.A.*, **45,** 115 (1959).
14. D. B. Judd, "Appraisal of Land's Work on Two-Primary Color Projection," *J. Opt. Soc. Am.*, **49,** 322 (1959).
15. W. N. Sproson, "The Range of Colours Excited by a Two-Colour Reproduction System," *J. Brit. IRE*, **21,** 537 (1961).
16. R. W. Burnham, R. M. Hanes, and C. J. Bartleson, *Color: A Guide to Basic Facts and Concepts*. John Wiley & Sons, Inc., New York, 1963.
17. L. E. DeMarsh and J. E. Pinney, "Studies of Some Colorimetric Problems in Color Television," *J. SMPTE*, **79,** 338–342 (Apr., 1970).
18. E. F. MacNichol Jr., "Retinal Mechanisms of Color Vision," *Vision Rec.*, **4,** 119 (1964).
19. F. W. Billmeyer Jr., "Precision of Color Measurement with the G.E. Spectrophotometer. I. Routine Industrial Performance," *J. Opt. Soc. Am.*, **55,** 707 (1965).
20. H. K. Hammond III, "Colorimeters," in *Applied Optics and Optical Engineering*, Vol. V, Part II. Edited by R. Kingslake. Academic Press, New York, 1969.
21. *IES Lighting Handbook*, Fourth Edition, Illuminating Engineering Society, New York, 1966.

Index